OUR OWN DEVICES

OUR OWN DEVICES

The Past and Future of Body Technology

EDWARD TENNER

Alfred A. Knopf New York 2003

Library of Congress Cataloging-in-Publication Data

Tenner, Edward.
Our own devices : the past and future of body technology / by Edward Tenner.
p.   cm.
Includes bibliographical references and index.
ISBN 0-375-40722-7 (alk. paper)
1. Technology—Social aspects.   2. Technological innovations—Social aspects.
3. Human beings—Effect of technological innovations on.
4. Body, Human—Social aspects.   5. Body, Human (Philosophy)   I. Title.

T14.5 .T4588 2003
303.48'3—dc21   2002040694

To the memory of my father

Contents

Preface

T ECHNOLOGY APPEARS to have become a synonym for electronic systems. It should not be so. Just because microprocessors are all machines does not mean that all machines, even all important machines, are built around chips and circuits. In fact, in the early 1990s the economist Edwin Mansfield found that improved thread and stain removers had done more for America's productivity than personal computers. The time is right to think again about technology, and to start with some of the simplest forms.[1]

In *Why Things Bite Back: Technology and the Revenge of Unintended Consequences,* I defined technology as the human modification of the natural world. This book is about the changes we have made in ourselves: how everyday things affect how we use our bodies—how we sit, stand, walk, and communicate. And it is about their symbolic side: how they affect our images of each other. They are, literally, our own devices. Though people have been able to do surprisingly well without them, these everyday things are today part of a technological treadmill from which it is difficult to step down.

I started to consider these themes after participating in an on-line forum on technology and design sponsored by the Discovery Channel's Web site in 1996. The other guests were the cognitive psychologist Donald Norman and the civil engineer and historian of technology Henry Petroski, both of whom had written important and insightful studies of everyday objects. It was the best kind of discussion, and I regret that it is no longer available on the Web. One of the things I learned from our agreements and disagreements was that nobody had yet looked at ordinary things as the outcomes of a ceaseless interplay of technology, economics, and values.[2]

An engineer can look at the proliferation of tableware in the nineteenth century, for example, and consider the mechanics of producing and using specialized knives, forks, and spoons. A food chemist can evaluate the appropriateness of glassware shapes for wines and spirits. But science-based analysis does not tell us why men and women of one period were paying for proliferating variety. Some social historians believe that the Victorian upper classes were making daily rituals as complex as possible to intimidate and entrap upstarts. Likewise, paper clips could probably have been made in the eighteenth century; it was not only new metalworking machinery but mountains of new paperwork that hastened their development. Technology, often regarded as a prime mover of change, is also a response to long-standing trends.

A psychologist can identify many ways to make objects easier to use. Many tools could still be improved by scientific study. But performance is not the only measure of value. Our needs are complex and sometimes paradoxical. One of the most popular fields for awards of U.S. patents is new golf equipment, yet the U.S. Golf Association maintains a large, professionally staffed laboratory that rejects many of these innovations, not because they fail to work or are difficult to use, but because they are too effective or too easy to operate. Golfers, like other sports participants, must balance the search for a technological advantage with the prestige that comes from learning a difficult technique. In *Why Things Bite Back* I considered how innovation can maintain or undermine the challenge, the artificial difficulty, that underlies sports. Even the most fervent technology enthusiasts have not (yet) proposed to turn golf and other sports over to robots whose progress can be observed on monitors at the clubhouse bar. Manual skills have not declined; they have been migrated from work to hobbies and sports.

The line between manufactured things and information is also arbitrary. Our knowledge is filtered through objects. A second inspiration for this book came during a question-and-answer session after a lecture I gave in 1997 in the Urania-Gesellschaft in Berlin, one of the world's oldest and liveliest adult education centers. One member of the audience, an older man who (to judge from my hosts' disapproving glances) liked to confront the speakers, challenged me to name the most important invention in history. I thought for a moment and replied, "Eyeglasses, so as many people as possible will buy my book," breaking the tension but also provoking myself to think more about the body as information system. The questioner, who himself was wearing glasses, joined in the approving response. Perhaps there is such a thing as constructive heckling.

This book looks at everyday objects through a pair of terms: *technology* and *technique*. The first consists of the structures, devices, and systems we use; the second, of our skills in using them. Historians of technology customarily focus on the devices themselves, if only because user habits are poorly documented. Designers—often rightly—aim to make some skills obsolete. What virtue was there in trimming a wick, cranking an automobile, tuning a crystal radio set, or even (for casual amateurs) focusing a camera manually?

In fact, technique is much more important in our lives than we realize. Many things seem to act almost magically on their own. We can sustain this illusion at length, but sooner or later we need an expert urgently to correct an "automatic" system gone awry. The zero-insertion-force floppy drives of older Macintosh computers—before Apple turned against the floppy disk on principle—were such a technology. They swallowed disks gracefully after only a gentle nudge. Unlike the IBM-compatibles, they released them again by software control from a pull-down menu on the screen, not with a vulgar spring-loaded button. But the drives were still machines, and the disks sometimes stuck. A simple technique was needed: straightening a paper clip and poking it in a discreet little hole with a recessed switch. Even users of well-designed objects need to know a few tricks.

The interplay of technology and technique is literally at our feet. The shoelace can encapsulate the themes of this book. For all the embrace of advanced materials in athletic shoes (which we will explore in Chapter Three), shoe fastening has changed little for two centuries or longer. The obsessive protagonist of Nicholson Baker's novel *The Mezzanine* observes, "Shoes are the first adult machines we are given to master." Generations of inventors have proposed alternative fasteners, and many children today begin with the interlocking fabrics trademarked as Velcro. But sooner or later, just about every child in the world's wealthier countries learns to tie his or her own shoes. Few of us remember studying the procedure, yet it is complex enough to justify manufactured learning aids for teachers and parents and extended Internet discussions of methods. Try, without manipulating an actual pair of laces on shoes, to write a paragraph explaining to a novice—and there must be at least a billion or two around the world—how to tie the knot. If you can do that in your mother tongue and are proficient in another, try to explain the procedure in your second language. You will probably find it easier to restate the ontological argument for the existence of God, or any other philosophical or theological idea. Tacit knowledge, what we can do but seldom explain, is essential. The special vocabulary of the everyday is obscure. How many people know the proper words for the plastic- or metal-bound ends of shoelaces, *aglets* or *tags*?[3]

Baker's narrator could have discovered even more about shoelace technique. Simple objects can call for a surprising level of skill. Henry David Thoreau, for all his acuteness as an observer of nature, took some time to understand that his leather laces were constantly coming undone because he was tying granny knots rather than square knots. Subcultures have their own shoelace techniques, sometimes highly refined. In his memoir *The Duke of Deception,* the writer Geoffrey Wolff recalls learning only one useful thing from his detested maternal grandfather, a mean-spirited, self-made officer: the navy method of tying shoelaces so they would not unravel. Soccer players knot their laces on the radial (outer) side of the foot to avoid interference with kicking. Experienced hikers can adjust lacing patterns to terrain, keeping the toe box loose and the ankle tight uphill (to prevent twisting) and reversing the pattern on the way down (to protect the Achilles tendon), using a double-twisted knot to separate the two parts of the lace. Still another lacing can seat the heel firmly in the back of the boot. A skilled hiker can skip eyelets and use other lacing tricks to improve performance and prevent blisters. Athletes in other sports too learn to adjust lacing patterns. Joe Ellis, a sports podiatrist, has published instructions for varying laces to stop heel slippage, improve midsole flexibility, and protect tender toes and insteps. Many manufacturers, beginning with Nike, have even introduced variable-width lacing with alternating inner and outer rows of eyelets to permit wider and narrower fittings of shoes made on the same last. (Most wearers still use all the eyelets.) Even for the less active elderly, orthopedists recommend laced shoes over slip-ons because they better accommodate orthotics and swollen feet. In fact, experienced shoe fitters can show people of all ages how to use lacing patterns borrowed from hiking to compensate for mismatches between shoe shapes and individual feet. They warn that "advanced" shoes with loops instead of eyelets and round laces may cut into the foot.[4]

Old technologies, then, stay alive by assimilating new techniques as well as materials. But they also confer symbolic meanings. Shoelaces have an important cultural side, even religious implications. In most traditional Islamic societies they were avoided as inconvenient in removing and replacing shoes at the entrances to mosques. Slip-on footwear is obviously easier to use wherever custom dictates removing shoes at entrances to private homes. Social class also helps define closures. In 1666, King Charles II of England began a new fashion when, according to the diarist John Evelyn, he "put himself solemnly into the Eastern fashion, changing . . . shoe strings [forerunners of laces] and garters into buckles, of which some were set with precious stones." Through the eighteenth century and until around 1820, gentlemen in Europe and America used buckles on their shoes, not laces. When Thomas Jefferson adopted shoelaces at the beginning of the nineteenth century, he was defying criticism to embrace a democratic fashion he had observed as a diplomat in Paris at the eve of the Revolution.

Laces also represent group identity. In North America, Europe, and Asia, school authorities occasionally ban colored laces for their association with gangs. Leaving laces threaded but with ends untied has been an emblem of youthful rebellion. Not wearing them at all with lace-up shoes has, like baggy trousers, been part of inner-city prison chic. (At the other extreme, Lord Baden-Powell established a "Scout's way" of tying laces that concealed both the knot and the end of the string.) And even among conservatively dressed adult males there are notable differences in shoelace technique. Europeans tend to tie parallel laces across the instep, using hidden offset zigzags; Americans prefer the crisscross like a series of X's. Nobody knows when and why these patterns diverged. In the British army in World War II, soldiers were taught parallel lacing to aid medics in cutting to remove the boots of the wounded, but the custom is probably much older. Some American commanders have ordered troops to lace each of two pairs of boots with a different style, as a reminder to alternate them. Once I observed that a visiting European colleague had used a different pattern on each shoe. (The American method uses slightly less lace, according to the mathematician Ian Stewart.)[5]

In social interactions laces are no mere symbols. They serve as invaluable props. A loose shoelace justifies a time-out, the interruption of a conversation, the chance to observe. A London journalist reported that a financier she was interviewing would begin to adjust his eyeglasses and untie and retie his shoelaces when questions became uncomfortably personal. Viewers of *Dr. Strangelove* will recall the Soviet ambassador stooping to retie his laces at the apocalyptic end for a last bit of clandestine photography of the War Room. In fact, the shoelace enables many a devious maneuver. Cheaters in marathon races wait on the sidelines, bend to tie their shoes when the pack passes, then rise to join it in the excitement. During the 1992 Barcelona Olympics, the Cuban baseball team used frequent shoelace-tying to slow down the game and defeat their American opponents—at least, so U.S. journalists saw it.[6]

Finally, shoelaces show the power of technological improvisation, our ability to find unexpected uses for familiar things. Flat, strong, braided thread is easy to knot. In its benign aspect, a shoelace is a wonderful tool wherever a flexible cord is needed, for example, looped to make it easier for people with disabilities to open doors with knobs, or as an emergency tourniquet. It also has sinister uses, ranging from the concealment of crib sheets by students to suicide and murder. Prison and mental hospital authorities the world over often confiscate shoelaces from new inmates. (Airport alarms are tripped by metal arch reinforcements, but laces raise no questions.) Like clothespins, paper clips, and duct tape, shoelaces are technologies that leave especially ample room for new techniques elaborated by our imagination.[7]

Laces show that a simple body technique can be so valuable that changes become incremental rather than revolutionary. Production of shoelaces has long been automated, down to the attachment of the plastic aglets—quite a technological feat because laces are braided from multiple strands that must be woven together much as hanks of hair are braided, but on a minute scale. The only important recent innovation has been elastic laces that eliminate the need for knotting and make removal and replacement of shoes easier for people with disabilities and for those living in societies where shoes are for street wear only.

But fashion can make unexpected demands on technique. When shoe manufacturers began to turn to neat-looking but slip-prone round synthetic laces for a sleeker appearance in the 1990s, wearers complained. A winning New York Marathon runner from Kenya sponsored by Nike failed to break a record after his round nylon Air Streak Vengeance laces came untied three times, costing him crucial seconds. (Speaking of technological revenge, Nike paid him the $10,000 bonus he would almost certainly have won had the laces held.) Yet some shoe executives insisted that with proper knotting skills the laces would stay put.[8]

The manufacturers of nineteenth-century laced footwear and shoelaces could scarcely have foreseen the variety of motor and symbolic skills their innovations would unleash. They no doubt were thinking of more comfortable, flexible, and fashionable alternatives to buckles, buttons, and slip-ons. Communities of users—athletes, young people, criminals—

developed new techniques for manipulating laces. Of course, new technology could in turn serve these trends: braiding machines with today's sophisticated controls can produce patterns in laces, including ways to identify corporate and school affiliation. Technique and technology reinforce and modify each other, coevolving unpredictably and endlessly. Building on the ideas of the architectural theorist Christopher Alexander, the writer and designer Stewart Brand has illustrated how structures "learn" by absorbing the additions, subtractions, innovations, and restorations of successive owners. Adaptability is equally vital in designing the smaller things in life. What Brand writes about buildings, that each is a prediction and a wrong one, applies to our tools as well.[9]

Our Own Devices is thus an exploration not only of inventive genius but also of user ingenuity. It explores why we experience so many positive as well as negative unintended consequences. In *Why Things Bite Back,* I identified a class of problems that I called revenge effects, the side of technology that undermines its reason for existence. Some computer programs designed to prevent crashes, for example, have been known to cause them. But technologies can have equally unexpected benefits. Apparently frivolous ideas can have productive consequences. The chemistry of attaching oxide particles to magnetic tape was a by-product of efforts in the Weimar Republic to bind gold leaf more effectively to fashionably decadent black Russian cigarettes. More recently, the control room of a major neutrino experiment at the particle research center Fermilab near Chicago was inspired by the round subterranean den of the *Teletubbies* children's television series, familiar to parents and grandparents among the laboratory's staff. Scientists at Silicon Graphics have trained digital cameras on the circulating blobs in Lava lamps to generate truly random numbers, which are essential to many computer tasks but cannot be produced by computer circuits alone.[10]

This book starts by looking at how objects complement body techniques (as the French anthropologist Marcel Mauss originally called them), arguing with examples from warfare to music that some of the most important innovations have been less in the invention of things than in the development of new usages.

The core of the book considers body techniques and technologies through history, starting with the first technology most people encounter, infant feeders. It then starts, literally at the ground, by examining the simplest form of foot covering, the sandal—specifically, the thong sandal familiar to antiquity and spread around the world from Asia after

World War II. It continues with the twentieth century's most characteristic footwear, the running shoe, and its association with styles of movement. The following chapters move to the back and its relationship to seating. The chair is an originally Mediterranean and Middle Eastern creation that has displaced older forms of seating only recently in most of the world. The idea of designing a chair for work emerged only in the nineteenth century and is still one of the greatest challenges for designers. Reclining chairs, the subject of the next chapter, are not always fashionable but are usually well designed for what they do. The chapter on them shows how a technology first associated with aristocratic leisure and then with high modern style and health became a symbol of self-indulgence and obesity in popular culture.

The breakthrough technology for the hands in the last two centuries has been the keyboard, which has not only displaced the pen and the pencil as a writing instrument but has changed human control of many other processes. The chapter on the musical keyboard looks at how styles of playing and the principles of making instruments interacted. Its counterpart on the text keyboard shows how typists contributed as much as the original inventors to the definition of the technology.

What the piano did for music and the hands in the nineteenth century, mass-produced spectacles did for the eyes. The chapter on eyeglasses explores how their existence depended less on the understanding of optics than on the needs of society's growing literacy; yet the explosion of reading material in the nineteenth and twentieth centuries has in turn made us increasingly dependent on eyeglasses. The helmet, which started the first arms race in the ancient Middle East, is now as much an athletic as a military object, and is even an orthopedic appliance for babies.

The epilogue considers trends in body technology. Wearable and implantable devices have lost some of their cachet since the late 1990s; even a pioneer of the cyborg lifestyle has misgivings. Direct modification of the body itself—through exercise machinery, diet, medication, and in the future even regeneration of teeth and organs—is a possibility. And mobile devices are promoting the humblest digit to a new dignity, suggesting that there can be happy unintended consequences, too.

Like *Why Things Bite Back*, this book is not mainly for specialists in the history, philosophy, or sociology of technology. It is an exploratory work for all who are curious about familiar things. My goal is to provoke new ways of looking at the commonplace. For I agree with Oscar Wilde that the true mystery of the world is not the invisible but the visible.

. . .

My first debt is to the Woodrow Wilson International Center for Scholars, which gave me the initial support for this work. I am grateful to the staff, to my colleagues there, and to the interns who helped me with so much intelligence and enthusiasm, Meryl Hooker and Christopher Garces. I also appreciate the hospitality of Princeton University, in particular John Suppe, S. George Philander, and Robert Phinney in the Department of Geosciences; Michael Wood, Nigel Smith, and William Howarth in the Department of English; and Beth Harrison, the English Department manager. At the Princeton University Library, I especially thank Mary George, Kiyoko Heineken, Elaine Zampini, Stephen Ferguson, and the Interlibrary Loan staff. More than once the Princeton Office of Information Technology (OIT) helped rescue my programs and data from the frailties of hardware. I have also appreciated the help of Roy Goodman at the American Philosophical Society, and the wonderful reference staffs of the Princeton and Plainsboro Public Libraries, the Library of Congress, the New York Public Library, the library of the Victoria and Albert Museum, and (not least) Mary Carter-Hepworth of the Albertson Library, Boise State University.

Though this is a new work, I have also appreciated the encouragement of the magazine and journal editors with whom I have worked on pieces related to it: Jay Tolson (now at *U.S. News & World Report*) and Steven Lagerfeld at the *Wilson Quarterly;* John Bethell, John Rosenberg, Christopher Reed, and Jean Martin at *Harvard Magazine;* Victoria Pope and Tim Appenzeller at *U.S. News & World Report;* Luke Mitchell at the former *Industry Standard* (now at *Harper's*); and Jerilou Hammett at *Designer/Builder.*

At the National Museum of American History, I have been helped greatly by Arthur Molella and Claudine Klose of the Lemelson Center, and by John Fleckner and the staff of the Smithsonian Archives Center. Stuart Pyhrr and Donald LaRocca of the Metropolitan Museum of Art shared their deep knowledge and the awesome heritage of the Arms and Armor Department. Rolf Fehlbaum, Alexander von Vegesack, and Andreas Nutz of Vitra were most generous with both material support and encouragement, as were Bruno Haldner, Matthias Götz, and Matthias Buschle of the Museum für Gestaltung in Basel and Kenneth Boff and his colleagues at Wright-Patterson Air Force Base. Special thanks are also due to Walter Bradford of the U.S. Army Historical Center, Gail Bardhan of the Corning Glass Museum, Barbara Williams of the Shoe Museum of the Pennsylvania

College of Podiatric Medicine, Marilyn Hofer and Travis Stewart of the New-York Historical Society, Christopher Mount of the Museum of Modern Art, Judith A. Carr of La-Z-Boy, Robert Viol of Herman Miller, Wayne T. Stephens of Barcalounger, Greg Lucchese of Domore, Scott Underwood of IDEO, and the staffs of Steelcase and the Bata Shoe Museum.

I am also grateful to the many experts who responded to my queries, by electronic mail, by telephone, and in person: John S. Allen, Kenneth L. Ames, Susan M. Andrews, Jed Ballard, Edward B. Becker, Russell Bienenstock, G. A. Bettcher, William J. Bodziak, John T. Bonner, C. Loring Brace, Joseph L. Bruneni, Randy Cassingham, Peter Cavanagh, Perry R. Cook, Galen Cranz, Marvin Dainoff, Katherine A. Dettwyler, James Dickinson, Niels Diffrient, Clive Edwards, Jay M. Enoch, Jerry Epperson, Darroll Erickson, M.D., Peter Fletcher, Brent Gillespie, Sander Gilman, Edwin M. Good, Richard Gradkowski, Katharine C. Grier, Vincent Ilardi, Tim Ingold, Karl Kroemer, Walter Karcheski, Jr., Charles Letocha, M.D., Hayden Letchworth, Linda Lewis, John Logan, John Ma, Alan Mann, Russell Maulitz, M.D., Phillip Nutt, Josiah Ober, Scot Ober, Ron Overs, Charles Perry, Trevor Pinch, Clark Rogers, Paul Saenger, Steve Scott, Yoshiaki Shimizu, Fred Chris Smith, Heinrich von Staden, Valerie Steele, John Steigerwald, Manfred Steinfeld, Ko Tada, Neal Taslitz, Lara Tauritz, Johan Ullman, M.D., Michael Volk, Joshua Wallman, Emily Wilkinson, and Frank Wilson, M.D. My friend Charles Rippin has helped me on many points with great generosity. This can only be a partial listing; I am humbled by how many busy people responded to my requests for help. The result owes much to them; any errors or omissions are my own responsibility.

Without the encouragement and insight of my editor, Ashbel Green, and my literary agent, Peter L. Ginsberg, this book would not have come about. And on the personal side, the support of my family and the friendship of Barbara and Robert Freidin and Charles Creesy and Gretchen Oberfranc have helped sustain me.

Finally, I want to remember three exceptional authors from my time as an editor at Princeton University Press, who became colleagues and friends: Thomas McMahon, Amos Funkenstein, and Gerald Geison. All three combined superb scholarship with deep humanity, and their loss is still felt by all who knew them.

Edward Tenner
Plainsboro, New Jersey
August 2002

OUR OWN DEVICES

CHAPTER ONE

Technology, Technique, and the Body

I N A GARY LARSON cartoon, a number of dogs are tinkering with building hardware at laboratory workbenches. The caption explains that they are striving to improve canine life by mastering the Doorknob Principle. What makes this funny is partly the idea of pooch scientists standing on their hind legs, manipulating screwdrivers and even microscopes. It recalls the long-discarded notion that we humans are the only tool-using animals. It points indirectly to the unique versatility of the human hand, with its range of grips, and the relative specialization of other creatures' paws and claws.

There is something even stranger than Larson's image, though. Other animals have a surprising ability to manipulate human technology. Not all understand what people do with things, but they develop ways to work with human-made objects, and they transmit this knowledge socially. At the dawn of industrialization, the rats of early-nineteenth-century London, with no direct auditory or visual cues, had long known that water flowed through the lead pipes servicing houses. When sufficiently thirsty, they gnawed right through them. Unfortunately, by the 1830s the pipes sometimes contained gas; the holes left by the disappointed rodents added to the risk of explosions. (Rats also loved the wax covering early matches, brought them back to their dens, and ignited the phosphorus with their teeth, causing still more fires.)[1]

Bears in Yosemite National Park have learned to twist open screw-top jars of peanut butter, to break into food lockers with a combination of paw and snout, and to raid Dumpsters through supposedly protective slots. Bears cooperate to defeat other human technology; sow bears appear to send cubs into branches to dislodge carefully cached food, and young bears learn from observation how to break open automobile doors and penetrate the flimsy barrier separating the backseat from the food the owners thought they were protecting in the trunk. According to park rangers, who call the practice clouting, bears recognize specific brands and models, for example Honda and Toyota sedans, that are most vulnerable to attack, and use similar techniques on each model. When a particular model and color yield a rich cache of food, bears begin to attack similar vehicles every night. Mother bears show cubs how to pry open rear side doors by bending the door frame with their claws until it becomes a platform for reaching the backseat and trunk partition. Bears also brace themselves against neighboring cars to break the windows of vans more readily.[2]

Zoo officials "never use the words *can't* and *orangutan* in the same sentence," according to the comparative psychologist Benjamin Beck, a specialist in animal survival skills. A young orangutan in the San Diego Zoo became famous for unbolting the screening of his crib, removing the wires, and moving through the zoo nursery, unscrewing lightbulbs. According to Beck, an orangutan, unlike other great apes, understands what a tool like a screwdriver means in the human world and given the opportunity could use it to take its cage apart and escape. Other orangutans have learned to distinguish the faint sounds made by electrified barrier systems and to escape when they detect that the power is off.[3]

It should not be surprising, then, that dogs also learn the Doorknob Principle. An innocent resident of Takoma Park, Maryland, was allegedly mauled by a Prince George's County police dog in her own bed after it was sent in to look for a burglar in her basement apartment. It had gained access by turning a doorknob, which the woman has since replaced by a latch. Bizarre as these anecdotes appear, they make an important point. Those who create things, whether doorknobs or gas pipes, can only begin to imagine how they will be used. If we define *technology* as a modification of the environment, then we must recognize the complementary principle of *technique:* how that modification is used in performance. New objects change behavior, but not always as inventors and manufacturers imagine. And changes in behavior of people, as of bears and dogs, inspire new hardware, which in turn engenders more innovations.[4]

TECHNOLOGY = TECHNIQUE?

In many European languages the same term—*la technique; die Technik*—describes both things and practices. One of the most incisive and influential critics of technology, Jacques Ellul, argued powerfully that modern humanity is enmeshed in such omnipresent, interlocking technological institutions that technology and technique are inseparable. Ellul begins his most important work, *The Technological Society,* by insisting, against Lewis Mumford, that the machine is only a result of technique, not its source. Mumford established historical periods based on energy—hydraulic, coal, electrical; Ellul sought to understand the spirit behind the power sources and machinery. For him, technology was the product of the ancient Near East; in Western antiquity and medieval Christianity, it was always subordinated to other principles. Even the Renaissance and the seventeenth century pursued humanistic universality rather than technical proficiency; many supposedly practical books of this period lose the modern reader in digression and speculation. The universal application of technique began only with the eighteenth century and the French Revolution. Mechanization was only one consequence of "systematization, unification, and clarification," equally reflected in the suppression of customary law by the Napoleonic Code. History and philosophy as we know them emerged as intellectual techniques. In fact, Ellul argues, industrialization followed rather than preceded these intellectual and cultural transformations. With the rapid industrial development of the nineteenth century—which Ellul does not attempt to explain—came a new relationship of technique to society: "a reality in itself, self-sufficient, with its own laws and determinations." Political control of technique is an illusion. It is irreversible, "beyond good and evil," not merely morally neutral but "the judge of what is moral, the creator of a new morality" and thus of "a new civilization." Humanity has become "a slug in a slot machine," setting it in motion without controlling its outcome.[5]

Ellul's arguments are trenchant, brilliantly argued, and often persuasively illustrated. His focus on mental processes and habits as the foundations of technology is especially apt. And Ellul probably would agree that other organisms, including bears and dogs, become subject to the same technological regime. But there is strong evidence for the power of technique well before the revolutionary era. In fact, for better or worse, the dominion of technique is an ancient one. For Ellul, the Greeks simply despised technique and exalted pure theory. He cites Archimedes' destruction of the models that he used to construct his theories, once he had demonstrated his proofs geometrically. He also suggests that Greek ideals of harmony and moderation held back the development of technique. But there is no opposition between aesthetic ideals and the development of techniques. Experience and skill are hardly neglected in the Hippocratic writings in favor of abstract thought.[6]

Ellul also underestimated the power, even "autonomy" in his terms, of military technology before the eighteenth century. Swordsmanship and other martial arts were cultivated and transmitted by generations of medieval and Renaissance masters. But only in early modern Europe did the distinction between technology and technique become apparent. Between 1594 and 1607 Maurice of Nassau, Prince of Orange, showed how technique could transform technology. Matchlock muskets were heavy and dangerous. Soldiers had to hold their weapon in one hand and a lighted fuse in the other. While some hunting weapons already had spiral grooves in their barrels for greater accuracy, this delicate rifling needed more careful loading and maintenance than rough and dirty military field conditions could afford. Thus the military musket was inaccurate as well as awkward to handle. Maurice's genius was to see that organization and synchronized motion could make the crude musket an effective weapon. Inspired by an idea of his cousin William Louis, he assigned another cousin, Maurice's adjutant Johann, to break down the cycle of preparing, loading, aiming, and firing muskets into discrete steps. Johann of Nassau had a series of forty-three plates engraved, each illustrating one stage. It was no longer enough for drillmasters to teach the operation of the musket. Soldiers now had to be able to march forward, rank on rank, as they prepared the weapons for firing when they advanced to the front rank, then countermarch to repeat the process. Only with precision and strict discipline could they avoid serious injury to themselves or their comrades. But once the process—the technique—was mastered, the troops could lay down repeated, formidable volleys. An intense field of simultaneous fire could be effective, even with many shots going wide of the mark. Maurice and Johann's manual, *Weapon Handling* (1607), with its elegant illustrations by Jacob de Gheyn, transformed battle throughout Europe, especially through the victories of Gustavus Adolphus of Sweden two decades later. The press was a technology, and the printer's art a technique, that accelerated the diffusion of countless others.[7]

Here was technological synchronization three hundred years before Henry Ford. Some readers will insist that the musket itself, as a technology, somehow dictated the technique of drill. Certainly, battlefield experience determines which innovations spread and which are abandoned. But techniques do not create themselves, and neither operators nor their supervisors initially understand the full possibilities of devices. Ellul wrote in the shadow of a modernism that still sought a single best way to do everything, and for which (as the architectural critic David Heathcote recently remarked of the British official-planning mentality of the 1960s) "design is the search for the Platonic ideal and . . . variety is symptomatic of an unsolved problem." Ellul was right to underscore the constraints of technique, but wrong to deny its creative and improvisational side. In fact the two complement each other. Just when a technique seems to have proved itself inevitable and universal, an individual may develop and spread an alternative method. Organizations and professional groups codify best practice. Gifted individuals from time to time challenge the textbooks, often failing but sometimes revising them.[8]

TECHNIQUES OF THE BODY: MARCEL MAUSS

Ellul, like most analysts of technology, believed in a radical discontinuity between the industrial world and nonindustrial societies. An older French contemporary, the anthropologist Marcel Mauss, illuminated a side of technique that Ellul thought was too restrictive and unrelated to modern societies: the role of physical habits, which Mauss called body techniques. Ellul was correct in arguing for a concept of technique that included mental and social practices, but he ignored how important simple body techniques can be even in complex societies.[9]

Mauss introduced the idea of body technique almost casually, in an article in a French psychology journal published in 1934. He identified a set of human practices as "effective" and "traditional," and at the same time "mechanical, physical, or physico-chemical." These were the ways people learned to do things with their bodies. These patterns of motion were not haphazard; they were produced and inculcated by an entire society. They formed a framework of conduct. He coined the word "habitus" for these socially produced behaviors, which varied systematically "between societies, educations, proprieties and fashions, prestiges." Prestige was, for Mauss, essential to human body technique. Children imitate the actions of their elders, especially those with formal authority. The learning of body techniques is at once social, psychological, and biological. Mauss gave examples from the dances and rituals of the native peoples of Australia and

New Zealand. But some of his most interesting cases are taken from the early-twentieth-century history of sports, and from his own experience.[10]

Mauss remembered learning to swim first, then to dive. He also recalled being taught to dive with eyes shut, opening them only after immersion. By the 1930s, when his study of body techniques was published, children were expected to accustom themselves as early as possible to keeping their eyes open in the water, controlling the instinctive tendency to shut them. (The aversion to opening one's eyes in the water is now considered to be learned; swimming authorities encourage parents to begin instruction in infancy, when moving in water is still innately delightful.) Swimming, like other techniques, was an apprenticeship, but not a static one. Mauss's generation had learned to swim the breaststroke, with the head above water; by the 1930s, variations of the crawl had prevailed. No longer did swimmers swallow water and spit it as though they were "a kind of steamboat." Yet Mauss acknowledged that he still could not swim otherwise.[11]

Walking was subject to more subtle but still discernible techniques. In the Great War, Mauss recalled, the Western Allies moved differently. The Worcester Regiment, as a token of its valor fighting beside the French infantry in the Battle of the Aisne (he probably meant that of 1914), received special permission to be accompanied by a band of French buglers and drummers. The desired panache failed. For almost half a year, "the regiment had preserved its English march but had set it to a French rhythm. . . . When they tried to march in step, the music would be out of step, with the result that the Worcester Regiment was forced to give up its French buglers." And as recently as World War II, an American observer found British troops "looser-jointed than we are, with freer knee action," and French soldiers with "a long loping swing that seems to use the shoulders to push ahead with." Were these stylized expressions of each nation's civilian steps or the choreography of long-forgotten sergeant-majors, enshrined in drill manuals?[12]

Of course, soldiers are drilled by noncommissioned officers who build morale partly by teaching these distinctive motions. Their exercises produce what William H. McNeill, in his study of dance and drill, has called "muscular bonding," the solidarity of rhythmic synchronization. Both national armies and civilians march to different drummers. Mauss began to notice that American and French women also walked differently, and that even in French society, upbringing influenced gait noticeably. Girls raised in convents, for example, walked with fists closed, and Mauss recalled his own third-form teacher "shouting at me: 'Idiot! why do you walk around the whole time with your hands flapping wide open?'" Mauss believed that many of these differences arose from physical traits. The height of Englishmen, as well as the distinctive gait they had learned from their drillmasters, made it impossible to follow the French military band. Even the goose step had a biological explanation, the German army's way of achieving the greatest possible extension of the (long) German leg, as distinct from the short, knock-kneed French limb.[13]

Mauss was probably exaggerating the role of physical differences. Surely all European conscript armies of World War I included a significant range of heights and body types. And the goose step, officially the parade step, is only an exaggerated form of the Prussian marching style called the *Gleichschritt* developed in the early eighteenth century. Troops were taught to swing their legs with stiffened knees not only to build morale and cohesion but to produce an ultra-erect posture signifying and building discipline. The gait was also called the ramrod step, as straighter and more efficient iron ramrods replaced wooden ones in the Prussian army in 1718. Striking the rifle barrel with the open hand, while hitting the pavement with the heel, created audible reinforcement. So difficult and constraining is the goose step that it would be disastrous on maneuvers or in combat; its point is virtuosic display of self-mastery, not biomechanical efficiency. Posture and synchronization rather than stride length govern its success. The French military always eschewed this step for a style emphasizing mobility and flexibility. It must have been the difference in marching instruction rather than body proportions that frustrated the Worcester Regiment.[14]

WALKING AS A TECHNIQUE

Mauss had no idea how rich in meaning even a single body technique could be. Cultures can suppress certain techniques; for instance, in Mali, as the anthropologist Katherine Dettwyler found during her fieldwork there, some groups do not allow children to crawl, apparently because of hazards on the ground. At the other extreme, Western technology to accelerate walking may actually impede development. Up to 92 percent of families with babies have infant walkers, wheeled seats that let children move about before they can even crawl. Yet experiments have shown that infants using them sit and crawl one month after those who do not use them, begin to walk two months later, and score lower in mental tests. The walkers are thought to restrict the ability to explore and interact with the infant's environment. That is certainly consistent with the reports of many creative adults that locomotion promotes reflection. The writer Evan S. Connell once observed that great ideas come to people in transit, especially walking; Joyce Carol Oates has celebrated the stimulation of running and walking, citing Wordsworth, Coleridge, Thoreau, and Dickens.[15]

In adults, the range of walking styles has a history that would have fascinated Mauss. In ancient Greece, bold steps were associated with warriors and rulers. In Homer's epics, Ajax, Odysseus, and other heroes move "with long strides." (Females, women and goddesses alike, take dainty steps.) For the Athenians of the fifth century B.C., the waning of aristocratic power brought a new and more moderate male ideal, neither swaggering nor timid but graceful and dignified. The freeborn man of leisure moved calmly. In the *Nicomachean Ethics,* Aristotle observes that the *megalopsychos,* the great-souled man, moves unhurriedly. Deliberate walking was the mark of leisured men also for the Romans. Strides were broad and both feet probably remained on the ground for longer intervals than they do now. In Roman comedy, slaves ran about, possibly in a scrambling, Keystone Kops fashion. The American orthopedist Steele F. Stewart recalled that the norm for himself and other presumably middle- and upper-class children about a hundred years ago was a "mild Chaplinesque shuffling gait." It might have been a reaction to the clomping, clodhopping stride of assertive lower-class youth, but we can't know for sure; the body techniques of the poor are even harder to reconstruct than those of the affluent.[16]

Westerners and non-Westerners have long noticed each other's distinctive gaits, not always admiringly. The Bengali Brahmin polymath Nirad Chaudhuri recalled in *The Continent of Circe* that his Indian contemporaries seemed to move "like swaying trees"; in turn, young and old compatriots mocked Chaudhuri's walking style, which he learned from the British, shouting "Left-Right!" and "Johnnie Walker!" as he came down the street. (Indeed, the immense success of the international whiskey brand may have owed much to the bold tread of its booted, ruddy-cheeked Regency figurehead and his invocation of masculine ambition and self-assurance.) Until the middle of the nineteenth century, according to the anthropologist Masaichi Nomura, Japanese children learned to walk with the "namba" gait. It can be seen in old woodblock prints. Arm swing was limited, and each arm moved forward with the corresponding leg instead of with the opposite leg, as is usual in both Japan and the West today. To assure a more level platform for archery, some horses were even trained to move namba-style, with right and left legs synchronized. Japanese ethnologists today find a wider variety of gaits than in the West, but note that the Japanese still use less arm motion and that fewer Japanese than Americans or Europeans strike the ground with their heels as opposed to their toes or entire foot surfaces. Another cultural anthropologist, Tim Ingold, has observed that before their country was opened to the West, Japanese children learned to "walk from the knees," moving the hips as little as possible, whereas European and North American children are trained to "walk from the hips," maintaining their culture's prized erect posture by keeping the legs straight. For Ingold, the skill of walking is not just the addition of culture on top of a genetic capacity; it is the result of a developmental program that includes an entire society's technologies and techniques. For example, traditional Japanese walking gave a good footing on the often uneven local landscape; it was linked with the Japanese method of transporting heavy things by tying them to shoulder-borne poles. Yet the Japanese also have adopted some Western walking techniques with delight. An American named Frances Caldwell Macauley is said to have introduced skipping to Japan in the late nineteenth century, when she demonstrated it to her class of prospective kindergarten teachers in Hiroshima. It spread so fast that she soon saw an aged couple skipping down the street. In Chapter Three we will see that even in our own times, Japan's distinctive footwear has continued to influence gait.[17]

Some traditional walking techniques, long considered merely exotic, are biomechanically stunning. Women of the Kikuyu and Luo tribes of Africa, according to an international team of scientists who studied their performance, can carry up to a fifth of their body weight balanced on their heads at no additional metabolic cost—in energetic terms, for nothing. With heavier loads, up to 70 percent of body weight, their method is still more efficient than a military backpack. In our normal gait, the body acts as an inverted pendulum, regularly converting potential energy to kinetic energy and back, with the foot as the pivot and the body itself as a moving bob. Somehow the women are able to move so that a moderate weight makes the transfer of energy more efficient, canceling out the extra energetic cost of moving the weight. Sherpas have a similar load-bearing style especially suited for heavier weights. It was not until the year 2000 that a new computer program showed how potential and kinetic energy were exchanged in the African women's walk. Still unknown is how certain Africans, Nepalese, and other peoples around the world acquire this skill, possibly in childhood.[18]

Mauss's insights into body technique can be extended well beyond gait and gesture to technological and medical procedures. When Mauss first published his study, he noted that French telephone linemen were still using only crampons to climb poles. Only a year or two later did they begin wearing belts that looped around the poles, a method Mauss believed was in universal use among "so-called primitives." The Greeks and Romans treated burns effectively with cold water, yet for thousands of years doctors were taught to use oils instead (or sometimes simply to apply bandages), especially after the introduction of petroleum jelly in the late nineteenth century, a technology that actually set back burn treatment. Only in 1953 did a Los Angeles surgeon, Alexander G. Schulman, impulsively plunge his own grease-burned arm into a tub of cold water and discover that after an hour he could remove it with little pain. His further research and publications brought back a technique abandoned for over a millennium.[19]

Other doctors of our own time have discovered apparently original techniques that needed no medical theory of technology. Consider the Heimlich maneuver. Because choking fatalities have been documented at least to the time of Emperor Claudius, and because airway obstruction can be fatal in five or six minutes, it is remarkable how long it took physicians to develop a standard emergency technique. Well into the 1980s, for example, the American Red Cross still recommended slapping the victim's back—probably a less effective treatment than the medieval method of rolling him or her over a barrel. It was a single Cincinnati physician, Dr. Henry Heimlich, who in 1974 proposed that the best way to dislodge food from the airway is to compress the diaphragm by standing behind the victim, placing a fist between the navel and the rib cage, and jerking it upward with both arms. The back slap, he maintained, could easily lodge food more deeply in the throat. While the Red Cross and the American Medical Association continued to recommend back slaps as a first resort (at least until the then Surgeon-General C. Everett Koop took Dr. Heimlich's side in 1985), the procedure was estimated to have saved more than 9,000 lives in its first decade and 20,000 by 1990. By the mid-1990s, it was credited with 4,000 to 5,000 rescues annually.[20]

The Heimlich maneuver is pure technique. It needs no medication or apparatus; the inventor decided against using special hardware because precious minutes would be lost in deploying it. (After World War II, Heimlich had developed and patented a now-standard chest drainage valve, for treatment of collapsed lungs, from a 25-cent rubber toy that sounded Bronx cheers.) Yet the method is simple enough that many people who have used it successfully probably have had no more training than seeing a demonstration on television or on a restaurant poster. Much of its power comes from the fact that in a public setting, at least one or two onlookers are likely to be familiar with it. Victims can apply it to themselves, and the method can be modified for very large people, pregnant women, and unconscious people. A four-year-old boy used it to rescue his two-year-old brother.[21]

Mauss's ethnological range as an anthropologist seemed a distraction to Ellul, but it had an important basis that Ellul overlooked. The industrial and postindustrial worlds have no monopoly on technique. A novice computer user can locate and order a set of stainless steel knives on the World Wide Web in ten minutes, but it takes a five-year apprenticeship to become an accomplished maker of stone axes and arrowheads. Australian Aboriginal throwing sticks repay aerodynamic study. Western troops using powerful microprocessor gear are also trained in martial arts techniques developed with no electronic assistance. In fact, the machine sometimes can be the enemy of sophistication in the execution of technique: the historian of technology Arnold Pacey has observed that in the nineteenth century, traditional Indian hand looms looked crude to Western observers but (in the hands of local artisans) produced far better cloth than the output of the massive, beautifully machined British power looms. Likewise, it was difficult to equal the better grades of handmade Japanese paper.[22]

AUTOMATION AND THE REVENGE OF TECHNIQUE

Is skill necessary? Should it be? We are constantly reminded first of how everything has become so much easier to use, and second of how life has grown more complex and requires more education.

The cognitive psychologist and computer scientist Donald Norman has observed that many of today's familiar things required surprising degrees of skill in their early years. One early manual cautioned: "The use of the Phonograph must be learned. . . . [One needs] a few days to learn everything about it and only a few weeks' practice to acquire all the dexterity in its use." Users of early tube radios had to master separate processes of intercepting, tuning, detecting, amplifying, and reproducing. Setting up the wiring and tube(s) alone could take a dozen or more steps. Early automobiles required potentially dangerous hand cranking and long-vanished adjustments like the choke. Photographers before George Eastman needed to learn to use some of the most toxic chemicals known.[23]

Other technologies follow a path from minimal to complex technique and back again. Late-nineteenth-century telegraphers could recognize each other by individual styles (the phrase "smooth operator" may have originated in their milieu), and the beautiful keys they used can still be admired in museums. But the telephone and automatic keyboarding equipment ultimately destroyed their culture, as we will see in Chapter Eight. The first telephones, in turn, needed cranking like the first automobiles, but had an early form of voice recognition—a human operator. The dial handsets that succeeded them had to be rotated smoothly and released when the finger reached the stop. In fact, AT&T delayed the dial's introduction to the Bell System until 1937, claiming it was too hard to use. (And subscribers had to learn to release their fingers at the stop and not try to force the dial back to speed up the process.) Now buttons with tones are almost universal; telephones, like toasters, hardly need instructions.[24]

Straight razors are still made, and enthusiasts may still compare them favorably with the latest triple-cutter disposable blades, but few barbers use them. Consumers who complain about programming a videocassette recorder probably have never tried to thread a reel-to-reel tape recorder, let alone a film projector, and cable television even eliminates the skill once needed to orient antennas. Our objects seem to be modularized in cassettes and sealed, swappable assemblies. Indeed, this was the appeal of the original Kodak Brownie camera in 1900. The slogan "You push a button, we do the rest" made Eastman's fortune.

But is technique really abolished, or only relocated? Think again of the automobile. Fewer American drivers are buying manual transmissions, not only because new automatic designs are smoother and more economical than their predecessors, but because motorists have embraced a new and potentially hazardous technique: traveling while cradling and speaking into a portable cellular telephone. An air bag seems to be the ultimate passive device, but even it has implications for how we use our bodies. The American Automobile Association now recommends keeping hands on the steering wheel at the 9 and 3 o'clock positions, or even as low as 7 and 5 o'clock, instead of at the former driver-education-course standard of 10 and 2 o'clock, to reduce the risk that the air bag will burn the driver's hands or propel them up at the face. Advocates also say the new grip allows a fuller rotation of the wheel in emergencies without removing the hands. And some experts now believe that traditional hand-over-hand technique, while efficient in tight cornering, not only exposes arms to air bag explosions but promotes dangerous oversteering under stress. Children in the backseat are to be directed to sit up straight and not lean to the side if their cars are equipped with the new side-impact air bags. And skill is required to install rear children's seats safely; the U.S. National Highway Traffic Safety Administration (NHTSA) estimates that in 90 percent of cars, either the seat or the child is improperly secured. New safety devices thus may require us to spend hours changing the very techniques that were once taught for safety, and acquiring entirely new ones.[25]

Few drivers receive formal training in emergency handling of their cars, especially in snow and ice, and places for safe practice of these maneuvers are rare. Yet antilock brake systems (ABS), traction control, and other innovations may encourage drivers to venture out in conditions where these special skills are necessary, even with the new technology. And there is an element of technique even in the use of automatic devices: because ABS is unlikely to be universal in the near future, every driver has to be aware of how to handle a skidding car: the pumping action formerly recommended for the traditional brake, or the steady pressure needed by ABS. (To make matters still more confusing, it is not clear whether pumping is a more effective way to halt a skid or merely a relic of the days before disk brakes, when fade was a serious problem.)

Making inherently dangerous things easier to use creates other problems. The 9-millimeter pistol shows that simpler construction may demand more skill. In the 1980s, many American police departments began to replace their service revolvers with these semiautomatic weapons, made by Beretta, Glock, Ruger, and other companies, for more rapid firing against drug dealers' automatics. Originally temperamental, the semiautomatics now are more reliable and easier to use and maintain than revolvers. But it is precisely this simplicity that makes technique more complex. When the city of Washington, D.C., adopted the Glock 17 in 1988, officials especially liked the absence of an external safety lock. The Glock 17 has three mechanisms to prevent accidental discharge, but squeezing the trigger releases each of them. One of the internal safeties is intended to add an additional pull on the trigger before firing, but in the streets many officers may activate it early to be ready to fire instantly.[26]

The military works hard to prevent premature engagement by achieving "fire discipline." In Washington, there were over 120 accidental discharges of the new pistols in the first ten years, leading to settlements of over $1.4 million in just three months of 1998 alone. The Glock, according to its many admirers in police circles, is inherently safe. One simply needs special training to know exactly when to begin fire when the finger is kept on the trigger. (Because it will fire a round from the chamber even after the magazine has been removed, the Glock 17 also demands more careful maintenance procedures.) The simplicity and ease of firing, then, demand more rather than less training in technique.

Other sophisticated, seemingly transparent technologies also call for unexpected manual skills. The magnetic-striped MetroCards used by the New York Metropolitan Transportation Authority must be passed through a slot at subway turnstiles in a firm, fluid, even manner not easy for all riders to master. (Bus fareboxes swallow the card and return it, provided it is inserted in the single correct direction out of four.) A failed "swipe," as the motion is called, must be repeated at the same turnstile; otherwise the system temporarily invalidates it. Sometimes several passes are required before the card is read.

In at least one of the advanced research laboratories I have visited, technicians were on call to untangle the serpents' nests of cords and wires that accumulate behind and beneath equipment. Their specialty has a name, cable management, and pays salaries high enough that a representative in one laboratory apologized for the mess with the explanation that untangling would have strained the unit's budget.

At the level of individual users, the mouse was heralded as an unproblematic intuitive device but often demands either reprogramming (for acceleration or double-click speed, for instance) or adjustment of the user's technique. It clogs readily with dust and only the most thorough cleaning restores its responsiveness. And some ergonomists believe that, used improperly, it can be a greater health hazard than the keyboard.

TECHNIQUE WITHOUT TECHNOLOGY

Technique has a history even without changes in equipment. It is too bad that Mauss did not pursue his analysis of swimming history, because the sport showed the importance of culture for performance. It was a kind of aquatic marching, at first taught mainly by military instructors. The breaststroke that Mauss and his contemporaries had learned was favored by European and North American swimming teachers in the nineteenth century despite its obvious disadvantages: as the name implies, the full breadth of the chest pushes through the water, and the arms and legs also increase resistance as they return to the beginning of a stroke. Yet it was an intuitive and natural motion, and it helped keep weapons dry. Europeans as early as the seventeenth century studied the frog kick to prepare themselves for the water, and live frogs were even kept at London's Serpentine for swimming instruction in the early nineteenth century. England was the center of Western swimming, and English athletes had refined the breaststroke so thoroughly that other styles could not compete for decades.[27]

The more efficient crawl did not originate with the masters of the breaststroke, who had no incentive to abandon the style they had perfected. Two Native American swimmers, Flying Gull and Tobacco, had demonstrated their overarm style in London in 1844, and Flying Gull was able to cover a 130-foot pool in only thirty seconds, but their apparently thrashing style was condemned as "un-European." In fact, many peoples of the Americas, Africa, and Asia had been swimming hand-over-hand. European techniques, the dog paddle and the froglike breaststroke, were the exceptions. But the American Indian style, effective as it was, needed a European smoothness if it was to compete. It was the most distant European outsiders of the nineteenth century, Australians, who supplied the missing bits of technique. Observing the stroke of South Sea Islanders, they made the kick more pleasing to Europeans. They needed no devices for this; they were pure innovators of technique, as were the Swedes who discovered that the springboard—an early-nineteenth-century innovation of the German founder of modern gymnastics, Friedrich Ludwig Jahn—could be not just an entry platform to the pool but the basis of an independent athletic performance. Throughout the twentieth century, gifted swimmers and coaches have been developing the crawl and other strokes. Often it is swimmers like the backstroke specialist Allen Stack and the butterfly expert Mary T. Meagher whose intuitively shaped motions transform practice. Today, athletes and coaches are also using imaging technology to study the techniques no longer of frogs, but of marine mammals, birds, and flying insects.[28]

SPORTS AND THE FRONTIERS OF TECHNIQUE

The transition from military to recreational swimming instruction illustrates another displacement of physical technique: from work to leisure. Learning any sport is a conscious apprenticeship, or trial-and-error self-tutorial, in the controlled use of the body. Even if technique really is becoming less important on the job, more of it than ever is apparent at play.

Playful does not mean inconsequential. To the contrary, part of growing up is learning graceful motion. Throwing a baseball is a surprisingly complex learned behavior. Consider the phrase "throwing like a girl," directed at Hillary Rodham Clinton after she ceremonially opened a Cubs baseball game at Wrigley Field in Chicago in April 1994. The writer James Fallows dissected the First Lady's offending stance and found three elements that distinguished it from proper baseball style: she faced the target rather than positioning her trunk perpendicular to it and then rotating back to amplify the pitch; she kept her elbow below her shoulder while throwing; and she held the wrist closer to her head than her elbow was ("inside the elbow"). A leading coach, Vic Braden, described the proper motion as a "kinetic chain" in which momentum is built up first from the lower body, continued through the waist to the shoulders, through the upper arm, forearm, and wrist, like the snapping of a whip. Tom Seaver said he pitched with his legs.[29]

The pitcher's fluid motion may look graceful, but it is hardly natural. Right-handed athletes instinctively throw incorrectly when they try it left-handed. In the earliest days of baseball, it was even illegal; the 1845 Knickerbocker Base Ball Club rules prohibited throwing and specified stiff, underhanded pitching. Today's pitching at over ninety miles per hour leads to elbow strains and sprains and (notoriously) tears the muscles of the shoulder's rotator cuff; one of the most effective variations, the slider, is an ergonomic minefield for the pitcher.[30]

Throwing, like other techniques, has evolved through the experimentation of elite athletes and their coaches. The equipment of cricket changed little in the nineteenth century, but because the ball takes a bounce before the batsman swings at it, improvements in the condition of the pitch (grass surface) tended to help batsmen. Balls bounced true and thus became easier to hit. Bowlers responded with changes in their manner of releasing the ball, reintroducing variety and surprise.[31]

In American baseball, a few pitchers in the 1860s and 1870s learned how to throw and control a curveball. It arose not accidentally or incrementally from other pitching styles, but from the insight that a different technique could produce a startling result—an insight followed by many hours of experimentation, self-training, and practice, until the new technique was ready for competitive play. Even in our own time, pure changes of technique continue to appear. Without any significant change in equipment except for safer landing pits, high jumpers experimented with a half-dozen styles: scissors, eastern cutoff, western roll, and belly roll. In the 1968 Olympics, the American Dick Fosbury revealed yet another maneuver, a twist from a frontal position, clearing the bar headfirst and backward. Also in the late 1960s, Jean-Claude Killy challenged the wisdom of coaches who taught skiers to lean forward for maximum speed; he became world champion by leaning back. Athletes and coaches soon adopted Fosbury's and Killy's techniques.[32]

It is also possible for techniques to disappear if a new technology leaves them with no competitive advantage. Today the large-wheeled bicycle, called the Ordinary or Pennyfarthing, appears literally the height of impracticality. It is hard to mount and control, even harder to stop, and tends to pitch the rider forward, headfirst, when it hits an obstacle. Without a freewheel, the rider can never rest, except with legs draped over the handlebars in a downhill maneuver called coasting. There were schools for cyclists, and even examinations in some European cities. Yet according to the sociologist Wiebe Bijker, the Ordinary's perils were positive for its enthusiasts. These athletic, rich urban young men were not looking for cheap transportation; the cycles were luxury craft products anyway. Their goal was showing off their courage and skill to young ladies, who do not seem to have even tried to ride them, and to their peers. Many other potential users, of course, saw the design as merely dangerous, not daring, but none of the early modifications could compete with the Ordinary in speed, and some offered doubtful advantages in safety. By the 1880s, inflatable tires, originally developed to counter vibration, made it possible for new low-slung designs to outrace the Ordinaries, helping end their sporting monopoly. Chain drives on the safety cycles, and their great aerodynamic advantage over the Ordinaries, finally stabilized (to use Bijker's expression) the bicycle as we know it. Because the skills of riding the Ordinary no longer bought superior speed, they disappeared.[33]

PULLING TOGETHER

The interaction of inventors (who may or may not be athletes) with participants (who may or may not have technical skills) allows technology and technique to produce striking results envisioned by neither designers nor athletes. Rowing, fencing, speed skating, bowling, and bicycling show five outcomes of the interaction of technology and technique. In rowing, an innovation forgotten since antiquity was independently revived in the nineteenth century, but it took outsiders nearly a hundred years to refine a style that exploited it fully. In fencing, hardware innovation at first upset traditional technique, then rapidly reached a new equilibrium with it as both equipment and behavior changed together. In speed skating, a design

from the 1890s was ignored for a century, then swept the field when athletes finally adjusted their technique. In recreational cycling, an alternative
design helped change the nature of the sport itself, appealing to new riders
with a different attitude. And in bowling, new equipment has altered the
definition of winning technique.

Technology and technique were linked in the early history of rowing,
when it was a vital military skill rather than a sport. The sliding seat was
first used by rowers of the ancient Greek triremes. Their leather cushions
and low-level seating, lubricated by fats, promoted more efficient motion
with a sliding stroke that exploited leg power. Most rowers were free citizens, and the sliding stroke, then as now, required practice, so twelve
thousand men were paid to train for eight months annually. The cushions,
so familiar to Athenians that Aristophanes made comic allusions to them
in his plays, disappeared with this style of rowing around 400 B.C. as a new
design, the Carthaginian quadrireme, arose. Where the trireme had used
but one man per oar, the quadrireme deployed the massive power of multiple rowers on a single oar. Warships became troop transports for grap-
pling and boarding rather than the skilled precision ramming that was the
trireme's specialty. The exercises to maintain slide stroke skills were no
longer cost effective. Whether or not the Romans could have used the sliding stroke in their own galleys, it was lost to them and their medieval and
early modern European successors. But the quest for technique was not
over. Europeans had a great range of equipment and corresponding rowing styles. Venetian gondoliers use an oscillating stroke of a single oar as a
fish swings its tail, while the oarsmen of medieval galleys stood upright to
move their massive oars back and forth in a "walking" stroke.[34]

From the beginnings of modern competitive sport in early-nineteenth-
century England, athletes and spectators have recognized that equipment
and style go together. And debates about their relationship could persist
for decades. In rowing, some design changes were relatively uncontrover-
sial, notably the use of narrow-beam boats with internal keels and outrig-
gers, which helped transform the old heavy craft into lightweight shells
that demanded far more finesse and balance. The sliding seat, despite its
antiquity, had a mixed reception. Professional rowers in the north of England were first to revive it. They realized that moving forward several
inches at the beginning of a stroke and returning at its end permitted a
longer and more powerful stroke, assisted by the legs, multiplying the
oarsman's efficiency. With substantial purses awaiting the victors, grease

on trouser seats was a small price to pay for an edge. A slide of just nine inches could add sixteen inches to a stroke. While rowing clubs began to follow this principle in the 1870s—using seat mountings first of bone sliding on brass and ultimately of wheels in vulcanite grooves—some conservatives resisted. Even if the innovation made rowing more efficient and speedier, they objected, it was a labor-saving idea of professional scullers, and put a premium on fast entry and leg power rather than a hard "catch" powered from the shoulders and upper body. Motion from the hips and lumbar region was no longer paramount. Letters objecting to the sliding seat and its associated technique appeared in the London *Times* as late as 1933. Because of this aesthetic model of proper use of the rower's body, coaches and athletes revised their style only slowly. At the most important English regattas, Henley and Oxford-Cambridge, speed records actually declined after its introduction. Even after the sliding seat's acceptance, the fixed seat remained a norm for training in England and the basis of a style called English Orthodox that persisted well after World War II. An observer in the 1880s reported that Oxford crews were not synchronizing slide and swing but delaying the use of the slide until their bodies were upright, as though they were still using fixed seats.[35]

As in swimming, it was an Australian who began a revolution of technique, and at about the same time. (The American Orthodox style was only a modification of English Orthodox.) In the early twentieth century, Steve Fairbairn (1862–1938), the coach of Jesus College, Cambridge, introduced a revolutionary call to use the body's weight, and especially the legs rather than the shoulders, to move the boat. His crews' successes were the first major challenges to Orthodoxy. The next great innovations, building on Fairbairn's ideas, came in the 1960s. A club in the north German town of Ratzeburg at last developed an alternative method, now known as International Modern. By accelerating the slide in its approach to the front stop, the coach Karl Adam was able to assure a steadier hull speed, help the West German crew win the gold medal at the 1960 Rome Olympics, and guarantee the technique's international influence. Adam also made significant changes in conditioning and in equipment, including a longer oar with a broader blade, and he was a ferocious morale builder, but his great revolution was in rowing style. Through their mastery of body motion, Fairbairn and Adam reinvented the sliding seat.[36]

GETTING THE POINT

In fencing as in rowing, technology and technique have evolved together. Users of weapons have always sought more effective maneuvers, and their tactics have in turn inspired modification of their instruments. The fencing historian Nick Evangelista has described medieval swords as " 'can openers,' hacking and whacking devices, whose sole purpose was to find a way through armor," but professionals at the Royal Armories in Leeds have been reconstructing early combat techniques with replica weapons and now believe that even these apparently ungainly implements required considerable skill as well as strength. Beginning in the mid-sixteenth century, a lighter sword called the rapier coevolved with new techniques. It started as a heavy, sharp-edged blade with a complex and bulky hand guard, used offensively and backed up by a piece of heavy clothing or a smaller weapon in the left hand. Used by soldiers, duelists, masters, and armorers, the rapier became a different weapon as the possibilities of various techniques were explored. Cutting was found to be inefficient; the rapier became a thrusting weapon without a sharpened edge. It grew lighter and simpler, and the secondary defensive weapon disappeared. Not that this was a smooth evolution. For a time, blades lengthened, reaching up to six feet before legislation restricted them to three. Ultimately the weapon became the eighteenth-century smallsword. Each change in form was linked with a new style of practice, which had to be learned from a master. Fencing masters rose in society as they developed a body of techniques (with an extensive specialized vocabulary) for teaching as well as dueling and, like Maurice of Nassau, published diagrams of their maneuvers in illustrated volumes.[37]

The foil, today's basic fencing instrument, is a notable example of a technology developed for the sake of technique. It was a modified version of a dueling sword to be used in instructional exercises according to a system of rules and is still called a "conventional" weapon. (The épée, with its stiff, triangular blade, was introduced in the nineteenth century as an "unconventional" weapon that could score with a hit anywhere to the opponent's body.) While an unsuccessful electric scoring system was introduced as early as 1896 and the épée was adapted for electric scoring in 1933, it was the electrification of the foil in the 1960s that had the most radical, if mainly temporary, effect on technique.[38]

In the interest of objective judging, fencing officials adopted electric scoring for the foil. To accommodate the contact at the tip of the blade, the weapon itself was modified; the wire was run through a more rigid, weightier blade. Equipment at first could shock the perspiring athletes wearing it, and specially trained technicians now check the wiring. But these were the least of the new challenges to good technique. More important, certain light touches that judges might not have noted now could set off the scoring buzzer if the foil was handled with a new rapid motion.

Just at this time, the fencers and coaches of the Soviet Union and Eastern Europe were emerging as an international force. Backed by generous state support, rejecting traditional techniques in favor of speed and mobility, they perfected new moves that took advantage of the tip's sensitivity. Others found the new equipment harder to master. Marvin Nelson, a veteran fencer, official, and coach, deplored the "sloppy, pig-sticking" style that flourished as officials allowed fencers to maintain attacks when procedural rules ("right of way") should have given the touch to the other side. The flashing light overrode their knowledge of the rules. Meanwhile, the flexible weight-tipped blades whipped away from targets. Nelson recalled that he, like others, "was forced to fence with absence of blade and reduce my game to much more simple actions."[39]

These problems passed. Suppliers introduced better points and stiffer blades. And technique changed as well. Competitive fencers practiced with foils simulating the feel of the electric models. In fact, by the mid-1970s Nelson noticed a more vigorous game: "Fencers are 'carrying' the weapon more effectively—showing an improvement in awareness of the different parts of the blade. Thrusts or actions are being made from positions not usual in standard foil or in the first period of electric foil. . . . Foot movements are increasingly efficient." Still, these more sophisticated techniques have brought other changes in technology, notably grips with small projections for more secure finger holds. Introduced as "orthopedic" grips for fencers with missing fingers or other disabilities, these are now widely used as "pistol" grips. They promote a more vigorous style at the expense of lightness and flexibility and have been labeled by one authority as "this monstrous brood." And electric scoring demands self-discipline, as some athletes focus their attention so much on the gratification of a light or buzzer that their minds wander from the bout itself, letting their opponent score the touch.[40]

NEW MUSCLES FOR OLD

A second group of sports changed significantly in the last century thanks to new materials and manufacturing processes. In these cases, too, the crucial change was not so much the new equipment as the development of body motions to optimize it. Most skating remains relatively conservative in equipment despite countless refinements of skates and great improvements in ice conditioning equipment in recent years. Speed skating, too, changed little until very recently. In 1488, Leonardo da Vinci studied it but lost interest after failing to create a new design. Four hundred years later, in 1890, a Canadian and a German independently developed and patented a skate hinged at the toe and spring loaded, so that the boot could separate temporarily from the blade. Neither model appears to have gone into commercial production.[41]

Gerrit Jan van Ingen Schenau, a professor of biomechanics at the Free University of Amsterdam, originally wanted to design a safer skate, not necessarily a faster one. He told a journalist from a Japanese news service that from the early 1980s, many skaters had complained to him of pain in their shins. "I then realized that skating is a uniquely unnatural movement—different from walking or running—because the heel does not rise freely in motion." The lever mechanism let the skater lift one leg at a time, the one not doing the pushing, while keeping the runner on the ice, in principle reducing strain on the calf muscles. At some point his team noticed that speed skaters kept their ankles locked and pushed off their heels—unlike jumpers, who extended their ankles and pushed off their toes. The hinged skate in principle allowed the ankle to move with the blade still fully on the ice, using the calf muscles for a longer stride.[42]

In execution, the idea was not so easy. Early hinged skates had mechanical problems. A post on the upper part now fits into a cylinder near the heel of the blade for stability. Even so the blades did not, and still do not, make learning easier for beginners. Expert skaters, who had refined their techniques on conventional fixed blades, saw no advantage that would offset the physical cost of retraining themselves to use new muscle groups. And the clacking sound of the blade snapping back to the shoe could be disconcerting. The best of the experienced Dutch skaters even considered the new equipment dangerous. The inventors' breakthrough came with the youngest skaters who had invested less in conventional technique and had most to gain from learning a new style. They found that the skates—with the new technique—improved their personal records. Then the women on the national team tried them and changed over. When the men saw improving results, they followed. German, Japanese, and American skaters resented the new equipment but by 1996–97 the main objection was not to the new design but to that fact that the patent licensee, the Dutch manufacturer Viking, was allocating its limited production to the Netherlands team and other regular customers.[43]

The records of 1997 confirmed the skates' value. They helped athletes cut a full second, a long time indeed in international winter sports competition, from each 400-meter lap of longer events. In November and December 1997 alone, adopters of the clap skates equaled or exceeded sixteen world records. In the 1998 Nagano Winter Olympics, the Dutch skater Gianni Romme won the 5,000-meter race clocked at 6 minutes 22.20 seconds; his previous Olympic record was 6:34.96.[44]

Even, or especially, with results like these, skaters' views of the new technique are divided. Already before Nagano, one American official compared using clap skates to doping and corking bats, and a U.S. team member proposed changing to a mountain bicycle with studded tires. But afterward, Nick Thometz, the program director of U.S. Speedskating, acknowledged that his team would need to learn the new style. According to Schenau, the new style of coordination "requires its own type of perfection." To defenders, clap skates mainly permit a more efficient but still demanding technique; to critics they substitute strength for skill. Interestingly, the speed skaters who depend most on bursts of power, male sprinters, are the only ones loyal to conventional equipment, believing the new skates impede explosive starts. As computer-assisted design (CAD) makes it possible to retool the mechanical components more and more easily, they too may switch. Equipment and skating style will probably continue to coevolve for better or worse. Meanwhile, the clap skate illustrates a paradox: what athletes at first rejected as too difficult now is criticized as too efficient.[45]

GOOD FORM AND BETTER MATERIALS

In bowling, the great innovations have been chemical rather than mechanical; dimensions of alleys, pins, and balls have changed little. But the techniques of the game have been transformed. While a strike is every bowler's goal, games among proficient bowlers once were won by converting spares, much as golf matches have been decided by putting rather than driving. More than a thousand different leaves, or combinations of pins, may appear after the first ball hits, and serious bowlers once spent years learning the proper approaches to configurations with names like the bucket and the washout.[46]

For nearly three quarters of a century, bowling's technology and body skills were stable, founded on the control of a hard rubber ball rolling on a surface of any of several kinds of wood finished with lacquer or shellac, and striking solid maple pins. By the 1970s, equipment began to change. To compensate for their new protective plastic coatings and to save on the cost of wood, pins now had hollow zones that some said made them livelier despite American Bowling Congress (ABC) test results to the contrary. New finishing materials gave the lanes a smoother surface, and some bowlers began to soften their balls in dangerously flammable solvents to increase the balls' gripping power. This discovery led to a series of balls that could be thrown to hook more strongly than ever as they reached the pins. The ABC set hardness standards, but new cover stock materials, such as urethane in the 1980s and reactive urethane in the 1990s, have continued to make balls more powerful. Meanwhile, with the help of CAD software, engineers have designed complex internal systems of weights for balls that let them hook as sharply as an older style of outlawed ball doctored with metallic salts, the "dodo ball." (Polymers and ceramics were unimagined as core materials when antidodo rules were written.) Because reactive urethane balls can retain more of their energy as they slide through their early trajectory and roll down the lane, they use their superior grip to generate a powerful, sharp hook in the pocket between the 1 and 3 pins. Perfect 300 games, once uncommon even for the leading pros and rare achievements for other strong bowlers, have increased about a hundredfold from the days of the hard rubber ball. In 1997 a college sophomore rolled the first perfect 900-score three-game series ever sanctioned by the ABA. But many of the pros who had developed the strongest hooks with conventional balls were dismayed to find that reactive urethane balls were letting lower-scoring colleagues catch up. In fact, "crankers" had to unlearn their formidable deliveries, and some leading pros had to retire from the tour.[47]

As with the clap skate, a new generation of competitors forced the established champions to adapt their technique to the new equipment. Not all succeeded. But to many veteran instructors and pros, the loss has been collective as well as individual. In their view the urethane ball threatens the game's integrity by upsetting the historic balance between strikes and spares. "With today's balls," says one, "you don't have to make a good shot to knock down 10 pins; you pay $160 [for a bowling ball], hit the pocket, and you strike." And another observes that "[l]ots of dedicated people practice. But most of them practice just to strike," instead of studying how to achieve spares as well.[48]

While the cultivation of some techniques has declined in the United States, other techniques have flourished as Asia has embraced the game. In the 1980s, players in Taiwan began to develop a radical new style of delivery. Using balls typically weighing eleven pounds rather than the sixteen pounds customary for Western men, they gripped them from above and used their thumbs to impart a rapid backward horizontal rotation to the ball, strong enough to continue down the lane like a top, scattering the pins with explosive force. (In the epilogue we will consider the thumb as the Cinderella of the hand.) With proper spin, hitting the 1–3 pocket precisely becomes unimportant. A ball aimed at the head pin can miss by as many as seven boards on either side, giving the spinner or helicopter ball a fourteen-board strike zone rather than the Western three-board zone. Some Taiwanese bowlers wager on their ability to roll "perfect" strikes, not just toppling all the pins but removing them from the deck. This technique of the lighter Asian players not only has surprised Western pros but has taken world titles, first when You-Tien Chu of Taiwan won the AMF World Cup in Mexico City in 1983 and most recently in November 1998, when Cheng-Ming Yang, also of Taiwan, won the AMF Bowling World Cup competition in Japan. Unlike other bowling techniques, the Taiwanese method is unaffected by the many variables of lane conditioning. In fact, as one of the Taiwan coaches put it, the style was "developed in self-defense" against lanes "swimming in oil." It is also considered easy to teach.[49]

The helicopter shot does not depend on the sophisticated cores and advanced cover stocks developed by ball manufacturers, although some new balls must be especially well suited to it. As a technique, it could have been introduced a hundred years ago. Obscure bowlers may have experimented with it, but it does not appear in Western bowling histories. Lighter balls have always been legal and available, yet even smaller male bowlers, many of whom could probably have improved their game by using them, shunned anything less than the sixteen-pound maximum. Asian athleticism, with its tradition of lightness and maneuverability, was a more promising cultural tradition.[50]

TECHNIQUE AND INNOVATION

Skaters, rowers, fencers, and bowlers form networks of athletes, coaches, manufacturers, and engineers. Technique changes as knowledge is shared within this diverse community. Innovation in technique, like other invention, demands the power to see beyond daunting early problems and to visualize ultimate benefits. Successful experimentation can be painful. Ying-Chien Ma, a pioneer of the helicopter shot, suffered serious pain and needed seven wrist-related operations. His injuries illustrated the autonomy and power of technique in Ellul's sense, its frightening ability to take over life and endanger health. But Ma's success signifies the creative and autonomous side of technique, the power of athletes to overcome pain in upsetting the orthodoxies of their sport. Neither negative nor positive unintended consequences are the full story.[51]

Technique is crucial for the evolution of technology. As the economist Paul Romer has observed, a change in athletic technique often paradoxically requires a temporary decrease in speed as the fine points of the new style are worked out. Many inventions in their early stages underperform the best conventional equipment; early medieval cannon were distinctly inferior to the well-developed stone-hurling trebuchet. It takes many hours of practice to determine the potential of any device or technique. Romer's point is that landmark inventions are only the beginning of a process of refinement in which many people invest time. We can take his idea further to say that technologies and techniques coevolve. Inventors cannot foresee the uses and abuses of a major invention. Especially in the world of electronics, inventors are often unlike the people who will apply their ideas. In a survey of computer professionals, the most frequently endorsed maxim for software interface authors was "Know the user, and you are not the user."[52]

We all know how technologies can rebound against their inventors when others acquire them and learn or improvise techniques for deploying them. The Maxim gun, favored by European powers for suppressing colonial resistance in the nineteenth century, begat the rugged and portable Kalashnikov brandished by guerrillas and terrorists, and the automatic-weapons arsenals of gangsters and narcotics traffickers, in the twentieth. In our day, deviant techniques flourish, from the shattering of spark plugs to produce ceramic fragments ("ninja rocks") to break into automobiles in the United States, to the "social engineering"—usually confidence tricks—that criminals use to obtain computer passwords and personal identification numbers for fraudulent electronic transactions.[53]

If it is a truism that any security hardware can eventually be defeated by criminal ingenuity, there should be a corresponding maxim that almost anything can be made better by user experimentation. There is a benign and positive side to the unintended. Inventors also can scarcely foresee the changes in values that will give new meanings and techniques to what they have produced. Some of the most common things in everyday life reflect not only the ingenuity of the original producers but the experiences of generations, sometimes millennia, of users. When we use simple devices to move, position, extend, or protect our bodies, our techniques change both objects and bodies. And by adopting devices we do more. We change our social selves. In other species, natural selection and social selection shape the appearance of the animal. In humanity, technology helps shape identity. Our material culture changes by an unpredictable, dialectical flux of instrument and performance, weapon and tactic.

CHAPTER TWO

The First Technology

Bottle-Feeding

EVOLUTION HAS GIVEN us an enormous advantage over other animals, and a corresponding burden. We are specialized in applying techniques to our needs, and in transmitting techniques more complex than those of other organisms. While the complexity of many but not all artifacts has grown over the millennia—and we will see in the next chapter that some artifacts of "simpler" societies actually reflected skills more sophisticated than those of contemporary industrial peoples—the human capacity for complex techniques appears to have been present from the outset.

Just as techniques (including the hardware now customarily called technology) were essential from the origin of our species, so they are present in human life from birth. Of course, there are prenatal techniques, too—not only the variety of practices followed by prospective parents but also technologies for assisting in conception and diagnosing and monitoring conditions of the fetus. Like other techniques, these probably will have unforeseen consequences for the shaping of the human body. Genetic manipulation may ultimately reduce the frequency, at least in certain populations, of genes responsible for certain birth defects. It may also promote other genes that are culturally held desirable, governing body type, appearance, and intelligence test scores. Neither most advocates nor most opponents of these techniques have fully thought out the implications of calling them "engineering." Civil engineers—as Henry Petroski has argued in a series of books—advance through mistakes and even disasters, but misdesigned people cannot be rebuilt as failed bridges, collapsed tunnels, and failing roads may be. At the very least, in vitro fertilization may actually help increase the incidence of infertility, as we have seen that genetic testing can inadvertently spread genes for hereditary diseases while preventing their expression in a given family. Over the coming century, for an increasing number of couples, parenthood might require medical assistance. Already one in six American couples receives some kind of fertility treatment. It is easy to imagine that this proportion might increase.[1]

As common as prenatal tests have become in industrial societies, they have a small effect on the human body compared with those of an older form of intervention in our development: bottle-feeding. For the majority of people in the developed world, infant formula and its physical apparatus of bottles and nipples is the first technology. And its effects, if usually subtle, are lasting.

THE SKILL TO BE NATURAL, THE WILL TO BE ARTIFICIAL

Human nursing is different from the lactation of other mammals, even of other primates, because (as part of the price of our hyperdeveloped brains) the human infant is uniquely dependent. Other primate infants are born as vigorous little individuals who know how to look out for themselves. They need no help in finding and latching on to the mother's nipple. Other primate mothers do have infant care skills to learn, and second infants are easier for them, but by the time they are ready to nurse they have learned what they need, and the infant does the rest. By contrast, the human infant needs its mother's body not just for nourishment and shelter but for immunological protection; it has been called an "external fetus."[2]

Once more, human performance requires technique, as breast-feeding advocates are the first to acknowledge. Being a baby can be hard work. While sucking is a reflex, an infant must use sixty-three different nerves to suck, swallow, and breathe, according to lactation specialists, and about one in ten has some difficulty. Maternal behavior, too, has to be learned. New mothers are not innately prepared for many contingencies: insufficient or excessive breast milk, cracked nipples, and a great variety of infant behaviors. In the words of an African infant nutrition activist: "We make the mistake of believing breast-feeding is natural, an intuitive thing. But it's a learned behavior passed on from generation to generation. In the old days, the older women would sit there and encourage and tell you to do this and that—it was part of education." Now breast-feeding advocacy organizations and lactation consultants have taken the place of grandmothers.[3]

Humanity has a long history of feeding arrangements for mothers unable or unwilling to nurse their own children. Of course, the great majority of mothers did so, if only because there was no available alternative. In Greece and Rome the affluent often employed wet nurses, free as well as slave. Wet nursing was a thriving profession in the Middle Ages, a specialty of whole districts like the Casentino and Valdarno valleys in Italy, whence young married women and their husbands traveled to the Florence fairs to advertise their skills in verse. Wet nursing remained the rule for most Europeans who could afford it well into the eighteenth century. But religious and medical authorities were turning against it. The Catholic church vigorously promoted maternal breast-feeding not only to assure infant health but, modern historians have argued, to limit women's growing cultural and political influence. And the authors of the *Encyclopédie,* the multivolume summary of the eighteenth-century Enlightenment, advocated maternal nursing as a duty that they also hoped would reduce the public influence of women. Late-eighteenth- and early-nineteenth-century medical writers all over Europe extolled the natural benefits of lactation for mothers and infants and warned of the dangers to bodily mechanisms from frustrating the natural flow of milk. And romanticism continued what religion and rationalism had begun: the women of the American antebellum South were among history's most fervent believers in maternal suckling, on the grounds of natural duty, despite what became the postbellum cliché of the slave wet nurse.[4]

Bottle-feeding did not begin in the nineteenth century. There is a long history in many societies of feeding infants animal milk or cereals. The practice was most common in cold regions with abundant and relatively easily preserved animal milk, but it was not confined to them. Feeding vessels for infants have been found at sites as much as six thousand years old, and the Romans used artificial nipples. In late medieval and early modern Europe, feeding technology was the norm in areas of Scandinavia, southern Germany, northern Italy, Austria, Switzerland, and Russia. The historian Valerie Fildes has found evidence of centuries-old debates on artificial feeding and breast-feeding. In Upper Bavaria, peers denounced a woman born in northern Germany as swinish and filthy for attempting to nurse her own baby. Their aversion appears to have no local environmental basis; the reasons for it are unclear. And the physician of Louis XV of France reported no ill effects on the health of Muscovites and Icelanders from their custom of letting the smallest babies suck on tubes placed in containers of milk or whey. A twentieth-century demographer has confirmed the effects of artificial feeding, in Bavaria at least. After a higher rate of infant mortality in the first year in districts where artificial feeding was usual, there was a far lower rate in the next four years of childhood, possibly because the surviving one-year-olds had developed resistance to organisms in their diet. Artificially fed children also remained with their parents, avoiding the medical problems associated elsewhere with wet nurses.[5]

The industrial revolution, then, did not introduce the feeding bottle. Whether to absorb milk surpluses or (consciously or unconsciously) to control family size through higher infant mortality, many earlier communities used crude expedients. But they were the exception. The nineteenth and twentieth centuries brought three changes that ultimately made many and sometimes most infants in Europe and North America dependent on artificial food: new devices for milk delivery; new scientific and medical attempts to equal the quality of breast milk; and the rise of national and international dairy food markets. These turned out to have serious unintended consequences for both infants and their mothers.

THE TINKERER IN THE NURSERY

None of the feeding devices for infants before the nineteenth century had a chance of becoming a technological and cultural icon. Many of the problems were functional: the vessels must have been difficult to keep clean. Some were technical: there was no satisfactory artificial nipple. (When infants were not presented directly with the neck of a ceramic, pewter, or tin vessel, or with a cow's horn punctured at the tip, they were given a piece of stitched parchment with a bit of sponge inside. Nipples were also made of wool, chamois leather, linen, and even alcohol-preserved cows' udders. The preferred material for nipples in early-nineteenth-century America was silver, not simply for display but for chemical stability.) But the greatest problem was economic. Feeding devices had to be hand produced at great expense. Since elite families often employed wet nurses, the market for feeding technology had to be limited. Still, inventors began to patent new styles of feeding bottle. The first in America was the Lacteal of Charles Winship in 1841, shaped like a human breast, contoured to fit on the mother, and claimed to persuade the baby that the milk was hers ("a useful deception," in the inventor's words). But the delivery system was literally a technological bottleneck: the child was to suck on a sponge-stuffed deerskin teat.[6]

Four years later, the real revolution in infant feeding began. Elijah Pratt of New York, using Charles Goodyear's new vulcanization process, patented the first rubber nipple. A series of patents followed, becoming a wave in the late 1860s and early 1870s—a sign of the market's growth. These devices in turn helped make possible more successful containers; one of these, the O'Donnel bottle patented in Great Britain in 1851, was popular in the United States between the Civil War and the 1890s. It featured a flasklike bottle with a long rubber tube that had a nipple on the end. The tube was prone to bacterial contamination. A British alternative design of the 1860s, called the Mamma, had a feeding end modeled, probably of rubber, from a human breast and secured to the glass body, shaped vaguely like a whale, with an elastic band. Toward the end of the century, other models appeared that could be strung over an infant's cot for overnight feeding on demand or deposited on its chest with small legs.[7]

While the U.S. Patent Office had recognized a minimum of 230 feeding bottles by 1945, by the 1920s most had taken the form they have

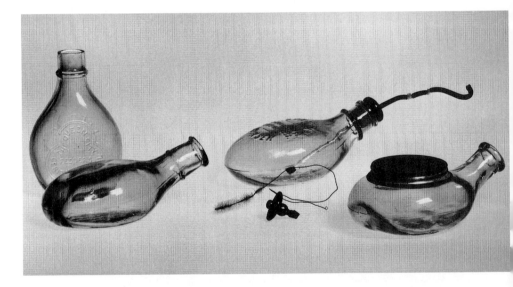

Before the twentieth century, hand-blown, narrow-mouthed bottles and long tubes were hygienic nightmares. The flattened Turtle was also called the Murder Bottle. Courtesy of Corning Museum of Glass, Corning, N.Y., left to right: nursing bottle, "Cerity & Morrel Feeder," after 1880 (58.4.16. United States. Colorless glass with an aquamarine tint; mold-blown), gift of Mrs. Thomas Wilmot; nursing bottle, "Tyrian Nurser," 1890–1910 (62.4.43. United States, Andover, Mass., Tyler Rubber Company. Colorless glass, cork, wood, brush, metal; mold-blown, assembled), gift of Mr. and Mrs. Paul Perrot; nursing bottle, 1870–1910 (66.4.53. United States. Colorless glass; mold-blown), gift of Arthur A. Houghton, Jr.; nursing bottle, patented 1890 (57.4.19. United States, New York, N.Y., McKinnon & Co. Colorless glass, metal, paper; mold-blown), gift of Hugh L. Kline.

today: wide-mouthed containers—originally, they were made of glass—simplified for sterilization and topped by rubber nipples. (The hygienic movement and wide acceptance of the germ theory did not eliminate other variations, including figural novelties shaped like shoes, turtles, and rabbits.) Early in the past century, rubber formulations were developed that avoided the cleaning problems, stiffness, and offensive taste and odor of the nineteenth-century product. Meanwhile, an automatic glassblowing machine invented by Michael J. Owens of Toledo, Ohio, and beginning commercial operation in 1903 was, by 1920, able to produce nearly 13,000 bottles daily, significantly reducing their cost—and also assisting distribution of bottled milk. (Even in the 1880s and 1890s, bottle production was only semiautomatic, and the lip had to be added by hand.) In 1919 the Corning Glass Works patented Pyrex, a heat-resistant glass that had grown out of its work with railroad signal lights; Pyrex nursing bottles appeared in 1922. Sterilizable, wide-mouthed bottles enhanced the hygienic aura of artificial feeding and remained the familiar standard until after World War II. Pediatricians and hospitals welcomed disposable, formula-filled bottles like the plastic Mead Johnson Beniflex of 1962 and later glass models, which replaced the unpopular and sometimes careless hospital formula technicians. Following early experiments with built-in thermometers, some plastic bottles for home use now change color to reflect the temperature of their contents.[8]

And a tradition of kitchen-table innovation endures. Advances in molding have recently permitted at least one variant recognized by a major cultural institution: in 1988, the Museum of Modern Art design collection added a plastic nursing bottle called the änsa, produced by an Oklahoma couple with no professional child development or design experience, shaped like a long doughnut for better infant gripping. Admiring as he was, the inventors' own pediatrician still had a concern: that babies would like the änsa so much their parents would let them go to bed with it. While there has apparently been no independent health evaluation of this design, the doctor was wary because bottles in general have an unfortunate effect on infants' feeding, for taking nourishment is also a technique. When a child sucks its mother's breast, it "latches on," taking the nipple into the mouth, clamping down with its jaws on the areola, and obtaining milk by inducing peristalsis with its tongue. In this process, the nipple can stretch to twice or three times its usual length, and the milk shoots from fifteen to thirty pores throughout the infant's mouth.[9]

Rubber, plastic, and silicone nipples function differently, with consequences for the infant's technique. When presented with a bottle, the jaws need not engage the nipple. Negative pressure alone transports the milk. The artificial nipple is more efficient; the baby has less work to do. For this very reason, unfortunately, once an infant begins to feed from a bottle, even soon after birth or as a supplement to breast-feeding, the new style of feeding is not easily abandoned. The infant has lost its innate tendency to open wide when feeding, expects to have a nipple solidly in its mouth, is used to a free flow of milk, and pushes its tongue forward: all behaviors that will frustrate its experience of breast-feeding. With reduced sucking, the mother's breasts will become engorged, reducing her milk production and making it more likely that the infant will continue to be bottle-fed. Infant feeding specialists call this effect nipple confusion or triple nipple syndrome. Breast-feeding and child welfare advocates are thus especially alarmed by hospital programs promoting formula-feeding for newborns and presenting mothers with free starter kits. An American hospital innovation of the 1970s, disposable sterile plastic bottles of formula for newborns, saved preparation time and improved quality but did nothing to overcome criticism. And advocates are concerned about "mixed feeding," the alternation of breast milk and formula.[10]

(Paradoxically, lactation consultants recommend a technological solution for nipple confusion and other feeding problems that combines ideas from earlier feeding vessels: placement on the mother's body and use of a long tube. The Supplemental Nursing System [SNS] is a plastic bottle filled with the mother's own expressed milk or formula, channeled through a tube that runs along the breast and is held in place with nonirritating tape. The tube is so narrow and flexible that the infant maintains or develops a normal breast-feeding technique and the mother's nipple receives the stimulation that assists lactation. A valve in the bottle releases the milk only when the infant begins to suck. For bottle-feeding infants, manufacturers also offer "physiologic" nipples that encourage natural tongue technique.)[11]

Besides the initial mechanical problems of bottle-feeding, there are consequences for the baby's mouth. Inexperienced parents may put children to bed with nursing bottles and let milk or juice build up around the teeth, causing decay. Bottle-fed children also have a higher rate of malocclusions. According to one study from 1981, 36.4 percent of children breast-fed not at all or for less than three months had occlusal anomalies, while 24.2 percent of children breast-fed for more than six months did— an increase of risk by 50 percent. A 1987 paper reported an 84 percent higher risk of malocclusion among children breast-fed three months or less or not at all, and the authors estimated that 44 percent of malocclusions among the children in the study were due to brief or nonexistent breast-feeding. One medical writer on breast-feeding believes that the pistonlike motion of the tongue in bottle-feeding—it has a rolling action in breast-feeding—may be responsible.[12]

CHEMISTRY TAKES COMMAND

The mechanical risks of the bottle are still small compared to its possible nutritional consequences. Pediatric researchers and nutritionists are fond of repeating that infant formula as a substitute for human breast milk is "the largest *in vivo* experiment without a control series."[13]

While artificial feeding methods had been folk practice for centuries, even millennia, scientific replacements for mother's milk were a nineteenth-century innovation. Early modern urban elites used some expedients we would find bizarre: only two of Mozart's six children, all fed mainly sugar water, survived beyond their third year. The cow and goat milk sometimes previously used as a substitute or supplement was too thick and alkaline. In 1741, the London physician Sir Hans Sloane established that at the Foundling Hospital, 53 percent of infants receiving animal milk or cereal mixtures died, as opposed to 19 percent of those with wet nurses. American nineteenth-century doctors documented that the milk of sick cows could transmit not only throat infections but life-threatening diseases like typhoid and tuberculosis. They warned about the deficiences of "swill milk" from cows consuming brewery slops. In late-nineteenth-century London, milk still might be adulterated with as much as 25 percent water. As railroad supply lines and chemical knowledge grew, preservatives—including truly dangerous ones like hydrogen peroxide—were marketed openly; by one estimate, they were used by half the dairy trade.[14]

Even as some manufacturers were producing dubious additives, more idealistic chemists were finding new ways to detect them. And still others had an even more ambitious and, it seemed, noble goal: developing scientifically optimal foods for infants. The problems of artificial feeding, like other technological dilemmas, began with the best of motives. In England's industrial cities, malnutrition of mothers and the feeding practices of untrained women who cared for infants during the working day had raised the mortality rate of young children as high as 55.4 percent in 1858. In the

1860s, chemists began to develop foods with patented or secret formulas that "humanized" cow's milk to make it suitable for human infants. Baron Justus von Liebig (1803–1873), a founder of nineteenth-century chemical research, education, and industry, and one of Europe's most influential scientists, tried to help his nanny's granddaughter, who was unable to nurse her own children. Earlier in the century, the family would have brought a wet nurse to live in, but the middle class was beginning to avoid what it perceived as a potential source of malnutrition and, linked with malnutrition in their minds, bad moral influence. Liebig developed an infant feed consisting of ten parts of cow's milk with one part each of wheat flour and malt flour, and a supplement of potassium bicarbonate. Liebig claimed that it offered nutrients in twice the concentration available in human milk, and that infants would thus need less of it. Unfortunately, his ratios were indirectly based on the methodologically flawed studies of a Berlin chemist named J. F. Simon. As a result the formula was too high in carbohydrates and had dangerously low levels of vitamin C, of other vitamins, and of amino acids.[15]

Liebig was an outstanding predecessor of today's academic scientist-entrepreneurs, not only a preeminent chemist but an esteemed mentor and the master of a flourishing professional network. He promoted his preparation vigorously in the popular science press, solicited testimonials, and licensed an English company that soon was producing a dried form with pea flour to be mixed with cow's milk. Meanwhile, a self-taught Texan inventor, Gail Borden, had turned from meal biscuits to milk after observing children die, apparently of malnutrition, on an ocean voyage. (He adapted a spherical copper vacuum condenser used by Shakers for condensing fruit juice.)[16]

Despite medical reports of the food's inadequacy and mixed responses from consumers, Liebig's prestige allowed him to maintain his claims for the superior nutritional value and digestive qualities of his product. At least one supporter wrote that because its composition had been scientifically shown to have "the very same ingredients" as human milk, "I cannot understand why they should be unable to digest Liebig's Food." It went on sale in the United States in 1869, advertised with the slogan "No More Wet Nurses!" Other entrepreneurs had already realized the product's economic potential—Liebig's Food sold for a dollar a bottle in New York at a time when that sum represented many workers' daily pay—and competition grew. In Switzerland, with its large dairy industry, the American diplomat Charles Page and his brother formed the Anglo-Swiss Condensed Milk Company in 1866 to use Gail Borden's 1856 patent for the production of condensed milk. The following year a Swiss merchant named Henri Nestlé developed an infant preparation that allegedly had saved the life of a baby that had refused all other food. It consisted of "good Swiss milk" and bread, "cooked after a new method of my invention" and marketed in tin packages, to be mixed with water. The market proved spectacular, and in 1873 Nestlé was selling half a million boxes annually in Europe and in North and South America. Meanwhile, an English chemist named Gustav Mellin developed a variation of Liebig's Food to which cow's milk as well as water were to be added. In the United States, the Borden Company was promoting its own Eagle Brand, and the pharmaceutical industry began to take notice of the potential market when Smith Kline & French bought the rights to a product called Albumenized Food. The manufacturers appealed to doctors and lay consumers alike. Some, like Nestlé, promoted the safety of using only (preferably boiled) water; others, like the makers of Mellin's Food, sought medical support for mixing their product with raw milk. Doctors could not agree on the optimum treatment.[17]

Between 1890 and World War I, some physicians tried to place substitute infant feeding on a scientific basis free from the manufacturers' commercialism. The leader of this movement, Thomas Morgan Rotch of Harvard Medical School, sought not merely an acceptable substitute for breast milk, but scientifically optimal nutrition for each child that could reduce the still-high rate of infant mortality. Using analyses by the Philadelphia physician A. V. Meigs, a significant advance on Liebig's work, he developed a series of tables instructing mothers on the preparation of a formula containing precisely correct proportions of fat, sugar, and proteins, to be compounded by the mother from milk, milk sugar, cream, and lime-water (a solution of calcium hydroxide in water). To avoid contaminated milk products, Rotch worked with scientific dairies to produce certified milk products with newly developed hygienic procedures. Other doctors in France, Great Britain, and the United States helped establish "depots" where pure milk could be provided under medical supervision. Contamination of commercial cow's milk nevertheless remained widespread through the 1920s, if British and American evidence is typical. Critics of the formula industry argue that even the best-managed programs did not contribute to the decline of infant mortality that commenced around 1905, but they did establish a disturbing and continuing link between medical clinics and artificial milk distribution.[18]

The complexity of Rotch's "percentage method," which turned the household into a small-scale chemical laboratory, led to its abandonment after 1915. (It had been influential mainly in the northeastern United States.) In its place, a new pattern of infant nutrition appeared: the marketing of infant formula to be administered under pediatricians' supervision. For these rising specialists, and for family practitioners, scientifically managed feeding was a medical crusade. While some authors present the medicalization of life as the imposition of professional judgment on an intimidated laity, the reality was more complex. Many doctors as well as mothers affirmed the superiority of breast milk. But older networks of support for nursing mothers encountering difficulties were declining in the early twentieth century. On the other hand, at least before about 1875, lay men and women were enthusiastic about the authority and capabilities of scientific medicine. Medicalization may have introduced new prejudices and errors, but it was not simply imposed by legislation. Women themselves turned, whenever possible, from the craft knowledge of midwives to the care of obstetricians for safer and less painful delivery. They believed in the movement called "scientific motherhood" as much as the physicians did. The technologies of bottle, nipple, and formula likewise appeared to mothers and physicians alike as a more modern replacement for the techniques of breast-feeding that had been transmitted informally. With hospital delivery (rising from 20 percent to 80 percent of American births between 1920 and 1950), artificial feeding was institutionalized. To prevent infection, hospitals limited the frequency and duration of contacts between mother and infant. Babies regularly received supplementary feedings, and were bottle-fed at night to let mothers sleep.[19]

GLOBALIZERS OF THE BOTTLE AND THEIR FOES

Confidence in science and aggressive marketing by manufacturers were not the only reasons for the success of infant formula. Beginning in the 1920s, the breast was sexualized in a way that made public feeding potentially more sensitive than it had formerly been. Bottle-feeding was associated with scientific motherhood and at the same time with freedom from domesticity. Some bourgeois European circles also welcomed bottle-feeding as a step against prolonged oral gratification and for the development of good habits. The result was a steady increase, though with many national and regional variations, in the proportion of bottle-fed infants between World War I and the 1960s. In one American study of 1958, 63 percent of infants returning home from the hospital were already consuming only formula, and 21 percent were fed only breast milk. Few employers accommodated working mothers who had nursing infants, but even in the Sweden of the 1960s, where new mothers remaining at home with their infants received 90 percent of their professional salaries, bottle-feeding prevailed. A revival of breast-feeding among middle- and upper-class women in North America and Europe began in the 1970s and remains a strong force, but it has delayed bottle-feeding rather than replaced it as a routine of upbringing.[20]

In Europe and North America, the health effects of infant formula are still debated. In the great age of expansion of bottle-feeding from 1890 to 1950, infant mortality also dropped markedly—from 140 to fewer than 40 deaths per 1,000 live births in New York City, for example. Reduction of digestive and respiratory ailments, notably diarrhea and pneumonia, was especially pronounced. In Sweden, an even more pronounced decline in mortality had begun in the late eighteenth century and continued through the nineteenth and early twentieth. In neither the United States nor Sweden did the trend appear to be affected by the spread of bottle-feeding or by the Depression of the 1930s. All this suggests that in affluent countries, formula-feeding was indeed a good alternative to the wet nursing that had been practiced so widely in early modern Europe.[21]

The great unintended consequence of artificial feeding has arisen not in the industrial countries but in the developing world. Especially since the late nineteenth century, North American and European farmers have produced abundant milk. Breeding and animal nutrition alone have raised the annual yield of a dairy cow from about 1,500 liters in the early nineteenth century to 6,500 liters—and for some breeds as much as 10,000 liters—today. Pasteurization has been commercialized since the 1890s, refrigerated trains have drastically reduced spoilage on the way to market or processing, and global beef imports have allowed more European farmers to specialize in dairying.

Meanwhile, the growth of cities and market economies in Asia and Africa made processed infant formula a valuable export. In these markets, infant formula remains costly for all but a small segment of families. The formula producers applied with great success the scientific appeals that had been effective in the West. Infant formula was also promoted as a sign of modernity and education; elites adopted it as a mark of their political and economic authority. In the rest of the population it was most influential in cities, where rapid migration and women's industrial labor helped disrupt the transmission of breast-feeding techniques. Urban slum life and disease can also interfere with lactation. And even low-income women came to

share the privileged classes' view of infant formula as a progressive and scientific alternative to breast-feeding, and the Westernized taboo on the public display of breasts. Advertising linked formula with infant health as well
as with prosperity and modernity. Intentionally or not, it persuaded many
mothers who could have established lactation successfully that they suffered from "insufficient milk" syndrome. In some countries, "milk nurses"
receiving sales commissions, some of them with real nursing credentials
and all easily confused with hospital staff, promoted manufacturers' products to new mothers in hospitals. Inadvertently, distribution of millions of
pounds of powdered milk for starving babies by the United Nations Children's Fund (UNICEF) and other agencies in the 1960s helped legitimize
substitute food in new markets. What helped the sick would surely benefit
the well.[22]

The result instead was malnutrition and death. The first prominent
crusader against formula-feeding in the Third World, the pediatrician Dr.
Cicely Williams, had promoted condensed milk in combating kwashior-
kor, a severe protein-calorie deficiency disease, in Africa. But in 1939, Dr.
Williams, then working in Singapore, was disturbed by the consequences
of feeding infants sweetened condensed milk. In those days the product
was not supplemented with vitamins D and A, so it had contributed to
many cases of rickets and blindness. Speaking on "Milk and Murder" to
the Singapore Rotary Club, Williams accused the producers of callous neg-
lect of infant life in the interest of profit.

Well-meaning agencies as well as commercial interests could work
against breast-feeding. Two leading pediatric public health specialists, Derrick and Patrice Jelliffe, call the distribution of powdered milk by feeding
programs in the 1940s and 1950s a "nutritional tragedy." In the absence
of health education programs, the product encouraged a shift to bottle-
feeding. The commercial distribution of formula, far from reducing the
rate of nutritional deficiences, increased them seriously. Bottle-fed babies
gain weight more slowly than breast-fed ones, and are more likely to suffer
from bacterial and viral infections and parasites in their second year.
Marasmus—a form of severe growth failure closely connected with the
lack of high-calorie foods—is linked with bottle-feeding. Dilution of cow's
milk formulas to reduce costs is a special risk factor for marasmus in many
poorer countries. Since mother's milk may be the only safe liquid in many
regions and feeding bottles may be impossible to keep clean without
refrigeration or sanitary storage areas, formula-feeding also promotes
infections, especially diarrheal diseases that inhibit appetite and lead to

malnutrition and more illness. Meanwhile, the bottle-fed infant receives none of the protective substances in mother's milk. In seven villages in the Punjab studied in the 1950s, mortality among infants bottle-fed from birth was fully 95 percent during the first eleven months, compared with 12 percent among infants breast-fed from birth.[23]

United Nations–sponsored efforts to encourage industry regulation faltered. In the 1970s, social activists armed with the damaging statistics and with Derrick Jelliffe's identification of "commerciogenic malnutrition" in Jamaica began to urge restrictions on the marketing of infant formula in the developing world. A boycott of Nestlé, begun by religious groups and others, was resolved with a vote of the World Health Organization's (WHO's) World Health Assembly in 1981 establishing a UNICEF code restricting advertising and the distribution of samples and intended to put proprietary formulas under strict medical supervision. Manufacturers' literature now extolled the virtues of mother's milk while encouraging an early transition to bottle-feeding. But activists, believing formula manufacturers were trying to circumvent the UNICEF code despite several amendments designed to close loopholes, renewed the boycott in 1988.[24]

HEALTH IN THE BALANCE?

In the First World, infant formula raises different but equally interesting issues. Like the WHO, the American Academy of Pediatrics (AAP) strongly supports breast-feeding. In a 1997 policy statement it declared that "breastfeeding ensures the best possible health as well as the best developmental and psychosocial outcomes for the infant." But the cohort born at the peak of formula-feeding, from 1946 to 1952, comprises the babies who started the boom, and they flourished, as the science writer Natalie Angier and others have reminded breast-feeding advocates. Yet whatever their present health—could they not be even healthier?—they did not necessarily have an easy time as babies. Breast milk is an elaborate package of chemicals developed over millions of years of primate evolution to promote the newborn's development and build up its defenses against infection. Some of its constituent molecules keep microbes from spreading from the digestive tract into the body's tissues; others reduce the availability of vitamins and minerals (especially iron) that disease-causing bacteria need; and still others help the work of immune cells and kill bacteria directly by attacking their cell walls. And a variety of white blood cells helps the infant produce antibodies and attack microbes directly. A pedia-

trician recalls how easily incorrect preparation of bottles by hospital formula rooms could caramelize the sugar and precipitate diarrhea. In fact,
many infections are more common in bottle-fed infants deprived of the
protective antibodies unique to mother's milk. Even now, pediatric
researchers estimate that 250 to 300 infants die each year from diarrheal
infections as a result of bottle-feeding; another 500 to 600 die from respiratory diseases. Middle ear infections, more frequent among formula-fed
children, have been treatable with antibiotics, but the high price of therapy
has included the rise of resistant strains of bacteria.[25]

Even more important and less well known is how formula-feeding
affects the long-term welfare of both infants and mothers. Whether to
breast-feed or bottle-feed has always been a cultural as well as a biological
decision. Mothers who choose one or the other method may well feed
their children differently after weaning, give them more or less encouragement in school. They may be more or less affluent than other parents, and
their children may have different peers and experiences. Bottle-fed infants
are also more likely to be given pacifiers, which in turn differ in design and
in their effects on the development of the mouth.

Despite all these possible biases there is evidence—sometimes limited
and controversial—of long-term benefits of breast-feeding. Breast-fed children may develop slightly higher intelligence than bottle-fed contemporaries. A number of chemicals that appear in human milk but not
in formula, including arachidonic acid (AA) and docosahexaenoic acid
(DHA), are known to promote brain growth. Longer-term breast-feeding
has been linked in one study to faster motor and cognitive development.
In another, premature infants tube-fed with human milk had an 8.3-point
IQ advantage over their formula-fed counterparts at the age of seven and a
half—this after downward adjustment to reflect their mothers' educational
and social status. Some studies show only weak or insigificant gains after
adjustment for family background, but the very first study of feeding
method and mental development, published in 1929, showed advantages
for breast-feeding at a time when formula-feeding was chic and breast-
feeding associated with immigrants and the working class.[26]

The immune system also may be affected by infant feeding. The nutrition researcher Alan Lucas has pointed out that studies of other animals
show effects of prenatal and postnatal nutrition on indicators of health
from size and metabolism to obesity, hardening of the arteries, and
longevity. There is evidence, if not yet proof, that there are critical periods
in human development, "programming windows" for stimuli that enable

or inhibit later behavior. Very early in their lives, chicks, mallard duck-lings, and some infant mammals become attached to the first thing they encounter, in nature nearly always a parent but in the laboratory some-times a human being, another animal, or even an object. Since the 1960s, scientists have been able to produce lifelong changes with brief early inter-ventions; rats receiving less nourishment in their first three weeks remain smaller than others. Baboons overfed in infancy gain weight in early adult-hood as though by a delayed causal mechanism. Lucas believes nutritional programming may have effects on human health.[27]

Are we not only what we eat, but what we ate? Fragmentary but intriguing evidence exists. Inflammatory bowel diseases may be linked to very early nutrition. Studies have shown that ulcerative colitis and Crohn's disease (a chronic inflammatory disease of the intestines) are less common among those raised on breast milk. Another study suggests that artificial feeding has been responsible for the long-term increase in juvenile dia-betes; the authors attribute a quarter of all cases to bottle-feeding. Because childhood infections of the ear and the lower respiratory tract may lead to chronic respiratory illness, the early benefits of breast-feeding can carry through to later life. There are also indications that breast milk may reduce the risk of malignant lymphomas, multiple sclerosis, and coronary artery disease. Exclusive breast-feeding for at least four months after birth has been shown to reduce the risk of childhood asthma significantly. The number of studies in each case is small, and breast-feeding advocates in medicine also acknowledge that human milk is no panacea; it probably only delays the onset of allergies, for example. But the studies have still contributed to medical organizations' overwhelming support for extended breast-feeding.[28]

Industrial and postindustrial culture also affects the timing of feeding. For most of human history, there was no supplementary feeding, and infants were allowed access to milk frequently, with short intervals between feeds. Most of today's breast-feeding mothers, especially in eco-nomically advanced nations, nourish their infants fewer times each day. The original style is healthier for babies, helps prevent jaundice, and it's less likely that overenthusiastic sucking by hungry infants will cause the mother pain. While the standard U.S. breast-feeding handbook now rec-ommends nursing "at least every two or three hours," this is not feasible for many employed mothers, though their stored milk may be fed to their infants by sitters.[29]

Breast-feeding may also protect mothers in ways that are not fully

understood because our culture—even among most breast-feeding mothers—departs from many of the patterns present during the evolution of
our species. Katherine Dettwyler has argued persuasively that human
infants follow a "hominid blueprint" for extended nursing. Studies of other
living primate species have established a formula linking weight at matu-
rity to age of weaning. For human weight, the formula predicts, conservatively, a natural age of weaning of from 2.8 to 3.7 years, varying with adult
female weight. Studies of gestation time and weaning suggest that in
humans, as in chimpanzees and gorillas, breast-feeding would last at least
six times as long as gestation rather than for nine months, as many medical
texts assert. Other primate studies link weaning strongly with the eruption
of the first permanent molar; this happens to human children, regardless
of nutritional status, at the age of 5.5 to 6 years, about the time they
acquire adult immune competence. Thus infants are biologically inclined
to continue nursing until their third or fourth year or even longer. Katherine Dettwyler believes there may be benefits for both mothers and infants
in prolonged lactation; she cites research that suggests that small but significant numbers of mothers in industrial societies continue suckling their
children discreetly well beyond the second and third year. Indeed, in 1999
a Chicago newspaper found a local four-year-old preschooler with his own
computer who scandalized guests at a parental party by taking such a milk
break, raising his fists "Muhammad Ali–style, and declar[ing], 'That was
sooooo good.' " And despite the reservations of some psychologists, there
seem to be no physical or mental ill effects among late-nursing children.
One of them in Chicago grew up to be a massive offensive linesman at
Texas A&M who says he has not been sick since eighth grade and has no
memories of nursing.[30]

Because this ancient pattern of breast-feeding is difficult to study in
Western societies, we still are not sure of its effects on maternal health.
There is some evidence that prolonged breast-feeding has protective value
against breast cancer, especially among premenopausal women. Societies
in which breast-feeding is widespread tend to have lower rates of breast
cancer, and when infants are fed unilaterally, the suckled breast is significantly less likely to develop cancer. Breast-feeding is also thought to
reduce ovarian and endometrial cancer. Since the choice of whether to
bottle-feed or breast-feed may be associated with so many differences in
diet and other practices, these findings are more suggestive than conclu-
sive. But it still seems likely that future clinical studies will confirm some

significant long-term differences in the health of breast-feeding and bottle-feeding women.[31]

There is a clear benefit for women in natural regulation of birth patterns. In developing countries where mechanical or chemical contraceptives may be unavailable or too costly, prolonged breast-feeding—because it often inhibits ovulation—can space children naturally three or four years apart, the pattern among early hunter-gatherers. There are good evolutionary reasons for this pattern: sibling competition for a mother's milk endangers the survival of both infants and hence of their genes. The hormone prolactin, which stimulates milk production, is produced by the hypothalamus in response to the infant's suckling. Prolactin also suppresses ovulation and menstruation (the medical term for this effect is *lactational amenorrhea*). Because they generally gave their children to wet nurses rather than breast-feeding them, upper-class women in early modern Europe had much larger families than poorer women did. Swedish peasant women, and no doubt many others, were aware of this contraceptive benefit. But even in cultures where breast-feeding remains the norm, ovulation sometimes resumes while a mother is still nursing. Suckling at longer intervals appears to allow prolactin levels to sink; it is more frequent, on-demand suckling that keeps prolactin levels elevated.[32]

THE OUTLOOK FOR THE BOTTLE

What seemed to be an elementary chemical and mechanical challenge—duplicating the constituents of mother's milk and making it available to infants—has become a ubiquitous technology that changes both the bodies and the social lives of most men and women in industrial societies. Nineteenth- and twentieth-century technology did not begin humanity's cultural tinkering with ancestral patterns of infant feeding; we have seen that wet nursing and feeders were widespread in Europe. A baby nourished today with the latest version of infant formula may lack many nutrients and hormones unique to mother's milk, but it is certainly better off than its ancestors might have been with the often deadly mixes of their time and has far better prospects for health than an eighteenth-century child farmed out to a wet nurse miles away. And even breast-feeding advocates have made some room for artificial feeders. Bottles and other feeding devices are now recommended for administering the stored milk of nursing mothers while they are at work. While the medicalization of infant feeding a hundred years ago introduced some dubious concepts and practices—for example, strictly scheduled breast-feeding—that interfere with infants' natural demands and the beneficial synchronization of mothers' and infants' sleeping cycles, scientific research has also corrected errors. For example, in Europe and elsewhere, yellow early milk, called colostrum, was once discarded as poisonous to infants. While until recently some hospitals gave newborns sugar water solutions instead, English physicians had discovered the excellent nutritional value of colostrum by the early eighteenth century. Some historians believe those physicians' publications helped shape modern forms of family affection with maternal breast-feeding at their core.[33]

By 2000, the bottle and formula industry had achieved a paradoxical relationship with medical leaders and breast-feeding advocates. The boycott of Nestlé for its Third World marketing remained in effect but was less visible in the press. The manufacturers, for their part, conceded the superiority of human milk for infants in principle but still promoted their products through hospitals and encouraged supplemental feeding. Medical authority, once firmly on the side of the bottle, now supports breast-feeding. After declining in the 1980s, the proportion of breast-fed newborns in the United States rose to 59.7 percent by 1995, nearly equal to its 1982 level, but now the rate of breast-feeding is highest among more affluent and older women. Even in 1995, only 21.6 percent of infants were still breast-feeding at six months. There is still a vast domestic and international market for bottles and infant formula, $3 billion annually in the United States alone.[34]

The infant formula industry has achieved a curious symbiosis with its foes, the breast-feeding advocates. Even with the best lactation counseling and banked human milk from volunteers, at least a small number of infants will need formula. Our knowledge of the new revival of breast-feeding in the 1990s comes not from the U.S. Census Bureau or the Public Health Service but from a survey conducted by the Ross Products Division of Abbott Laboratories, a major producer of formula. Formula companies not only sponsor conferences on breast-feeding but fill a gap in consumer information: in Britain one formula maker sent out to parents 300,000 articles in support of breast-feeding. Companies openly acknowledge the superiority of breast milk and try to approach its composition more closely, though they obviously cannot replace the mother's enzymes and antibodies. But this research and improvement are also controversial, and not just because they may encourage bottle-feeding.[35]

The long-chain fatty acid DHA, present naturally in mother's milk, is known to help develop infants' brain cells. The WHO and the Commission of the European Community have endorsed it, and premium formulas overseas include it. The U.S. Food and Drug Administration approved it for inclusion in formula in 2001. Yet there are more than 160 other fatty acids in breast milk that do not appear in formula, and scientists are still unsure whether synthetic and natural nutrients are equivalent. The formula makers' goal has changed; where Liebig aimed to duplicate human milk, manufacturers now seek to match the "performance" of breast-fed babies. The position of the U.S. Food and Drug Administration appears in a headline in its consumer magazine: "Second Best but Good Enough."[36]

And as formula makers extol and emulate breast milk, even scientific advocates of breast-feeding are adding cautions. Alan Lucas, the infant nutrition scientist whose work on nutritional programming is some of the most powerful evidence against the bottle, recently called attention to studies concerning the long-term effects of early feeding on heart disease. When adult baboons were fed a typical Western high-fat "unphysiologic" diet, those that had been breast-fed in infancy developed more fatty arterial streaks than those that had been formula-fed. Human studies also suggest that prolonged breast-feeding increases the risk of cardiovascular disease in men. And environmental hazards may create new risks. In the 1960s, strontium-90 was found in Americans' breast milk, and dioxin was found in the 1980s. A Dutch study in the late 1990s revealed that polychlorinated biphenyls (PCBs), a neurotoxic threat to intelligence and learning, appeared in the bloodstreams of breast-fed children at a concentration 3.6 times higher than that found in the bloodstreams of their bottle-fed counterparts. Though these results are tentative and Lucas cautions against drawing any inference against breast-feeding, they do show that as we change our environment inadvertently, we modify ourselves unexpectedly as well.[37]

Despite all the evidence against it, the bottle is likely to remain the first technology of a substantial part of humanity. It gives fathers a nurturing opportunity denied them by nature. (In the 1980s, a prominent American male physician even invented the Baby Bonder, a bib of terrycloth and fleece with nipple openings through which men could nourish their children.) Formula is certainly superior to the animal milk that traditional societies used as a breast-milk substitute. Because American physicians still have limited training in helping with breast-feeding technique, but extensive information on prescribing formula, and because American insurers are reluctant to pay for lactation consultants, mothers are likely to breast-feed for a matter of months rather than the year or more that medical societies as well as evolutionary biologists and epidemiologists recommend.[38]

One reason is that breast-feeding, like many other techniques, conveys conflicting values. Many of its advocates consider it a feminist act, a rejection of the contemporary sexualization of the breast. It is an affirmation of women's autonomy and unique capacity in the face of medical and commercial interests that initially sponsored "scientific" feeding, from the patriarchal Dr. Liebig onward. But breast-feeding, as we have seen, has also attracted male (and conservative female) campaigns to restrict married women to a nurturant domestic sphere. The health benefits of nursing for mothers appeal to liberals' greener side, but if expressed too forcefully they can also appear to some progressives as holistic fascism. The bottle, and modern infant formula, are much easier to oppose in the developing world than in Western societies that do not seem prepared to pass legislation guaranteeing nursing mothers equality in careers, or even decent accommodation in public spaces. Even where breast-feeding could save money for governments, as in U.S. income-supplement programs for women, infants, and children, authorities have not promoted it vigorously. Bottle-feeding may exact a high price in medical costs, but as long as medications exist to treat the illnesses it increases, the technological treadmill will roll on.[39]

On balance, then, the twentieth century gave a scientific and moral victory to breast-feeding but a de facto social and economic victory to bottle-feeding after the first few months of life. But breast-feeding advocates, scientific and lay, have made remarkable gains in transmitting and improving a sometimes challenging technique over a technology that, for all its progress, has remained merely adequate. Now there is the prospect, equally heartening and disturbing, of cows genetically engineered to produce human milk. What Liebig's chemistry was to the nineteenth century, molecular biology will surely be to the twenty-first. But even if the children are healthy and the cloned cows contented, and even if women could take safe medication that conferred on them the benefits nursing provides, could the transmission of one of the most profound human skills become threatened?[40]

CHAPTER THREE

Slow Motion

Zori

WHILE NURSING IS our first technique—and, for many of us, the bottle our first technology—the first rite of technical passage for the greater part of humanity is the use of footwear. We are not the only animal that uses external things to modify its body: hermit crabs, for example, inhabit discarded shells of other creatures. Nor are we the only animal that makes things to modify its body: chimpanzees sometimes use large leaves for walking, as they do for eating. But we are the only animal that both makes and needs these objects for its well-being. None changes us more significantly than shoes. Footwear alters the foot itself, sometimes disastrously, but it helps prevent conditions that may have even worse effects. It changes not only our contact with the world but our perception of it. And through the meanings which have been assigned it, it affects how we relate to others socially.

The footwear consultant and columnist Dr. William Rossi has identified seven and only seven styles of shoe: moccasin, sandal, boot, clog, pump, mule, oxford. The moccasin, a piece of leather wrapped around the foot and fastened with a leather thong or other material, is thought to be the oldest foot covering. But sandals—pieces of leather or other material held on the foot with straps or other devices—are almost as ancient. Other footwear forms are combinations or extensions of the moccasin and sandal. Our word *sole* is derived from the Latin *solea,* a sandal made of heavy leather or woven straw with loops for straps of plaited straw or rawhide, and many of our other shoe styles are moccasins sewn or bonded to a sandal platform. North American Indians had three styles: moccasins with continuous soft soles, hard-soled moccasins, and (especially in what is now the U.S. Southwest and parts of Central America) sandals of hide or plant materials.[1]

Because true moccasins are unsuitable for contemporary paved roads and sidewalks, the sandal remains the most ancient form of footwear in common use. In fact, rubber and plastic sandals are among the most plentiful manufactured objects on the earth's surface. No official international statistics are available, but the Canadian shoe industry specialist Phillip Nutt estimates that in the last sixty years, the earth's factories have produced fifteen billion to twenty billion pairs. The thong sandal, with two straps anchored at three points, forming a **V** with its apex between the first and second toes, is known to nearly every nation and social class. The variety of its names in English alone reflects its ubiquity: zori (*zories* is the official term of the U.S. Customs Service), thongs (originally Australian), flip-flops (originally New Zealand), slippers (Hawaii), slaps, flaps, beach walkers, and go-aheads. On one Caribbean luxury cruise, the writer David Foster Wallace counted over twenty makes. Used apparently independently by societies around the world for millennia, the thong, or zori, went into mass peacetime production more than fifty years ago thanks to new materials and manufacturing techniques. Like other body technologies, it helps shape body techniques. But to understand the zori, we must start by asking why people need even this simplest of footwear.[2]

BARE FEET

Sandals, like other footwear, are a compromise. The naked foot functions best without them. While bipedalism, as the performance of apprehended drunken drivers trying to walk a straight line suggests, becomes hazardous whenever sensation and balance are compromised, the structure of the foot is not to blame. The foot is one of a complex series of adaptations to bipedal locomotion, and one of the last to achieve its current configuration. It is a masterpiece of 26 bones, 33 joints, 107 ligaments, and 19 muscles that can withstand the repeated forces of up to 600 pounds while running. Yet each square inch of the sole has 1,300 nerve endings to make possible the constant adjustments needed for its role in weight bearing, balancing, and propulsion. The human foot is unique in the animal world for its two arches, transverse (side to side) and longitudinal (lengthwise). These let us stand upright by spreading the weight on the plantar surface (sole) of the foot as broadly as possible and by forming a tripod, absent in other primates. (Chimpanzees and gorillas can walk on their hind legs for a limited distance, but they cannot stand still bipedally on the ground without an object for support.) The first digit of the human foot, the big toe or hallux, has grown to bear the stress of locomotion, and the ball of the foot cushions us and helps us push off. The human foot also retains surprising potential dexterity. People born without normal hands have been known to develop complex manipulative skills in their feet, transferring the brain's motor control from one set of limbs to the other. A French physician, commenting on the mediocre work of the armless painter César Ducornet (1806–1856), once widely exhibited in provincial museums, noted that the mind that conceived the work was inferior, but that the foot had executed it correctly.[3]

People in industrial societies may assume that the tender sole would soon be ravaged by its environment without the protection of footwear. In fact, wearing shoes creates this sensitivity. After a week of barefoot living, the foot forms a protective thickened layer that, unlike calluses, does not generally block the pleasurable sensations of contact with the earth. A billion people on earth still go barefoot—some of them in harsh conditions, like the Seri of the Sonoran coast of Mexico, opposite Baja California, whose feet are toughened by the sand until they form a "Seri boot." In the rough, hilly regions of the Congo, unshod indigenous people developed a keratin coating on their soles so thick that they were able to walk on live coals without feeling pain. Some urban workers have managed without shoes. One investigation of several hundred barefoot rickshaw men of Shanghai, whose work took them over twenty miles of cobblestones and pavement every day, found that arch deformities and foot ailments were rare.[4]

Even in the industrial West, a surprising number of enthusiasts for barefoot living remain, despite a cultural bias dating from Roman times, when Greek barefoot customs disappeared, sandals represented imperial power, and bare feet became a mark of slavery or the most extreme poverty. In Coptic Christianity and later in some Western monastic orders, abandonment of footwear signified deep humility and poverty. St. Francis embraced the custom as a way of imitating Christ, though the order he founded later turned to sandals. But many other men and women, and especially children, have avoided footwear not to mortify the flesh but to indulge it. Well into the twentieth century, country children often went barefoot, especially in Ireland. So strong was the custom that Friedrich

Engels, in *The Condition of the Working Class in England,* reported that Irish immigrants were spreading it, especially to "the poorer native women and children of the factory towns." At the turn of the twentieth century, many Irish children still wore no shoes, even in strictly uniformed schools. In fact, some schools barred shoes as ostentatious. To urban Americans of the nineteenth and early twentieth centuries, John Greenleaf Whittier's "barefoot boy with cheek of tan" represented rural innocence, and now signifies nostalgia. In the sleepy county seat of Belvidere in northern New Jersey, for example, I have seen a plaque marking the "shoe tree," where rural children were said to have put on their shoes before entering a nearby church, reversing the injunction to Moses to "put off thy shoes" in Exodus 3:5. (Even in the United States, bare feet are still permitted in some Amish and Hawaiian classrooms.)[5]

In the early twenty-first century there is a vigorous barefoot hiking movement; one of its most active groups is led by an English-born theology graduate living in Connecticut, Richard Keith Frazine, who literally seeks to tread lightly on the earth. "Walking barefoot, as Nature intended," he explains, "humans hardly disturb even the most delicate ground cover and can delight in the soft, carpet-like feel of moss in good conscience." His fascinating book shows how the technique of barefoot walking must be relearned. Months of conditioning may not be necessary, as skeptics believe, but it takes several miles of practice on well-maintained trails a few days a week for two or three weeks to accustom the feet to what was once their natural activity. The rush of new sensations from the environment, while ultimately pleasurable, can at first be overwhelming in their unfamiliarity, as though we had grown up with our ears covered. And barefoot hiking requires learning a somewhat different technique of walking, not just for enjoying contact with the forest floor, but for noticing and avoiding obstructions and poison ivy and dealing with insects. "You must choose each footfall," Frazine explained to a local reporter as they crossed a stream. "Anticipate each step. If you kick, shuffle or drag your feet, your chance of injury multiplies enormously." Proper barefoot hiking demands a vertical step in which the foot conforms itself to the forest floor. A Canadian barefoot hiker describes a maneuver for reacting to sharp objects: rolling the foot and rolling off to the side. Even the return to intimacy with nature demands learning technique—and using technology like insect repellent, and lanolin as a foot conditioner.[6]

SHAPING HEALTHIER FEET?

Despite the still flourishing, innocuous fetishism of the baby shoe, the unshod child's foot remains an exemplar of the body's uncorrupted state. Bare feet appear to be safer for children than many shoes, according to one study of injuries seen in the emergency room of the Children's Hospital of Philadelphia in 1988. Falls cause half of all childhood accidents, and the study showed that injuries from loss of footing were more common among children with smooth-soled shoes than among the barefoot.[7]

Two generations ago, medical and popular opinion recommended children's shoes not only for safety but to prevent flat feet. The stigma of the flat foot appears to date from nineteenth-century military medicine, preoccupied with foot irregularities as signs of physical and mental degeneration and unfitness for service. (In German-speaking Europe, Jewish feet were especially suspect.) Medical authorities believed in early mechanical intervention for healthy foot development. As late as 1920, a medical text warned that "infants should be taught to creep, and early walking discouraged." Orthopedists and pediatricians favored stiff shoes with corrective inserts, some still used in special cases and others long forgotten. By the 1930s, no fewer than fifty "doctor's" shoes and one hundred types of arch support were available in the United Kingdom. Manufacturers in the United States (following the lead of the physician, shoemaker, and master salesman William Mathias Scholl) were equally zealous in producing "corrective" juvenile shoes. As William Rossi later observed, these shoes had no scientific basis. Even now there is no standard for a healthy or unhealthy, high or low arch.[8]

And at least in the United States, the military itself has changed its stance. For decades, North American army doctors rejected conscripts and volunteers alike for flat feet, even star athletes like Satchel Paige (who thereby enjoyed a lucrative civilian career during World War II) and accomplished hikers like the professional hunter who was turned down after walking two hundred miles to Edmonton to enlist. A 1947 Canadian army survey of foot problems warned that training recruits with low arches was a "useless waste of time, effort and money." This opinion was challenged only in 1989, when scientists at the U.S. Army Institute of Environmental Medicine in Natick, Massachusetts, found that soldiers with low arches had half as many foot problems as those with high insteps. High arches may be more rigid and unstable; flat feet, unless they are painfully so, are more flexible and better equipped to absorb the shocks of exercise. One of the study's investigators told a reporter: "I've seen drill ser-geants with arches as convex as the bottoms of rocking chairs, who are active and successful."[9]

Beginning in the 1970s, medical researchers began to question the value of many corrective shoes prescribed for children. They saw great variation in normal, healthy young feet. A study of the feet of 2,000 chil-dren, published in 1971, revealed that of those who were wearing special shoes for flat feet, only 43 percent had true pes valgus (the technical term for an abnormally low longitudinal arch). The "corrective" shoes that these children had been wearing had not had any effect on their arches. In the early 1990s, orthopedists in India examined the static footprints of 2,300 children and found that the incidence of flat feet among those who used footwear, especially closed shoes, was three times higher than among chil-dren who were exclusively barefoot, although the longitudinal arches of the great majority of the flat-footed children recovered naturally after the age of eleven. Studies elsewhere found that 80 percent of two- to four-year-old children were beginning to show malformation of the toes as a result of wearing shoes; that nearly three-quarters of children's shoes are between a half size and 3.6 sizes too small; and that the lasts (wooden or plastic forms on which shoes are built) of children's shoes differ from the dimensions of normal children's feet. After a review of the literature, the orthopedist Lynn T. Staheli concluded that the best criterion for children's shoes is how well they approach the barefoot state, including their spa-ciousness, lightness, flexibility, and porosity. Except in cases of congenital deformity—some flat feet and other conditions can be painful and do ben-efit from treatment—the dream of molding ideal feet through stiffly engi-neered shoes is as dated as the idea of the superiority of formula to mother's milk.[10]

Children's and adults' feet still need protection, and not just from hos-tile terrain and climate. The domestication of dogs, for example, has exposed humanity to the risks of the bloodsucking hookworm, endemic in the tropics and subtropics. In the glorious days of shoeless youth in the American South before World War I, there were nearly four million cases of hookworm infection, and even one barefoot enthusiast in Charleston, South Carolina, acknowledged that he needed hospital treatment for it while growing up. Hookworm larvae usually enter the body through exposed toes and multiply in the intestine; the parasite can rob children of up to a quarter of their normal growth and drain the energy of adults. Elsewhere in the world, many of the most dangerous parasites are waterborne, notably the flatworms that cause schistosomiasis (bilharzia), which affects up to 200 million people globally. Footwear can be a barrier against the larvae, which are transmitted by snails spread inadvertently since the nineteenth century by migration, colonization, and irrigation projects. The alternative to covering one's feet may be chronic damage to internal organs from the massive release of eggs by the females. Next to sanitation, footwear has been such a major theme of antiparasite campaigns that Southern critics of the 1909 (John D.) Rockefeller Sanitary Commission for the Eradication of Hookworm Disease accused the philanthropist of planning to enter shoe manufacturing. These critics were overlooking the drop in the rate of infection once children reached the "shoe age" of fourteen. And during Henry Ford's unsuccessful attempt to grow rubber in the Brazilian Amazon, the *seringueiros* (rubber gatherers) balked at the company doctor's campaign to combat hookworm by replacing their sandals with shoes.[11]

There are other risks in modern life. Ubiquitous rusty nails can transmit tetanus. Even the original habitat of barefoot hiking, the woodlands of the northeastern United States, are infested by deer ticks bearing Lyme disease. Urban environments present risks as well. In the 1950s, physicians at the University of Hong Kong found the feet of the city's unshod fishermen and -women more mobile and structurally far healthier than those of a group of their shoe-wearing patients. Impressed as they were with the fishers' ability to grasp lines and nets with their toes, they also found in their keratinized soles "many minor lacerations due to traumata" and observed that striking stones and other objects on the ground had produced toenails that were "thick, cornified, and short with uneven and jagged edges." Footwear, then, may not correct but it can protect. Each modification of the environment increases the pace of the technological treadmill. Once we have stepped on it, we have stepped into footwear.[12]

THE SANDAL: ANCIENT SOPHISTICATION

The simplest solution is the sandal: open footwear, secured by leather, fiber, or other material or, especially in India, grasped by a knob between the first and second toes. Sandals may have appeared independently in widely different societies, or they may have been diffused and elaborated during the thousands of years of migrations by which the earth was peopled. One of the first depictions of sandals, on a five-thousand-year-old Egyptian palette, is of the barefoot King Narmer followed by a servant bearing a pair of them. The Egyptians of the Eighteenth Dynasty (1567–1304 B.C.) wore sandals of leather and of woven palm-leaf strips little different in appearance from some of today's beachwear. By the end of this period, there was a hieroglyph for *sandal*, a long oval with an inscribed, upside-down **V** representing the thong, that would be universally recognizable today. We know more about sandals and shoes than about most other items of apparel from prehistory through modern times. Many of the plant fibers used to make the sandals of desert peoples were unappealing to insects, while tannin has prevented the biodegradation of leather deposited in bogs and even in the remains of the *Titanic* on the North Atlantic seafloor.[13]

Today sandals evoke primitive simplicity and a return to the natural. And they do affect our feet far less than closed shoes. They may be the most gender-neutral footwear of all, removed from the swagger of boots and the swaying walk of high-heeled shoes—even though recent medical research has suggested that the latter do not have nearly the radical effects on posture that laypeople and professionals alike had supposed. But sandals, while connoting naturalness, are far from simple. They have inspired not only elaborate techniques of manufacture but unexpected changes both in the feet and in the act of walking.[14]

One of the first memorable descriptions of these differences occurs in Herman Melville's *Typee* (1846): the toes of the Marquesas Islanders are "like the radiating lines of the mariner's compass, pointed to every quarter of the horizon." Steele F. Stewart, an orthopedist who studied barefoot peoples around the world a century after Melville's voyage to the South Seas, underscored the differences between these and the shoe-wearing nations. The fourth and fifth toes have a prehensile curl toward the midline of the foot, and barefoot peoples grasp things with their toes. Their normal gait is smooth and restful, a rolling motion beginning on the heel, continuing with the outer edge of the foot, and ending with the ball of the foot and the toes, which extend and contract as the foot makes contact, helping give a final push while walking. While Stewart did not believe that sandals interfered with the foot—at least, not if they were cut to the foot's natural shape—later studies suggest that even the simplest footwear starts to rearrange the bones of those who habitually use it. A team of Japanese medical researchers compared the feet of barefoot East Javanese, sandal-wearing Filipinas, and shoe-wearing Japanese, and discovered that the second were in some ways closer to the third than to the first. In the Filipinas, the ratio of foot breadth to foot length in proportion to their body mass was similar to that of (shod) Japanese women. Sandals also interfere with the roll that Stewart observed. And while they do not deform the small toes as most closed shoes do and preserve and even develop the toes' natural gripping abilities, thong sandals do increase the separation between the big and second toes of children who grow up wearing them. Japanese authorities, it is said, were able to distinguish assimilated Koreans from ethnic Japanese during World War II by inspecting their bare feet, as Koreans generally did not wear zori. On the other hand, expatriate Westerners are sometimes bemused to find their children growing up with such a gap. Those who take to zori later find the adjustment less natural. A young American teacher who came to Hawaii in the 1960s, Victoria Nelson, took months to develop the necessary "zori callus" between her toes, alternating several pairs so that the blister, whose location varied from pair to pair, had time to heal.[15]

To judge from extensive remains that have been found, early sandal technology makes much of today's mass-market footwear appear crude, no matter how elaborate the machinery that produces it. Armed with a new carbon dating method, accelerator mass spectrometry, which spares all but a tiny sample of ancient textiles, archaeologists have been discovering that complex and beautifully produced sandals were older than they had thought. Some sagebrush-bark shoes from Oregon turn out to be nine thousand years old, and various remains from a single cave in Missouri have been dated as between eight hundred and eight thousand years old. The earliest was a padded sandal of plant fiber with a pointed toe, a slingback formed from twisted lengthwise elements, and a cord zigzagging through loops across the foot, tied down at the ankle. (Some sandals were made from *Eryngium yuccifolium,* or rattlesnake master, considered an antidote to snake venom.) From the Anasazi people of the desert Southwest, the Utah Museum of Natural History has hundreds of sandals made from yucca leaves and cordage, dating from seven hundred to two thousand years ago, and also exhibiting an impressive variety of forms, weaves, and tying systems that are surprisingly like the technology of today's open footwear.[16]

Impressive as these sandal collections are by today's standards, they are crude compared to the examples in another noted collection from the Southwest, 188 sandals from about fourteen hundred years ago, collected and first analyzed by the early-twentieth-century archaeologists Ann and

Earl Morris at a site in northeastern Arizona that had been occupied by ancestors of today's Pueblo peoples. The Puebloans' descendants had turned to leather moccasins by the time of the Spanish conquest of the sixteenth century, but whatever the reasons, more sophisticated workmanship was certainly not among them. The examples in the Morris collection are unique and nearly flawless; learning to make them must have required careful observation of master weavers. While the Puebloans' agriculture, houses, and pottery were unremarkable, the complexity of the sandals' construction startled the archaeologists who rediscovered them in the 1990s. One of the investigators, Kelly Ann Hays-Gilpin, has identified at least twenty-six textile techniques used to make them, and a graduate student of another investigator prepared a master's thesis on the toe area of the sandals alone. The investigators believe the variety and complexity of the geometric patterns, more elaborate than the designs of any of today's footwear, were associated with communities, families, and individuals, their raised patterned soles leaving distinctive footprints to be read by friend and foe. People were living together in larger units, and the decorations may have helped to assert or maintain the identities of groups. But the sandals were functionally as well as symbolically ingenious. They had doubled-warp toes (with twice as many lengthwise fibers) for extra protection. Twill twining was used under the toes and the ball of the foot for a flexibility matched by few sandals today. Raised sole designs under the ball of the foot also improved the sandals' grip on stony or wet surfaces. The depth of their treads, rounded and patterned heels, ridges oriented in multiple directions, edges pointing outward to expel water, all meet the latest specifications for rubber safety footwear published by the American Society for Testing and Materials. The hide moccasins that replaced the sandals by Columbus's time were functionally and aesthetically more primitive than the earlier footgear. New materials may have unintentionally broken the transmission of refined and beautiful techniques centuries before the age of plastics.[17]

THE SANDAL: JAPANESE FRUGALITY

Steele Stewart wrote of "pan-Pacific" sandals worn for protection of the sole by the inhabitants of the volcanic cordilleras of Central and South America and of the volcanic Pacific islands whose shores are lined with sharp coral. Japan did not have the same rough terrain as other Pacific islands, but it did develop footwear well suited to a climate in which hot, humid summers alternated with cold, snowy winters. Japan is a densely populated country, with much of its land unsuitable for agriculture, and its people have had to use its limited supplies of food, timber, and other resources efficiently.[18]

Traditional Japanese footwear suited this environment admirably. It also responded to religious and cultural influences. In the sixth century, under the influence of Buddhist teachings against animal slaughter, wooden clogs called geta replaced leather shoes and boots outdoors. (Even now, many leather workers are descendants of the former pariah caste of burakamin, to whom this trade was reserved.) Geta, unlike Western clogs, are flat platforms supported by two transverse blocks between two and four inches high placed far enough back so that the front end can tip forward. They were more elegant versions of the wooden planks fastened by rice straw straps that rice paddy workers had already been using for hundreds of years. Two cloth strips forming a V are fastened at the sides toward the rear of the geta and meet between the first and second toes. Geta raise the wearer above mud and puddles, originally protecting the hem of the kimono, the universal outer garment of men and women alike. Sizes were standard for men and women, and even left and right geta were interchangeable, as the forward point of the V was in the center of each platform, not toward the body's center where the toes would naturally lie. Men's zori and geta were (and are) square, while women's had rounded corners. While some women's geta were luxuriously fitted with sole coverings of plaited rushes and rich fabric thongs, male counterparts could be so similar to one another in size and appearance that drunken dinner guests often walked off with the wrong pair. The best geta were carved from a single block of pawlonia wood, sometimes covered with tatami (bamboo) insoles; others were assembled with tongues and grooves. Far from declining with industrialization, geta became affordable for most poor Japanese with nineteenth-century machine production. But artisanal techniques survived. Well into the twentieth century, craft shop owners traveled deep into Japan's mountain forests to buy whole pawlonia trees for the best-quality geta.[19]

Other footwear needed no quest for rare materials; it recycled common ones. The original Japanese sandals, *waraji,* made of rice straw, appeared even earlier than geta, about two thousand years ago, as rice was becoming a staple of the Japanese diet. Like their yucca counterparts in the New

World, they were one of history's most stunningly economical items of costume. Rice straw, a waste product of food production, became both a practical and a religious item. Shinto shrines have sacred straw ropes. Straw was the material of the roofs and mats of traditional houses, raincoats, and rain hats. Woven straw matting, bound with straw rope, secured food products in transit. And the same renewable straw was woven into *waraji* sandals with a twisted straw cord passing through loops at either side of the foot and heel and secured at the ankle. (An immense straw *waraji*, supported by more than twenty bearers, was recently dedicated to Tokyo's Sensoji temple, as an embodiment of the protection of the nation.) In the ninth century, a new kind of straw sandal was introduced, the zori, following the V pattern of geta. Rush, bamboo sheaths, and other economical materials were often used instead of straw. While *waraji* remained country work footwear, zori ultimately were produced with many variations of design and workmanship, including luxurious white brocaded silk wedding models and others with double rice-straw soles presented by the fiancé's family to the bride-to-be on their betrothal.[20]

Straw sandals, along with geta, formed part of an economical and hygienic way of life that ultimately helped set the stage for Japan's economic rise and world cultural influence in the Meiji era. Easily removable footwear have always been preferred in societies living at mat level rather than elevated in chairs. While part of the floor of country houses was bare earth and considered an extension of the outdoors, sandals and clogs were left at the edge of the raised wooden platform that marked a family's living space. Even when zori were used as house slippers, they were not allowed to touch the sensitive surfaces of tatami mats. *Tabi,* mittenlike socks with separate room for the big toe, were always worn with sandals outdoors, and sometimes with geta. (Some Romans wore a similar woolen or leather foot covering—called a *soccus,* whence the English word *sock*—with a leather counterpart of the zori.) Untainted by the dirt of the street, *tabi* were ideal for indoor wear. After the introduction of indoor plumbing, users of bathrooms stepped into special zori to protect the feet (and living areas) from contamination, a custom that still prevails. The design of the sandals and clogs permitted people to change pairs rapidly without touching the footwear. So strong was this custom that, according to tradition—or perhaps it is a Japanese urban legend—when the first trains opened service in Japan in the late nineteenth century, travelers left their sandals on the platform before boarding and were surprised to find them gone when they returned.[21]

Luxurious or simple, geta and zori were hygienic, and not just because they were left outside. Climate as well as belief long inhibited the development of closed leather shoes in Japan. Just as heat and humidity fostered the open-plan post-and-beam Japanese house with its sliding partitions, they encouraged open footwear. Leather shoes rot easily, and they also trap warmth and moisture. Even all-leather sandals may fall apart in the humid Pacific, as Victoria Nelson discovered when she moved from northern California to Hawaii. As closed leather shoes became more common in Japan after World War II, dermatologists reported a surge in athlete's foot, from 31 percent of all fungal infections in 1945 to 76.8 percent in 1955, according to one study. In one factory, nearly three-quarters of workers wearing closed shoes had athlete's foot, while only about a quarter of those with sandals and slippers suffered from it.[22]

Closed shoes also reshaped the feet of Japanese in the postwar era, especially in the 1960s and 1970s, when the wearing of geta and zori with *tabi* declined even more sharply than in the immediate postwar years. Hallux valgus, the often painful inward deformation of the big toe, was unknown in early Japan, though it has occurred among other barefoot peoples. There were no signs of it in footprints from Japan's Jomon period (a time usually defined as from 10,000 or 6000 B.C. to about 300 B.C.) that were discovered in the north of the island of Kyushu. Two orthopedists at Kyorin University in Tokyo reported that they never had to perform surgery for hallux valgus until after 1972; then they saw eighty-five patients in less than ten years at the university orthopedic clinic. Only after children were allowed to wear fashionable shoes, beginning at age fourteen, did signs of hallux valgus appear.[23]

SANDALS, SHOES, AND TECHNIQUES OF MOTION

While zori helped protect the foot from infection and kept it closer to its natural shape than closed shoes do, they also affected how the Japanese walked. Of course, traditional Japanese differed in their walking styles even as people of all nations do today. Edward S. Morse, an American zoology professor who came to Japan in 1877, observed that the geta in a household closet displayed "the same idiosyncrasies of walking as with us,—some were down at the heel, others were worn at the sides." But footwear still influenced gait.

The centered thong may have encouraged the pigeon-toed gait that foreign visitors noticed among Japanese women; kimonos, apparently shapeless when laid out but worn tightly on the body, also affected women's walking style. When the hourglass figure, with its full bosom and hips, began to influence Japanese fashion in the 1890s, women needed walking lessons to use the new foreign shoes. A satirical Japanese cartoon of the time shows one such group with the caption: "Follow me, ladies. Feet pointed straight ahead and chest out. Good! Now smile."[24]

For men, the influence of zori and Western shoes has taken a different course. In the Meiji era (1867–1912), when Japanese offices came under Western influence, male garb was much more strongly influenced by European and American fashions than women's clothing was. Shoes went along with Western hats and coats, inconvenient as it was to remove and replace them. But boys as well as girls continued to wear clogs and sandals as they grew up; Japanese schools still require this style of footwear. The results fascinated nineteenth-century Western visitors. One of them, the American Alice Mabel Bacon, observed that geta worn from an early age strengthened young children's feet, so that those who grew up to be artisans were able to grasp work with them: "Each toe knows its work and does it, and they are not reduced to the dull uniformity of motion that characterizes the toes of a leather-shod nation."[25]

The usual Japanese description of the difference between Japanese and Western male walking uses the contrasting metaphors of "piston" and "swing." At least one Japanese academic who has studied gait, Michiyoshi Ae, even detected a different sound made by the walk of an American woman studying in Japan, a sharper "Ka-Ka-Ka" as opposed to the dragging pattern of the Japanese ("Klan-Klan-Klan"). Growing up wearing zori appears to be at least partly responsible for the Japanese style. A Japanese graduate student at Western Michigan University, Ko Tada, used electrical measurements of muscle contraction, videotape, and force platforms to compare the walking of ten other students, five Americans and five Japanese who had been born in Japan and grew up there. The gaits were measured both when the subjects were barefoot and when they were wearing flip-flop sandals, contemporary versions of the zori. Tada found that Japanese and American young men have different, unconscious techniques for walking in flip-flops. Americans, unaccustomed to them, tend to grip them more firmly and to insert the foot farther into them, keeping a smaller angle between foot and sandal. The American walk coordinates the motion of ankle and knee to keep the center of gravity vertically steady during motion, promoting forward acceleration into the next step. The Japanese subjects dragged the heels of the sandals as they walked. Both barefoot and with flip-flops, the Japanese landed not with the heel strike characteristic of the American subjects but with the forefoot or flat-footed, a technique that does not absorb the shock as effectively. (*Ashinaka,* half-footed sandals worn by medieval Japanese soldiers in wartime, saved weight by eliminating the heel altogether.) The risk of injury leads in turn to shorter strides that reduce the braking force. The Japanese stride, even barefoot, restrains the full motion of the thigh, a muscular pattern that Tada and her thesis supervisor, Mary Dawson, believe originated in the need to keep the sandal from flying off the foot like a projectile as it advances. The Japanese also tend to lean forward as they walk with rounded shoulders, while Americans have a straighter carriage, which Tada attributed to "a long history of wearing shoes with flexible soles."[26]

Ko Tada cites studies suggesting that neuromuscular patterns fixed in childhood shape not only the mature walking gait but also the style of running and jumping. Childhood footwear thus can ultimately shape sports performance. Tada notes that holders of Japanese running records, including one finalist in the 1992 Olympic Games, are athletes who developed the swing rather than the piston walk. Otherwise Japan has comparatively few world-class runners.

Of course, as Tada recognizes, other biological and cultural differences besides footwear might affect gait, and her work is only a beginning. But anecdotal evidence from Japanese and Western sources tends to support her findings. The son of the Japanese sports scientist Michiyoshi Ae spent his first years in the United States and was not given zori or geta to wear. (He was not even allowed to ride a tricycle, as the pedaling position develops a muscle on the front of the leg associated with the piston walk.) By the first grade he was walking with the Western swing style. As he grew, he retained this stride even when wearing geta. His sister also learned to walk in the Western manner. On the other hand, some Japanese-American children who grow up in the United States wearing traditional Japanese footwear develop the piston walk. What we wear evidently helps determine how we move.[27]

THE SPREAD OF ZORI: MIGRATION, WAR, AND PEACE

More than a hundred years after the opening of Japan to the West, then, the Japanese have a dual system of footwear: traditional thong sandals for children, closed shoes for adults. Each type of footwear has positive and negative consequences for health and athletic performance. But just as the

Japanese were turning gradually but decisively to Western-style shoes for adults in the postwar years, the zori was finding its way around the globe. It was becoming the first world shoe.

As they migrated across the Pacific in the later nineteenth century, Japanese men and women brought their national footwear with them. Even an early Japanese Christian minister wore burlap sandals. The simplicity of materials—wood and fiber—meant that immigrants could fashion their own geta, zori, and *waraji* in nearly any warm climate. And they did. Before World War II, Japanese working on the sugar and pineapple plantations of Hawaii wore geta, if not every day, then for walking in damp fields or doing laundry in concrete-floored plantation washhouses. There were also lacquered dress geta for dancing. While the decline of rice farming cut off the straw for sandals, the saltwater marshes of Hawaii abounded in a bulrush like the rush that rural Japanese used for weaving tatami mats and footwear. A few Japanese-American artisans continue the craft to this day.[28]

The manufacture of thong sandals on American soil appears to have begun in Hawaii during World War II. The firm of Scott Hawaii, established in 1932, had made rubber plantation boots for local workers. During the war, shortages of raw materials led to a shift to rubber sandals—it is not clear how many of these were zori style—and other open footwear. These also proved popular among sailors of the U.S. Navy on their way to the Pacific theater. On the hot steel decks of warships in tropical climates, rubber sandals were superior to government-issue footwear. In submarines they were a must. Indeed, even during the Vietnam War, shipboard American officers and men arranged to buy locally produced flip-flops.[29]

In Asia, war also helped spread Japanese footwear. Like their American counterparts, Japanese soldiers often found original-issue equipment unsuitable for local conditions. Once more, the technique of making straw goods proved its versatility. In a photograph from the 1937–38 China campaign, a group of Japanese infantrymen sit together on a straw-covered ground, looping cords of straw through their big toes as a first step in making *waraji,* and surrounded by well-made finished pairs. But straw was not the only material Japanese soldiers used. In Southeast Asia, they also began to cut up worn-out tires that they made into zori. During the battles for Malaya in 1941, thousands of British troops died in prisoner-of-war camps, prevented from escaping by boot-inflicted foot ailments, while Japanese troops were able to move through the jungle with what seemed eerie silence.[30]

The crucial time for the worldwide spread of the zori was the decade after the war. The details are not clear, but it is likely that rubber footwear was one of the early products of a recovering Japan. Japan no longer controlled the rubber plantations of Malaya, but the plantations and other Asian rubber producers still needed a market for their product. The manufacture of closed leather or even plastic shoes requires dozens of specialized machines; thong sandals need only a few. Extremely light, they can be shipped without individual boxes. Natural, synthetic, or recycled rubber can be used with fillers and blowing agents to create different grades and densities. The better models do require individual molds, but cheap sandals can be die-cut by semiskilled workers from a large sheet with tread patterns embossed on either side. For the most common thong sandals, the soles are drilled with a narrower hole at the top and wider ones at the bottom, and the molded plastic thongs are inserted into the three holes with large tools resembling crochet hooks.[31]

Best of all, the new zori could be made of materials other than rubber. A growing number of synthetic polymers became available to the industry: synthetic rubber, polyvinyl chloride (PVC), and ethyl vinyl acetate (EVA). Where the straw zori and *waraji* had represented self-sufficient frugality

More than a thousand years old and now worn by poor and rich around the globe, the zori is endlessly adaptable to new materials. This experimental sandal, called "Cold Feet," has a neoprene outsole overmolded on a polypropylene insert containing refreezable, cooling "blue ice." Designer: Gretchen Barnes. (Courtesy of IDEO)

and use of agricultural waste, these nonrenewable materials symbolized the cosmopolitan and technologically advanced postwar Japanese economy.

THE NEW PAN-PACIFIC SANDAL

Rubber and plastic zori were not limited to Japan. In fact, as interest in closed footwear grew after the war, the Japanese considered them neither traditional nor Western. In the memory of one Japanese-American scholar who was a young man in the 1950s, it was Harry Belafonte who popularized rubber sandals simply by walking down the main street of Kyoto wearing them during a tour in the middle of the decade. Belafonte's style impressed crowds and began a national fad. But the new zori was more than a Japanese event. It was a technological reincarnation of what Steele Stewart, as we have seen, described as the pan-Pacific fiber sandal.[32]

In the United States, rubber sandals began to appear in California in the early fifties. Allan Seymour, a surfing entrepreneur and historian of the sport, recalls seeing the first cheap models in Laguna Beach, where he was growing up, worn by soldiers returning from the Korean War. Bins of them at 29 cents a pair soon appeared outside liquor stores and markets in Laguna; they became wildly popular on the beaches, and standard attire for surfers. They were badly made. The straps often broke or "blew out"— the phrase is an unconscious reminder of the role of automobile tires in the early history of rubber sandals. They lasted for a month of surfing, and broken and discarded sandals littered the beach at low tide, but they were priced as disposables. In northern California the Beats are said to have adopted zori as part of the culture of Zen. By the early 1960s rubber sandals had made their way down the coast to northern Mexico, and were selling under the trade name Havaianas (Hawaiians) in Brazil. In Mexico and South America, they were originally considered plebeian rather than chic. Mexico City residents disdained them; in Brazil, they inspired the phrase pé de chinelo—literally, "slipper foot"—for the downtrodden. The Brazilian government included them in its working-class price index, along with milk, bread, and beans.[33]

All the while, zori were spreading slowly to the American heartland as an exotic novelty. A small advertisement in the May 1958 issue of House Beautiful presented sandals in men's, women's, and children's sizes "[f]rom an Oriental Teahouse . . . For Beach, Pool, Shower or Street." The text lauded the sponge-rubber product as the "[n]ewest style-rage inspired by the mystic East and imported direct from Japan—yet practical and modern as Miami. Strong, sturdy, lightweight, comfortable, long wearing . . . skid-proof, silent," and only $1.95 postpaid. Six years later, they were no longer a novelty. Another advertiser was offering women's Japanese-made "Dress Zori Sandals" with leather insoles and straps for $2.95.[34]

In Australia, too, postwar beach (if not bohemian) culture created a ready market. Australia's giant tire manufacturer Dunlop was already beginning to make sandals in 1954, using inflating rubber with air bubbles in molds and letting it shrink to size in steam rooms. The real boom started with the arrival of the Japanese swimming team at the 1956 Melbourne Olympics, wearing rubber zori as part of their uniform. By 1957, the David Jones department store in Sydney was selling them to delighted customers from a bin at the door. Soon local makers were cutting them out of sponge rubber sheets, three pairs at a time. Even then, demand seemed insatiable. A Dunlop executive ordered 300,000 cut-rate pairs on a single short trip in 1959. But while in largely chillier North America they remained beach accessories, in Australia they became everyday mass footwear. Dunlop alone sold a pair a year for every man, woman, and child in Australia. A local joke was that you could tell the bride at an Australian wedding because she was the one wearing new thongs. By the late 1980s the sandals were not so funny, and certainly not chic. One newspaper critic, Leo Schofield, called them "the national symbol of bad taste," on a level with furry toilet seat covers, and "[t]he ugliest and least practical kind of footwear. Also dangerous." In 1995 the director of the Melbourne School of Fashion, Miriam Cuna, declared them "vulgar" and "awful." David Jones had dropped them, and even a spokeswoman for Kmart told a journalist, "We don't advertise them." But they remained popular. In 1995, a single large distributor imported 2.5 million pairs of high-fashion models annually, and Kmart itself sold 240,000 pairs in the state of Victoria alone.[35]

Meanwhile, another style of sandal was making news. The North Vietnamese army and Vietcong, perhaps taking a leaf from the Japanese army's book, made their own sandals from rubber tires. The soles had multiple slits for straps—the design resembled the sandals of the Romans—and, according to an American physical anthropologist who examined recovered specimens later, could last for years. Ho Chi Minh, who had worn white sandals during the independence movement against the French, now himself set an even more austere example by keeping the same pair of rubber-tire sandals for years, refusing to replace them. But this austerity lost its prestige after victory. The civil servants of Ho Chi Minh City were under orders to shed their sandals for closed shoes at work as a symbol of their country's new economic realism and aspirations.[36]

In Hawaii during the 1960s, zori showed how the simplest of technologies could evolve into a complex system of ethnic symbols. The first civilian sandals produced by Scott Hawaii after the war had crossed straps over the instep, but zori emerged as favorites by the early 1950s. Steve Scott, a son of the founders of Scott Hawaii, recalls his family making the early models from molded pieces of EVA from Japan, fitting them with leather insoles and thongs. By the time Victoria Nelson arrived in the late 1960s, Japanese students at the University of Hawaii wore *lauhala,* a local version of the classic zori with a bottom made from native plant leaves and a velvet thong, also known as Jap slaps. Haole (mainland European-American) surfers wore thick black rubber zori with a racing stripe down the side and a thong of parachute fabric. For the fashionable there were "cork-soled plastic-thonged steamboats [presumably local slang for bulky footwear] made in Taiwan or Brazil" and the "Hollywood forties-style wedgie with bejewelled thong favored by Dragon Lady types." Some had holes running through them for draining water; others had no thongs at all but came with a preparation for sticking them directly to the wearer's sole. Stores served locals with seemingly unending varieties to match the tacit dress codes. If Hawaii was a cradle of multiculturalism, zori were its baby shoes. And even today, zori are as ubiquitous in Hawaii as in Australia. Their secret, according to Steve Scott, is the variety of materials—including EVA soles injection-molded with a new technique borrowed from the athletic shoe industry—that keep production costs low and styles and colors easy to change.[37]

THE ZORI'S WORLD MARCH

The zori did not remain confined to the Pacific Rim, though many of the world's sandals are still made there. It became one of humanity's first universal manufactured goods, much more so than the branded soft drinks and cartoon-character merchandise that in many places could be bought only by the relatively well off. During Nelson's tropical idyll of the 1960s, Mitsubishi and other giant Japanese concerns took over the small factories of Kobe, where many zori were made, and decided to shift production to Taiwan, where costs were lower. Taiwanese manufacturers had experience in producing synthetic goods like plastic tablecloths, and it was not hard to convert their machinery to the production of footwear uppers. Existing technology, some of it located in duck huts and hog houses, needed mainly the spread of new techniques. The Japanese trading company inspectors, as one of them later put it in an interview, "moved around like bees spreading pollen among separated manufacturers so that innovations in manufacturing spread quickly." At the same time, the Republic of China (Taiwan) was beginning a drive to expand its export industries because it expected American aid to be phased out. Despite or because of the presence of the Japanese giants and the skills they brought, opportunities opened for smaller companies. Especially for the supply of PVC, which was winning favor over rubber (because of the latter's high vulcanization cost), Taiwan had an excellent base: the plastic industry had burgeoned from only two factories in 1948 to more than four hundred in 1966. Sandal and shoe plants followed. (Zori probably were not the only sandal forms made but almost certainly were the most common.) From 1969 to 1988, the number of registered footwear firms increased from 75 to a peak of 1,245. Even after a decline in production in the 1990s as cheaper labor costs in mainland China, India, and elsewhere led to a shift in production, Taiwan remains a major center of capital, design, machinery, and supplies.[38]

A 1982 publication of the International Labour Organization in Geneva provides a clue to the spread of sandal manufacture. While zori manufacture of course requires far more capital than artisanal leather sandal making, it can be done efficiently in a relatively small shop. The report lists thirty-one stages of closed-footwear making and only two of the injection molding of flip-flop sandals. A large shoe manufacturer can gain an advantage from adding additional equipment and workers at many steps. It pays to be big. The maker of PVC-injected sandals needs no skilled employee, only two or three operators and a packer. There is no equipment that would reduce costs substantially for high-volume plants. Because opportunities for automation were so limited, Taiwan government statistics from 1976 showed that in the plastic footwear industry, the few very large factories of a thousand or more employees actually added only two-thirds the value per employee that the many plants with under thirty employees added.[39]

With a global market in machines and variety of materials, sandal production could move wherever there was a market and capital. For international giants like the Canada-based, originally Czech manufacturer Bata, the largely barefoot parts of the developing world were a challenge to promote the "shoe-wearing habit." Thong sandals, at an international price of 30 to 40 cents, cost less than an unskilled worker's daily wage, and even the higher-quality models could sell at retail for only 65 cents. In Kenya, for example, Bata was able to build its sales from 972,000 pairs in 1964 to two million in 1968. (Even in 1997, many villagers in Kenya were still barefoot, risking threadworm infection transmitted in outhouses; a resourceful Japanese community development specialist was teaching rural women to make straw zori.) But there was room for many local entrepreneurs, too. Because plastic zori wear out in as little as three months and are not readily repairable, the market is steady. Urbanization makes footwear increasingly necessary. There is evidence that a number of prominent families in developing countries formed at least part of their capital with zori factories. And sandal making can become big business indeed; in less than forty years the Brazilian company Alpargatas, once a manufacturer of rope-soled canvas moccasins, has sold two billion pairs of its Havaianas brand flip-flops. Sales are now at 125 million pairs a year, and Alpargatas is producing high-fashion lines for sale at luxury outlets overseas.[40]

Sandals have built not only wealth but nations. In the early 1980s, overseas Eritreans presented the Eritrean Popular Liberation Front (EPLF) with an Italian PVC injection molding machine that could produce a hundred pairs of black (for camouflage) plastic sandals per hour. The rebels imported base granules of PVC but supplemented it by recycling the worn-out footwear of their troops. Just as family companies later diversified from sandals into higher-margin activities, the EPLF later was able to open pharmaceutical factories using foreign feedstocks to make 40 percent of the basic drugs Eritrea needs. (EPLF sandals had straps across the forefoot rather than thongs.)[41]

THE UNDERSIDE OF A STYLE

The zori sandal appears to be wonderfully universal, improving the health of people in the developing world yet adaptable to the high fashion of the West. The president of Alpargatas calls his Havaianas "the most democratic product made in Brazil. The guy who owns a mansion wears Havaianas and the guy that cleans his swimming pool wears them, too." Queen Silvia of Sweden delighted her hosts on a recent state visit by wearing the company's Havaianas Brazil model, with its little national flag. In the United

States, a leading magazine for young women urges its readers to "let the air breeze between your toes—and . . . show off your sexy feet." The illustrated sandals closest in appearance to the original Japanese zori are made by Calvin Klein of wood with leather thongs and cost $245. They probably are not much more or less comfortable than many of their discount-store plastic counterparts.[42]

In their global reach, though, sandals unfortunately represent more than affluence and glamour. As the footwear of the world's poor, they also stand for the replacement of the frugal environmental ethic of early Japan with a worldwide throwaway society. Occasionally old flip-flops are recycled; in fact, an enterprising Liberian artisan, Saarenald T. S. Yaawaisan, has fashioned entire working toy helicopters, blade tips and all, from a rainbow of discarded sandals. The throwaway society still has a way of prevailing, even in Monrovia, where Yaawaisan ultimately had to buy brand-new sandals from a local factory to meet growing demand for his delightful playthings. Unfortunately, it is probably more usual to burn sandals with other refuse, potentially generating dangerous dioxin. And dumping them may not be much better for wildlife. It is not only a few surfers who abandon broken sandals on the beach today. Discarded plastic footwear is a major component of the world's flotsam, washing up in surprising places. In 1996 the Australian Cocos and Keeling Islands, home to a number of endangered species, were assaulted by hundreds of thousands of discarded flip-flop sandals, rejects of Indonesian manufacturers. After touring the islands, an Australian member of parliament, Julian McGauran, declared: "The beaches are the home of the green sea turtle and the blue rubber thong. One of them has to go."[43]

Meanwhile, on other sands, thousands of miles away along the Arabian Sea in the Indian state of Gujarat, 35,000 men from remote villages were earning as little as $1.50 a day for breaking up the world's rusty, unwanted ships, riddled with asbestos and toxic chemicals. The men wore only plastic sandals on their feet. In New York harbor, Brazilian merchant seamen embark on the North Atlantic wearing shorts and sandals. In Calcutta, children make flip-flops with scissors nearly the length of their arms. In Jakarta, Indonesia, shacks are built over a surface of old rubber sandals retrieved from refuse piles. And in formerly prosperous developing countries, the plastic sandal is a harbinger of the return of bad times; in Iraq in 1998, a former computer science student was selling them in a marketplace.[44]

The zori's universality has a double meaning. To billions of people it is a foothold on an arduous climb to sharing the prosperity of the developed nations. But it is an insecure foothold, and where parasites or other natural hazards are not serious issues, it may not be an improvement over the pleasure of walking the earth on broad, bare feet. But it is too late to return. The pioneer surfer Allan Seymour put it aptly: "We called them go-aheads, because you can't walk backward in them."[45]

CHAPTER FOUR

Double Time

Athletic Shoes

I F THE FLIP-FLOP SANDAL is the revival of a primeval classic with new materials, the athletic shoe is a rare innovation in costume. Richard Wharton, a London retail shoe executive and sneaker connoisseur, has called sports shoes "the first new kind of footwear in the past three hundred years." The novelty is as much in attitude as in construction.[1]

For most purchasers, the athletic shoe is yet another example of what could be called potential consumption. Home fireplaces and apartment balconies are seldom used, yet convey warmth or fresh air by inspiring fantasies of actually lighting a fire or sitting outside. Few owners of swimming pools actually do laps or even spend more than occasional minutes in the water; the possibility of plunging in is what cools them. Likewise the hypothetical jog, or pickup basketball game, or distance walk invigorates the athletic-shoe wearer. Even boat shoes, disconcertingly grippy on urban pavement, are always ready to step aboard somebody else's yacht. Unlike sandals, which usually constrain the gait, sport footwear with its stripes and swooshes holds out the possibility of speed without exertion.

SLOW TIME, FAST TIME, HARD TIME

The environmental educator David Orr has contrasted the "fast knowledge" of Promethean problem-solving Western societies with the "slow knowledge" of communities living together with other species in their habitats. The athletic shoe was born with the chemical transformation of the nineteenth century and may be the leading material sign of the enthusiasm for speed that has marked the last 150 years so strongly. If the original rice-straw zori stood for slow knowledge, the athletic shoe reflects and proclaims the glory of speed itself.[2]

Sandals and sneakers represent a contract between the two experiences of time. One form tends to limit speed, the other to promote it, a distinction apparent even on the streets. A federal police officer and sociologist who has done fieldwork on the crack cocaine traffic in Honolulu reports that the runners—homeless men and women who shuttle between dealers' safe houses and drug buyers—are usually able to recognize police by their black tennis shoes, worn even undercover "to be ready to fight, chase a suspect, make an arrest, etc." The runners, lacking "tactical training," wear the most comfortable local garb, sandals or flip-flops. These observations suggest that the boundary between trained officer and low-level addicted criminal is more than just the gap between enforcement and transgression. It is the difference between two ways of moving, between fast time and slow time.[3]

As early as the 1960s, black prisoners in South Africa's infamous Robben Island penitentiary were allowed only rubber sandals as footwear. More recently, prison authorities worldwide have issued sandals, often flip-flops, not only to save money but to compel a slower, controllable style of movement. The bright orange or red jumpsuit and zori has become ubiquitous as the prescribed garb of the accused in televised arraignments. In court, sneaker-wearing convicts—especially wearing civilian-style clothes—can be an escape risk not only because they can run faster, but because they can blend more easily into a crowd. Conversely, sandals can impede mobility even in warm climates; one South Carolina escapee was apprehended trying to buy a pair of shoes to replace them, while another fugitive in Arizona was caught by a discount store security guard after switching his sandals for a pair of Reeboks. And the growing complexity of athletic shoes has become a security risk in itself. The American Correctional Association (ACA) security manual for prison administrators cautions that normally "tennis shoes from the community and shoes with air pockets or pumps should not be authorized, as they provide a particularly convenient hiding place for contraband."[4]

On the street, especially in the 1980s and 1990s, one synonym for sneakers was "felony shoes": a means of escape for offenders more ambitious than Honolulu crack runners. Perpetrators like the same showy styles as other young people despite the added risk of identification. Police tracked and caught at least one convenience store robber by following the blinking lights built into the heels of his running shoes, earning him an appearance on a new television series, *America's Dumbest Criminals.* Other officers identified a young bank robber by the unusual thick-soled shoes he was wearing when arrested on a minor traffic charge. America's most proficient cat burglar of the 1990s, Blane David Nordahl, routinely discarded his shoes (as well as clothing and tools) after his sneaker print on a countertop led to arrest and a three-year jail sentence. The FBI is able to identify 70 percent of the footwear impressions it receives, according to the former head of its footwear unit, William J. Bodziak.[5]

Sneakers are as suited for self-assertion as for flight. The bouncing and dragging "pimp walk" was a staple of black exploitation films of the 1960s and early 1970s, just before the sneaker explosion. In Tom Wolfe's 1987 novel *Bonfire of the Vanities,* youthful black defendants in the Bronx County Courthouse defy officialdom with the pimp walk's successor, an insolent gait called the "pimp roll," a "pumping swagger" that another writer, a London journalist, has described as "that swaying, strutting way you walk when your soles are bouncy and your ankles supported"—a gait less feasible in sandals than in running shoes.

As usual in conflicts, the police have body techniques of their own. Many younger officers are bodybuilding fitness enthusiasts far from the old doughnut-munching stereotype, and athletic shoes fit their new image. One of Wolfe's white protagonists, the assistant district attorney Larry Kramer, begins to wear sneakers on the subway hoping he will be mistaken for one of the plainclothes policemen who have adopted them en masse. The Chicago Police Department recently converted from standard issue black oxfords to a leather shoe close in construction to running shoes.[6]

The athletic shoe embodies speed, dynamism, ambition, and the search for a technological advantage. Apart from relatively high-priced specialized sports models, such as the Teva brand developed by a white-water-rafting enthusiast, the ordinary flapping sandal, whether in flimsy plastic or luxurious materials, connotes voluntary or enforced idleness. The story of the sport shoe is as remarkable as that of sandals. And it, too, shows how new techniques of motion interact with new materials and processes to create endless variations on a classic theme.

RUNNING WITHOUT SHOES

The specialized running shoe is remarkably recent in Western history. Even before shoes were made by machine, it would not have been difficult to use lasts and materials optimized for running. When Romanticism brought an ethereal spirit to ballet in the early nineteenth century, shoe-makers responded with the pointe shoe. The new dance style, made famous by the ballerina Maria Taglioni, was and remains one of the most demanding of all Western body techniques. The shoe, with its reinforced toe box modified and conditioned by each dancer according to her personal formula, was not the hardware that made the style possible; the technique brought the new form of construction into being. (And technique is so important that most ballet dancers have so far rejected high-technology pointe shoes designed to protect toes and cushion impacts, partly because the new models are stiffer and partly because dancers are so attached to their idiosyncratic customization rituals.) It took far longer for purpose-built athletic shoes to emerge. The reason for the delay lies not just in the limits of technology and materials but in attitudes toward running and toward shoes.[7]

Running may be the oldest sport of the West. It played an important part in ancient Egyptian festivals. Egyptians recorded times for distance running that compare with those of the best nineteenth-century European athletes, and the Egyptians were almost certainly barefoot. In Greece, at the first Olympics in 776 B.C., races of from about 192 meters to 5,000 meters were run without shoes. Although Greek message carriers ran in soft leather boots called *endromides,* and although the Greeks were aware of spiked shoes used in the icy Caucasus, they had no interest in footwear for athletics. The original Olympic starting line had grooves for runners' toes, and in the absence of cleats even sprinters are pictured standing upright at the start rather than leaning forward on their hands. Bare feet were, of course, part of male athletic nudity, a custom that scandalized foreigners but served the Greeks themselves as proof of their superiority to barbarians. Even "hoplite" foot races based on warfare, run with shields and at first with helmets, omitted sandals. (Women, who ran wearing light clothing at special events for them, at least sometimes wore sandals.) The Greeks applied their ingenuity to running terrain rather than foot coverings. They went to some expense to prepare the track for bare feet. The soil was weeded and dug up before the events, and covered with fine sand. At Delphi the cost of the sand alone was considerable: over 83 staters, the equivalent of more than ten months' wages for a laborer. Runners trained for this surface, practicing on deeper sand. Today's experts recommend grass as the best running surface and warn that running on beaches endangers the Achilles tendon, but in antiquity, a carefully prepared layer of sand over dirt may have been the state of the art. Late in the Olympics, during Roman rule, Greek athletes did begin to wear a type of sandal called the *krepis,* probably owing either to the influence of Roman customs or to changing standards of track maintenance. By the time of Diocletian (A.D. 245–313) there is a reference to shoes called *gallica* used for running. But, having been born with the Romans, the athletic shoe appears to have vanished in the West with the end of the Olympics and the disappearance of Greco-Roman physical culture under Christianity.[8]

Unlike the ancient Greeks, who prized moderation too much to train for anything like the modern marathon or, indeed, any race longer than about three miles, many New World peoples were superb distance runners. Running was not only part of a spiritual regime, as it was for the Greeks; it was often essential in hunting and for communication among dispersed settlements. Native Americans developed sophisticated shoe forms like the moccasin that continue to influence the footwear industry today, and some athletes and couriers wore sandals or moccasins. The Tarahumaras of the Sierra Madre in Chihuahua, northern Mexico, were famous for a kickball race lasting up to sixty hours, run with sandals. Today their descendants still earn a living as corn farmers, their feet protected only by tire-tread sandals with leather thongs (huaraches), routinely covering ten miles or more a day at work or going to school in their harsh climate. In the 1990s they defeated some of the strongest U.S. athletes in hundred-mile ultramarathons in the Southwest, on at least one occasion discarding the running shoes donated by footwear companies that had sponsored them. Complaining that their feet were sweating, they replaced the shoes with new huaraches they had fashioned soon after their arrival in the United States from dumped tires that by Mexican standards had ample tread left. (Their U.S. supporters failed to find a tire company to replace the backing of the shoe manufacturers.) Moccasin-wearing Mohave couriers covered a hundred miles a day, at least in the times before the martial tribe's defeat by U.S. troops in 1859. Many Native Americans were able to run scores of miles daily without any foot protection at all. The Hopi Louis Tewanima (1879–1935), a 1912 Olympic silver medalist, was said as a youth to have run 120 miles from his village to Winslow, Arizona, and back, barefoot, just to see the trains coming through. Photographs of Native American races from the early twentieth century suggest that bare feet were still the norm.[9]

FROM PEDESTRIANS TO MARATHONERS

Europeans and European Americans took a different course. As the very term *chivalry* implies, European combat arts were rooted in an equestrian society that left walking and running to laborers, servants, and common soldiers. There was no courtly counterpart to the armed footraces of the Greek Olympics. Real knights did not run. Competitive Western footracing reemerged in early modern England in the lowest domestic ranks. Footmen ran at the sides of coaches to protect them from obstacles and raced to the inn at the next destination to procure lodgings. This strenuous pace required constant training, and a first-rate footman could cover sixty miles a day. In the early seventeenth century footmen began to race each other, but without the glory of Greek or Indian runners, despite the new term *celeripedians*. Servile most remained, although some artisans and even a few highborn young men also took up running, and footmen's employers bet on their servants' matches. Footmen wore pumps, light shoes with thin soles and no fasteners, kept on the foot at the toe and heel—a style now best known in women's dress shoes and men's formal shoes, not as athletic gear. One of the early aristocratic runners, the Duke of Monmouth, Charles II's eldest son, once competed in boots as a handicap and won.[10]

Competitive walking and running took the public imagination only in the nineteenth century. Paradoxically, it was improved coach transportation and, later, railroads that promoted them. When the roads were dangerous and disagreeable, only rare eccentrics and occasional foreign travelers joined the poor and criminals on foot. Once even workers could afford cheap railroad tickets, walking was no longer a necessity but increasingly a recreation, especially since trains also served destinations with picturesque paths. As recreational walking emerged, contests flourished. Beginning in the 1830s, walking footraces in England, the United States, and Continental Europe became mass spectator sports that sometimes attracted over 25,000 people. Usually run for ten to fifteen miles, the smaller contests took place in enclosures built by the tavernkeepers who sponsored them, but the larger ones were held at horse tracks. The competing pedestrians, as the contestants were called, sometimes wore gaudy clothing like jockeys' silks. In fact, the footrace was a kind of poor person's horse race without the expense of the horses. But the "peds" themselves could earn handsome purses and, along with jockeys, were America's first professional athletes.[11]

If the pedestrians used specially built shoes, they did not call attention to them. They were the sports stars of the time—some of the first modern celebrity athletes—and shoe manufacturers would have been natural sponsors. Edward Payson Weston, the most famous of the later nineteenth-century American long-distance walkers, was supported on one walk by a sewing-machine maker, a druggist, a photographer, and a clothing store in return for promotion of their products. But there was no shoe manufacturer. Weston did advise his many admirers who were taking up walking to get comfortable, low-heeled shoes. But his technique, almost a shuffle with little bending of the knees—to conserve energy—made minimal demands on footwear. He wore sturdy shoes with high red tops. The leading mid-century distance runner of England and America, William Howitt, billed as "The American Deer," is shown in a contemporary drawing wearing low, slipperlike shoes similar in shape to the former footmen's pumps. Only in the 1880s, just before the decline of pedestrianism amid corruption and drug scandals, did shoemakers start to advertise. One firm, McSwyny of New York, boasted of "Bengal Tiger, Alligator, Porpoise, and Seal Skins, Tanned especially for my use," with the endorsement of Daniel O'Leary, who had walked 520 miles in 139 hours. But in design these were probably conventional mid-length leather boots.[12]

Specialized footgear seems to have appeared first in the 1870s, when interest began to turn from marathon walking to speed running, and amateur associations were beginning to promote track and field for exercise and health. The main innovation, besides the use of lighter leathers, was spikes. Patented for cricket in 1861, they were soon applied to running. Peter R. Cavanagh, a leading biomechanics scientist and athletic footwear analyst, believes that running shoes first diverged from street shoes only in 1865, with a pair of spiked running shoes made for Lord Spencer. Probably about the same time, a New York shoemaker named John Welsher ran an advertisement before a celebrated race, offering walking shoes for $6.50. For $7.50, they were available with springs: one of the first energy-return systems in athletic shoes. Despite these beginnings, the later nineteenth century was not a great time for running footwear. Team sports, such as football, cricket, and baseball, as well as participant sports, such as croquet, lawn tennis, and bicycling, were eclipsing pedestrianism. The spiked shoes sold by the A. G. Spalding Company were costly by contemporary standards. In Bolton, England, an amateur runner named Joseph William Foster designed and made his own spiked running pumps in 1892 after finding available models too heavy; they were the foundation of an athletic shoe company that later became Reebok. By 1897 the Sears, Roebuck catalogue featured many specialized sports shoes, including calf-skin track shoes with spikes, "lace nearly to the toe," men's "Kangaroo calf" bicycle shoes, canvas baseball shoes with leather soles and counters, and canvas "gymnasium shoes" with canvas soles. There were "tennis oxfords" in black sateen and "tennis bals" ("[a]lso used for yachting, baseball and gymnasium") in white duck, a light grade of canvas. Missing were any shoes for road running or long-distance walking. In fact, most of the men's and women's street shoes shown in the catalogue were built on needle lasts that appear grotesquely pointed a century later. But while Sears sold "Feel Ezy" and "Corn Cure" shoes and police models with wide toe boxes, most of their products must have been agonizing over long distances. The same catalogue featured an assortment of corn and bunion plasters.[13]

Without help from the shoe industry, runners of the early Boston Marathons of the late 1890s seem to have improvised their shoes from other sports. The 1898 winner, Ronald MacDonald, wore bicycle racing shoes. His successor in 1899, Lawrence Brignola, used a model with "light leather uppers, lacing nearly to the toes, light leather soles and rubber heels," according to a newspaper account. Runners of the time used chamois skin inserts covering only the front part of the foot; these probably aggravated blisters. (They also clenched cork grips tightly in their hands, exactly contrary to today's practice of avoiding tension in other muscle groups.) Nor were there great improvements for distance running before World War II. The second-prize winner in 1932, John A. Kelley, used indoor high-jump shoes. Despite openings slit in the tight shoes to relieve pressure, Kelley had to drop back with blistered feet. Other runners used de-spiked track shoes and bowling shoes, but these variants do not seem to have gone into production. "The poorest running shoes manufactured today would be far better than anything we had years ago," Kelley later remarked. Although a retired English shoemaker named Ritchings custom-made excellent white kid leather shoes for competitors, marathons were agonizing. Runners would soak their feet in beef brine to harden them and rub their shoes with neat's-foot oil to soften them.[14]

THE SNEAKER ERA

Athletic shoes as we know them did not originate with the street shoe industry of a hundred years ago that supplied the pedestrians and most of the early marathoners. They began with the rubber business and a largely different set of manufacturers and techniques. Native peoples of South America had for centuries used the milky sap of the Hevea tree to coat their cloaks against rain and even to make shoes and bottles. They filled bowls with *caoutchouc*, as they called the material, and molded it by immersing their feet and then exposing them to flames that solidified the sap, yielding custom-fitting and waterproof if bloblike footwear. In the eighteenth century, French scientists began to study the tree and the process, and to advocate new industrial uses. By 1770 the first erasers appeared, "[West] India rubbers." For Northerners as for the Indians, shoes were a natural application for the material, given the rain and mud of both Europe and New England, but processing coagulated rubber was a challenge. Turpentine as a solvent made rubber goods sticky. Newly discovered ether produced better results but was so costly that only personages like Frederick the Great of Prussia could wear riding boots produced by submerging lasts in a solution of rubber and ether. Nevertheless, by the early eighteenth century consumers were buying footwear of cut sheet-rubber, and (nearly a century before L. L. Bean) a Boston merchant was importing 500,000 pairs of gum shoes made in Brazil on lasts he supplied. In 1832 a U.S. patent was granted for a way of "attaching India Rubber soles to boots and shoes." But the fashion was brief. In North America, unlike in England, natural rubber hardened in the cold winters and softened and stuck in the hot summers. By the 1830s the nascent U.S. rubber footwear industry was collapsing.[15]

In 1839, Charles Goodyear of Massachusetts discovered a combination of sulfur and heat that at last made rubber a serviceable material. One of his first licensees was the Goodyear Metallic Rubber Shoe Company in Naugatuck, Connecticut, later merged into the giant holding company U.S. Rubber, formed out of nine manufacturers in 1892. Today we think of rubber mainly as a tire industry, but even as recently as 1901, half the crude rubber production of the United States was used in footwear, and U.S. Rubber moved into the automotive market only to diversify and offset fluctuations in the shoe business. Its plants overwhelmingly produced overshoes, galoshes, and boots for miners, fishermen, policemen, and others who worked outdoors or in hazardous conditions. The first sport shoes with rubber were probably the beach slippers first worn in Europe in the mid-nineteenth century, when Europeans and Americans discovered the seashore—previously considered a desolate, marginal zone of salt-marsh farming—as a summer playground made invigorating by sunlight and salt air. As the landscape historian John Stilgoe has observed, the heat and shifting movement of sand make it an inviting but exasperating surface. Sand scorches the bare foot but its grains spill into shoes. While Charles Goodyear was developing vulcanization in the United States, John Boyd Dunlop was finding a way to bond rubber soles to canvas uppers, marketing them as "sand shoes."[16]

In the mid-nineteenth century, vulcanized rubber-soled shoes were best known as croquet sandals, specialty footwear selling to the affluent starting in 1868 at $6 per pair, several times the cost of good leather shoes. A few elite pedestrians, at least in England, also used the new material. Sir John Dugdale Astley, the sport's reigning arbiter, referred to "india rubber shoes which fitted him like gloves." In the later nineteenth century, rubber soles spread for special purposes. Thieves and prison guards alike appreciated their silence, hence the enduring nickname "sneakers," which first appeared in 1873. The rubber joint, called the foxing, between sole and upper reminded others of the Plimsoll line around ships' hulls, marking the limit of safe loading. The popularity of the nickname Plimsolls for the footwear suggests that rubber recreational shoes were growing in popularity as early as the later 1870s. New models were introduced for boating, tennis, and cycling.[17]

The rise of the rubber-soled shoe in the early twentieth century was connected with basketball rather than with track. Spalding introduced a gum rubber marathon shoe in 1908 but returned to leather in 1913 after complaints of excessive wear. Meanwhile, basketball had originated as a winter sport at a school for YMCA instructors in 1891; by the early twentieth century it was swelling YMCA memberships. It was partly for this market that U.S. Rubber developed its Keds brand, introduced in 1916–17. Several companies had trademark claims on *Peds*, and *Keds* was less a play on *kids* than a neutral word inspired by the recent success of Kodak. But the name did attract children and parents and helped give the company an identity; until then, U.S. Rubber had remained more of a trust than a modern corporation, with thirty plants and thirty rival brands. Converse All-Stars also appeared in 1917. With B. F. Goodrich, Dunlop, and Spalding, these competed for the growing sneaker market. Converse, kept out of other outlets by the rubber companies, became a specialist in athletic footwear and began to publish a basketball yearbook that has become a standard reference on the game. Until World War II, the athletic shoe industry remained divided between the makers of leather sports shoes and the sneaker companies rooted, with a few exceptions like Converse, in the rubber industry. Sneakers of all kinds remained largely limited to sports

W. SCHNEIDER.
SHOE.

No. 564,767.   Patented July 28, 1896.

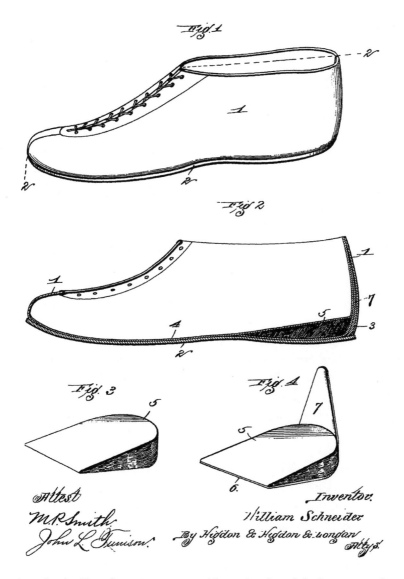

Fig. 1

Fig. 2

Fig. 3   *Fig. 4*

Attest

M. R. Smith.

John L. Tunison.

Inventor.

William Schneider

By Higdon & Higdon & Longan

Attys.

Specialized athletic footwear was not widely produced until the late nineteenth century. The Schneider shoe, designed for the increasing number of indoor gymnasiums with wooden floors, has "flexible elastic pads" in its heels for absorbing shocks. Whether or not this design ever went into production, the search for optimal resilience in sports shoes has continued.

and children's wear; only the Duke of Windsor experimented with them as street shoes.[18]

The 1930s were important for athletic footwear because of the growing understanding of the relationship between the skills of athletes and the construction of their shoes: technology following technique. Innovations often came from sports rather than from manufacturers' laboratories. It was in this period that Vitale Bramani, an Italian mountaineer, developed a new sole material after six of his climbing companions perished because they had discarded their heavy approach boots before being surprised by a storm and had to spend the night in mid-climb in lightweight boots. Vibram made possible a single boot with both strength and grip. Around the same time, the American Paul Sperry introduced a new kind of deck shoe with a superior grip after nearly dying in a sailing accident. The young Bavarian shoe manufacturer Adolf Dassler, founder of the later Adidas, was also an athlete. Dassler's great achievement of the 1950s was a new kind of soccer shoe that matched a new style of play. The dominant British-made models were boots of heavy vegetable-tanned leather. Their studs of layered leather were attached with bent-back iron nails that could penetrate the insole when the studs shrank after repeated wearings and exposure to moisture. Painful as they could be, they were well suited to the prevailing English "kick and rush" method. Dassler worked with West Germany's international team to make possible a new style of play, based on mobility and control of the ball. To reduce weight and aid passing, he lowered the shoe's profile and softened the toes, replacing heavy materials with soft kangaroo leather. Interchangeable spikes replaced the built-up leather studs. To maintain spacing of the eyelets he used stiffer leather trim, stabilized with two and later three stripes across the instep. In 1954, these innovations combined to help the German team win the World Cup from the Hungarians, thanks to their ability to change from short to long studs after rain had drenched the field.[19]

The classic age of the sneaker was the baby boom era of the 1950s. Life and dress became more informal after World War II, especially in the United States. Adult men and women wore hats less often, and sneakers gained favor with children over leather shoes. In fact, they were among the first products sold explicitly for youth appeal. School authorities dropped their initial resistance. When the New York State legislature abolished its public school dress code in 1957, other states followed. While total footwear sales remained flat at 600 million pairs per year, sales of sneakers grew from 35 million in the early 1950s to 150 million by 1962. Just as the

1,304,915.

Patented May 27, 1919.

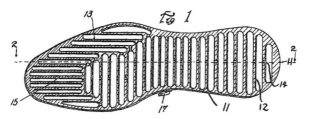

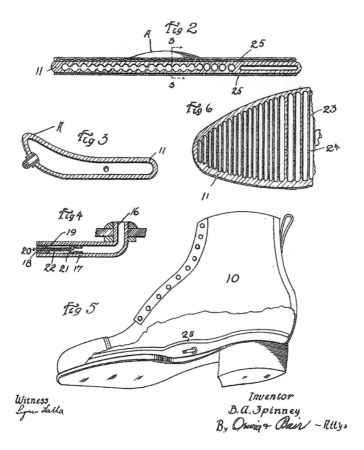

Witness
Lynn Latta

Inventor
B. A. Spinney
By Owing & Bair ~ Attys

B. A. Spinney was one of a number of now-forgotten inventors experimenting with air-cushioned insoles early in the twentieth century. Leaks doomed early models. The idea had to wait for new gases and containment systems.

Hat Council of America (now long defunct) unsuccessfully promoted the health benefits of headgear in the same era, the Leather Industries of America warned in magazine advertisements, press releases, and even cartoons of the dangers rubber-soled sneakers posed to growing feet. Although sneakers had been available with heel and arch supports since the 1930s, one advertisement in *Parents'* magazine cautioned: "Madam, only leather shoes assure your child proper support, correct fit and protection." So powerful was the campaign that even an overview of sneakers published in 1978 recalled mothers' warnings of "flat feet, fat feet, swollen ankles, bulbous heels, ingrown toenails . . ." To the future newspaper columnist Richard Cohen, growing up amid this propaganda, something was not right. Sneaker-wearing children of low-income families were becoming formidable runners. By the 1960s, medical opinion had already begun to turn. Only 11 percent of doctors disapproved of sneakers, according to a study commissioned by one manufacturer—for Cohen, a warning against medical alarmism.[20]

THE TUBE AND THE SHOE

Just as medical opinion was turning in favor of sneakers, a technological and social transformation was beginning. The boundary between the competitive athletic shoe and the sneaker, once sharply defined except in basketball and tennis, was starting to break down. And international markets, once strictly segmented by continent, were combining. The makers of leather shoes were losing not only children but adults to the diverse group of rubber companies and athletic suppliers.

What changed the sneaker into the modern running shoe was a series of new techniques. The first was essential: jogging. It originated in New Zealand in the early 1960s as a public fitness movement that sent dozens of men, women, and children trotting through public parks. Bill Bowerman, a founder of Blue Ribbon Sports (later Nike) and a nationally famous track coach, was amazed at the speed and endurance even of elderly New Zealand runners. In 1963, he introduced jogging classes in Oregon; YMCAs across the country used his materials to promote their own programs. Running was starting to influence style: in August 1966, *Footwear News* noted that the film *The Loneliness of the Long-Distance Runner* was helping sales of "track shoe fabric casuals." In 1967 Bowerman coauthored the best-selling *Jogging: A Fitness Program for All Ages*. Track or marathon shoes were not really necessary, the authors said: "Probably the shoes you wear for gardening, working in the shop or around the house will do just fine." But Bowerman's consulting orthopedist concluded, after reviewing medical articles on an epidemic of foot and leg injuries among runners, that shoes were indeed the problem and that special shoes with more elevated heels were needed.[21]

While Nike and its competitors made their reputations on the endorsements of many of the world's greatest athletes, it was really the out-of-shape, casual runners who inadvertently triggered a technological revolution. The nineteenth century's great pedestrians had contributed almost nothing, though their mix of running and walking was not so different from jogging. The shoe renaissance was a favorable unintended technological consequence of the worldwide spread of a technique. Seldom in the history of any apparel industry have so many new ideas emerged in only two decades, from the mid-1960s to the mid-1980s. Already in 1966 the U.S. branch of the Bata Shoe Organization (originally a Czech company, reestablished in Canada) introduced the two-color, injection-molded sole on its basketball model the Bullet. Thus began the decline of vulcanizing (processes derived from Goodyear's original purification of rubber with heat and sulfur). From Europe came the "training shoe," so called because it was designed not for any one sport like track or football but for conditioning programs to prepare for competing. From America emerged the heel wedge, developed for runners in 1962 by New Balance, a Boston orthopedic shoe company that applied other lessons from corrective footwear, widening the toe box and eliminating seams. Americans also were the great tread experimenters. New Balance introduced a ripple sole, and Nike's first model featured a sole molded on the household waffle iron of Bill Bowerman, bringing a cleatlike surface to the rubber sole itself. (Bowerman intended it for tracks; road runners, not the inventor, first promoted it as a training shoe.) And Japan was an important early innovator. Sneakers were already beginning to supplant zori among children in 1955, although the influence of traditional footwear remained so strong that one marathon shoe of the same year actually had a tabilike divided toe box. The manufacturer, Tiger, introduced the now ubiquitous nylon uppers in the 1960s. It was also using arch supports and continuous midsole cushioning on different shoes; Nike, still under its original name of Blue Ribbon Sports, combined these innovations in a single shoe, originally made for the new company by Tiger.[22]

The technology that did the most for athletic footwear starting in the late 1960s had nothing to do with shoemaking. It was television. Athletic shoe promotion had a long history, going back at least to A.G. Spalding's out-fitting of the track and field events of the 1904 Olympics in St. Louis, and his assistance to U.S. marathoners in the 1908 games in London. The basketball star Chuck Taylor joined Converse in the 1920s and added his signature to a line of sneakers, still a classic brand with 600 million pairs sold through the twentieth century. The favorite model of the 1932 Montgomery Ward mail-order catalogue was the Dutch Lonborg Basketball Shoe, designed by a championship-winning Northwestern University coach and featuring a "non-heat insole" and sponge-rubber cushion heel. Still, most print accounts of sporting events did not mention the winners' equipment, and even specta-tors might need coaching to recognize the Lonborg by its khaki-colored duck and double toe cap. Television meant that recognizably branded shoes, espe-cially when worn by champions, could have instant national and global recognition. Even before the 1954 World Cup match, Adi Dassler had been producing his shoes with contrasting stripes because young customers were requesting them not by their brand name but as the shoes with this feature. The Adidas trefoil appeared during the broadcast of the German team's World Cup breakthrough in 1954. The German baby boom generation was soon clamoring for them. In Britain, too, enthusiasm for Adidas was strong: the shoe consultant Phillip Nutt recalls that his father, who made traditional models, abandoned the soccer market and converted his factory to children's shoes. The power of the visual endorsement was born.[23]

Paradoxically, the greatest American media testimonial was made by athletes who removed their shoes. After finishing first and third in the 200-meter event in the 1968 Olympics in Mexico City, the black sprinters Tommie Smith and John Carlos took off their shoes before climbing the medal stand, bowed their heads, and raised their fists in black gloves as a protest against racism. Appearing in their stockinged feet to symbolize black poverty, they inadvertently drew attention to their footwear, Suede model Pumas with a distinctive pattern of swirling and crossing strips, made by another branch of the Dassler family that had become Adidas's fratricidal archrival. Some shoe industry veterans believe the incident was crucial in accelerating the black youth market, which in turn was shaping national popular culture. (The writer Stephen Talty has remarked that "black New York teenagers . . . are to sneakers what the Académie Française is to the French language.") The European market followed a parallel course. Soccer enthusiasts in England began to wear Adidas Sam-bas and even crossed the Channel to find the latest sport shoe models. The 1972 Olympics was the high point of the Adidas strategy: more than three out of four athletes and all officials wore their shoes, including thirty-one of thirty-eight gold medalists in track. Since then, most athletic shoes have been designed with a trademark prominently displayed as an element of stitching. Visual marketing made team adoption and celebrity endorsement more important than ever.[24]

The controversy surrounding payments to Olympic athletes and others has only occasionally touched on either the technology of shoes or the style of play. The technical innovations of the 1970s and 1980s were an explosion of creativity, of new styles and processes, sometimes cosmetic or faddish but often meticulously researched. The contemporary footwear industry developed few entirely new materials. A standard handbook published in 1987 notes that by 1959, all "basic polymers important to footwear manufacture" were already commercially available. Natural and synthetic rubber, modified by a number of agents for optimal durability, elasticity, and appearance, are still the favored materials for outsoles.[25]

It is the upper of today's athletic shoes that determines not only much of their look but also the international organization of their production. Today's upper is an intricate weave of synthetics and often leathers, setting it off from the usually black or white canvas of the 1950s sneaker. The change amazed veteran workers in the industry. When Nike began U.S. production in a former sneaker plant in Saco, Maine, in 1978, the cutters they hired had been accustomed to producing just three pieces, a vamp (for the forefoot) and two quarters (for the sides and heel). Now there was a rainbow of nylon, vinyl, transparent mesh, and suede, each with different properties, that had to be cut precisely and stitched together to form a rugged unit. Performance, comfort, looks, cost, and visual marketing—not necessarily in that order—have determined the geometry of the upper. Combinations of parts can interact in unexpected ways. In 1979, the Nike Tailwind was acclaimed by runners for its new cushioning under the Air trademark, only to fall apart in the middle of races as bits of metal in the silver dye slit their way through the mesh.[26]

Assembling the upper does not necessarily require as much labor as the complexity of the project suggests. In the mid-1990s, reporters discovered that of the $70.00 retail price of a pair of Nike Air Pegasus shoes, production labor accounted for only $2.75, less than a third of the $9.00 needed for the materials. That does not mean that labor is unimportant. The difference in labor costs has slashed the number of athletic shoes produced in North America, and even those labeled "Made in U.S.A." have components, sometimes including stitched uppers, from around the world. Asia is especially attractive because of its dense network of suppliers. The Nike Air Max Penny of the mid-1990s had fifty-two components from the United States, Taiwan, South Korea, Indonesia, and Japan. Using competing sources around the world can provide powerful leverage in pricing and quality control.[27]

The global footwear market that emerged in the 1970s did more than spread manufacturing. When factories in the United States closed, it was not only production-line workers who lost their jobs but skilled technicians like the last makers and pattern cutters. Many expert workers went overseas as foremen in the factories of Taiwan and Korea, until local supervisors were trained to replace them. Essential skills supporting design slipped away. The last-making industry had declined from forty-five U.S. firms producing 6 million pairs of lasts annually in 1920 to three firms making 1.3 million pairs in 1980. Fax machines, spreading from Japan around the world in the 1970s and 1980s, were making it possible to coordinate the assembly and marketing of rapidly changing footwear models for the first time; their electromechanical predecessors, telex machines, had been useless for graphics. Traditional designers had worked on actual shoe lasts covered with paper and masking tape, which was cut off in sections to create patterns. The inside and outside of the shoe displayed the same structure, following the same paper sections. Now the industry turned to a new set of specialists, industrial designers with no footwear experience. At their disposal was a new technology, computer-assisted design and manufacturing (CAD/CAM), hardware and software that can rapidly translate ideas into prototypes and marketable products. In the 1980s, a number of American and European companies developed CAD/CAM systems for footwear. As usual, the process began with the shoe's last. But now the design was digitized, painted with a grid representing surface areas and rotated on its lengthwise axis, each square being converted to computer code. (Bata worked with a Cambridge University professor who developed the algorithm.) Instead of drawing on the actual last with paper and masking tape, a designer could instead transform the electronic last, viewing it from any of more than two dozen positions and experimenting freely with color. Now industrial designers without shoe experience could work with software models of lasts. The outside of the shoe no longer had to reflect the interior. It was open to fashion.[28]

Athletic-shoe making could now attract talent from other professions. Tinker Hatfield became Nike's vice president of design and special projects in 1984, three years after joining the company as architect of its display rooms. A former college athlete with no previous shoe industry experience, Hatfield specialized not only in collaborating with outstanding sports figures but in introducing motifs from popular culture. Working with Michael Jordan, he designed the original Air Jordans in 1985. They were inspired by the fiery nose ornamentation of World War II bombers. Jordan's 1990s shoes included jagged Afro-Pop designs with what Hatfield called "an aggressive caricaturelike quality." Hatfield also has claimed inspiration from Mighty Mouse cartoons (for their affinity with footballer Bo Jackson), Native American moccasins, and more recently automobile styling. And the sneaker industry was one of the first to turn to nature for design ideas. In the 1980s, Nike sponsored research by a zoologist, Ned Frederick, who had written his doctoral thesis on the musculature of the striped skunk's feet. Frederick became Nike's research director in the early 1980s. Years later in England, the first prize of a 1998 Royal Society of Arts contest to design an ideal running shoe went to a transport designer at Coventry University who was inspired by animal footprints to equip his model with small claws for a better grip in wet weather. Reebok hired him.[29]

The new materials were made not only to perform—the athletic endorsers of Nike and other companies needed, after all, to win in them—but to dramatize performance. Color television and especially cable continued the process that had begun with the 1954 World Cup. Thanks to television and to new design and manufacturing techniques and materials, the shoe itself could now tell a story not just to a few close-up spectators but to national and international audiences. As Hatfield explained: "When Michael comes down the court, we want you to see him coming—in shoes that you recognize from a hundred yards. That is the philosophy behind the Air line." Strong design told a story to the all-important retailer as well as the consumer. And as dramatists, shoe designers assumed some of the aura of film directors. Tinker Hatfield became a world celebrity in his own right. In 1997, Reebok featured its own director of research engineering, Spencer White, as "The DMX™ Design Guy" in national television advertising with Shaquille O'Neal among others, and even as a full-length cardboard retail display figure.[30]

THE KERNEL

The shimmering textures of the contemporary running shoe, essential as they are to marketing, are intended by makers and buyers alike as only the outward expressions of a complex technological product. The focus of athletic-shoe development for the last twenty-five years has been largely elsewhere: in the midsole region, which once did not even exist. Rarely has so much money been invested for so long in a space of a few cubic centimeters in an item of apparel.

The midsole grew in importance in the 1970s, as biologists, physicians, and sports scientists began to pay greater attention to the physics of running. One of their first surprises was the cost of wearing any shoes— the price in energy, that is, not in dollars. Peter Cavanagh has observed that the typical running shoe not only makes the foot 25 percent heavier but also adds that weight at the rapidly moving point where the body needs the most energy. A few noted athletes have been able to profit from the savings. The record-breaking English track star and biologist Bruce Tulloh, who wrote the British best-seller on recreational running, raced barefoot while still a university student and thereafter competed brilliantly with nothing more on his feet than adhesive bandage wrapped around the big toes. Griffiths Pugh, a physiologist, measured Tulloh's shod and unshod performance and found that wearing footwear while running requires between 1 and 2 percent more energy than running barefoot, an immense margin for any elite athlete. Thanks to an exceptional light but rugged build and meticulous planning, Tulloh had run 3,500 miles without injury by the late 1970s. Meanwhile, the Ethiopian Abebe Bikila, who normally ran in shoes, electrified the 1960 Rome Olympics by discarding them when they became uncomfortable and winning the marathon event barefoot. In 1985, the South African–born Zola Budd, running barefoot, established a record for the 5,000-meter race. Even two NFL placekickers, Tony Franklin and Rich Karlis, have performed without shoes. While few other athletes can forgo protection, the dissenters found support in the late 1980s in research by R. McNeill Alexander, a biomechanics pioneer. The biologists Giovanni Cavagna and Rodolfo Margaria had already shown that runners move by storing and releasing elastic strain energy, like bouncing balls. Alexander found that the tendon returns 93 percent of its energy as it recoils and the arch about 78 percent. The midsoles of the running shoes he tested, including most of the leading brands, returned between 55 and 65 percent of the energy needed to deform them.[31]

While the foot theoretically remains the ultimate running technology, shoes are still a must. Since high-performance surfaces for conventional running, "tuned tracks," exist, it should be feasible to design one especially for use without shoes. But because many surfaces are unfriendly to the foot, it is the midsole that absorbs much of the stress of hitting the ground. Each time we touch down while running, our feet sustain an impact as though we were taking a step with two grown people sitting on our back, in Peter Cavanagh's metaphor. A 150-pound man sustains a half-million pounds of pressure in a mile. Fortunately the flexing of the knees and the stretching of the gluteus, hip, and ankle muscles dissipate most of this shock. But especially as millions of people around the world began to run for health in the late 1970s, running-shoe makers began trying to assist this natural cushioning. This is not as simple as it seemed to the late-nineteenth-century shoemaker who offered springs as an option. If the knee does not flex enough, harmful forces may be transmitted to bones and muscles. Human feet and legs form a complex system with many possible interactions of parts and running surfaces.[32]

The breakthrough in midsoles came in 1975, when Brooks Shoe Company introduced a shoe called the Villanova with a midsole and heel wedge made of ethylene vinyl acetate (EVA). This polymer, which we have already encountered as a sandal material, had been developed by a chemical engineer, David J. Schwaber, at the Monarch Rubber Company of Baltimore, a major aerospace contractor. While natural and synthetic rubber can also be made elastic by using a blowing agent to fill the product with bubbles, EVA offered an excellent combination of light weight and resilience; soon it displaced rubber. In five years in the mid to late 1970s, midsole foams lost three-quarters of their weight. EVA has helped runners go longer distances with fewer injuries, but it has also shortened the useful lives of many shoes. It goes flat like sparkling water and champagne, although with use rather than time. Soft EVA provides excellent cushioning at the cost of some stability, but it also loses its shock absorbency, becoming (in the words of one manufacturer's spokesman) "almost a hard plastic." Firmer EVA degrades more slowly and provides more stability, but also affords less protection against injuries. Polyurethane, the major alternative and complement to EVA, lasts longer but does not offer the same cushioning and may be half again as dense.[33]

The 1980s were the great decade of the midsole, with a profusion of exotic shapes and materials. A series of cushioning systems transformed the athletic-shoe market again. Heavier but more durable than EVA and polyurethane, these systems did not completely eliminate the old materials, but they did absorb impact more effectively and helped the shoe last longer. The most famous and commercially successful was Nike's Air. Air bags in footwear are an old idea; about 150 patents were filed between

1882 and 1970. Until the 1970s, none was satisfactory. Because of the very forces that make cushioning so desirable, the air leaked and the shoes went flat. Just as architects began literally revamping shoe uppers, engineers were breathing life into the midsoles. Frank Rudy, who had resigned a senior aerospace technology management position in 1969, codeveloped a liner for ski boots and applied a similar principle to a polyurethane film sac that transferred gas among compartments to cushion athletes' movements. Diffusion pumping, as he and his coinventor Bob Bogert called it, shifted material more efficiently than alternative cushions and springs. Of several companies that offered the invention, only Nike saw beyond its earlier failures, developing a flexible but nonleaking air liner and finding gases with molecules that, unlike common air, would not be forced out. (When the Tailwinds started to fail in 1979, consumers erroneously thought the air compartments had deflated, probably misinterpreting the expression "blowout," shoe manufacturers' slang for falling apart. Later metallic finishes created no such problems.) As a rookie with the Chicago Bulls in 1984, Michael Jordan linked the red-and-black Air shoe with legendary jump shots that made it the most rapidly profitable and influential patented technology in the history of sports, vindicating Nike's record-shattering $2.5 million contract with the player. As important as Jordan's endorsement was Nike's development of "Visible Air" models, including the Air Max, that made the gas-filled chambers visible through the sides of the shoes. These made it easy to explain the system to the sales force, retailers, and consumers. The "Visible Air" principle worked as the picture windows in American car washes and translucent Macintosh computer cases do, selling a process by turning it into theater. In their first six years, sales of the fifty-odd models of Air shoes exceeded $2 billion and helped Nike increase its market share to 36 percent.[34]

Not everybody applauded. Surprisingly, there are no published tests of the durability of air inserts, but one orthopedist has estimated they extend a shoe's life by a quarter to a half. (*Runner's World* magazine still notes the insert can deflate.) The original Rudy patents expired in 1997, but meanwhile other makers had developed their own midsole systems: Reebok's Pump (user-inflatable air bladder) and Hexalite (thermoplastic honeycomb created by the aerospace company Hexcel), ASICS Tigers' gel (silicone), Brooks's Hydroflow (liquid silicone–filled, featuring a dual-chambered heel), Etonic's Stable Air (encapsulated air), and Puma's and Adidas's synthetic gels. A company called KangaROOS USA introduced a resilient midsole with a NASA-developed material called Dynacoil in 1985.

There are other ways to control the properties of shoes. The inner heel can be reinforced to combat that ubiquitous source of running discomfort and injuries, overpronation, the tendency to roll the foot excessively inward after initial contact. Some shoes have composite roll bars for stability.[35]

Both the weakness and the strength of running-shoe research has been that none of these approaches, even the Nike Air system, has displaced the others. There are so many variations of human legs, feet, running, and walking styles, and so many surfaces and conditions, that *Runner's World* long ago abandoned its early system of ranking shoes according to scientific tests. While models fall into a number of types for overpronators and supinators (runners who turn their feet outward), each buyer—whether athletic or casual—has to search for the right match. Some brands use straight lasts, others curved "banana" lasts. The difference is palpable. And a shoe that feels comfortable in a showroom may be painful after actual exercise. Because of all the variables in real-world conditions, laboratory tests also have limited value. The Tailwind seemed to offer impressive gains in efficiency, yet no elite runners were able to improve their times with that model. The genius of the industry has been not any single invention but the variety of equipment to suit not only the techniques of different sports but the bodies and running styles of so many people. Without this diversity, far fewer people would have been able to run competitively. Many more would have given up after injuries.[36]

POSTMODERN MATURITY

By the late 1990s the logic of running shoe development was under strain. Thanks to CAD/CAM, short development cycles were flooding the market with new models. By early 1997, Nike designers were producing up to four new shoe models each year, introduced seasonally, for a total of more than eight hundred new designs annually, more than three for every business day. Obviously these numbers would have been unthinkable before computerized logistics for coordinating international shipment of parts and materials and advanced systems for tracking sales and inventories. While sneaker publicity had become ubiquitous in popular culture, it was for that very reason extremely expensive; according to a *Washington Post* investigative article in 1995, the $70.00 price of a pair of Nike Air Pegasus shoes included $4.00 for advertising, 16 times the $0.25 research and development costs. Even with features readily visible and identifiable, consumers could be easily confused, and only a limited number of specialist retailers could give buyers useful advice. The search for optimal fit and performance could make shopping Sisyphean, as even recent models were discontinued to make way for yet newer ones. And the prices of the most coveted models remained daunting. A high-performance running shoe selling for $140.00 has a useful life of 300 to 500 miles, according to shoe experts. Accepting 400 miles, the cost per mile would be $0.35, more than the $0.325 allowed by the U.S. Internal Revenue Service in 1999 for automobile use as a business expense, which includes not only tire wear but depreciation, insurance, fuel, motor oil, and maintenance.[37]

The sport shoe market began to mature in the early 1990s. After more than doubling in the 1980s, to reach 384.2 million pairs in 1990, unit sales fluctuated in the 1990s. According to the Athletic Footwear Association (AFA), U.S. sales were 325.4 million pairs in 1998, barely more than the 325.1 million pairs recorded in 1994, though dollar volume had grown from $11.66 million to $13.8 billion because prices had crept up by about 3 percent annually. Still, between 1997 and 1998, the market declined by 6 percent. The AFA's report found signs of hope but also noted "more stores with more space for more shoes than consumers want" as a big problem. By the end of 1999, one American athletic-shoe chain was preparing to close more than a quarter of its 236 outlets in a bankruptcy reorganization, and the big sneaker manufacturers were pledging expanded advertising budgets to help the sporting goods chains.[38]

Among the industry's problems was saturation. Once sneakers had replaced street shoes in all but formal business and professional settings, there was no more room for growth. As the podiatrist and industry analyst William Rossi observed in 1998, annual consumption of shoes per capita had doubled from 2.5 pairs in 1920 to 5 pairs in 1978 but had remained steady since then; athletic shoes had taken market share from dress footwear. And contrary to the images of advertising and popular culture, people were actually spending significantly less of their income on footwear: from 1.4 percent in 1960 to 0.6 percent in 1997. Athletic-shoe marketers could find no major new categories to offer after the aerobics, cross-training, and walking shoe introductions of the 1980s. Besides, in 1998, 54 percent of shoes were bought only for casual use, 17 percent for walking for exercise, and only 27 percent for sports participation. In 1997, the youth market began to shift to "brown shoes," including chunky hiking boots, more durable for city commuting and often costing no more than sneakers. Some manufacturers began to adapt construction techniques from athletic-shoe making to produce leather street shoes; but these underscored the loss of interest in the sneaker look. Meanwhile, claims of technical innovation by sneaker manufacturers were sounding less credible. Rossi believed they were exaggerating their technological innovations, observing that while other industries spend an average of 3.7 percent of sales on research and development, the leading athletic shoe companies together spent only a fourteenth of a percent and had made no significant innovation over the last five or ten years.

The very organizational and technical successes of the manufacturers introduce new problems. When the same model is produced by multiple international sources, differences arise not only in appearance but in fit. And familiar, favorite designs disappear in the quest for novelty; their rapid changes recall the annual automobile redesigns of the 1950s that now even Detroit has long since abandoned. No wonder that since the early 1990s many consumers have rediscovered Keds and other sneakers of old-fashioned canvas and rubber with the familiar "fox banding" stripe where the upper and sole join.

For all this, the athletic-shoe industry still has a formidable base of designers and biomechanics specialists and may transform itself again. Nike, for example, is continuing to extend the Air sole to cushion more and more of the foot with less foam, and has been developing models with a springy plastic structure that eliminates the need for any midsole. But since most shoes are still bought for mainly nonathletic use, it is uncertain whether the 1980s boom can be repeated. The big future market may be in the sandal wearers of developing countries: in 1998 even Nike announced a new line of low-priced canvas shoes at $15, for distribution only in the Third World.[39]

Most seriously for the sneaker industry, it began to lose its cachet with the press in the late 1990s. Without important new "operational" ideas like the Air sole and the Pump, and with the retirement of Michael Jordan from the Chicago Bulls, the industry was losing its dramatic appeal even as individual companies slumped or bounced back. British *Vogue* proclaimed the "Death of the Trainer" (running shoe), which was turning into "the antithesis of cool . . . a prop to bluff [one's] way into the world of hip." Now the athletic shoe was stigmatized as "bourgeois." "Fashionistas," as the fashion historian Valerie Steele has observed, are always seeking something new in sports footwear yet are seldom athletic themselves. One result is the rise of sneaker-inspired high-fashion street shoes that defy the ideology of the technical performance shoe. A designer for the Acupuncture brand disavows sports, proclaiming "vanity and fashion" as his goals.

One specimen at a recent Fashion Institute of Technology exhibition that Steele organized was a decidedly nontechnical Prada Sport model with an elastic cord threaded through a pull-through fastener. On the same theme, the French company No Name produces a shoe with a canvas upper mounted on a thick ripple sole like a row of cylinders, weighted to let the wearer experience new techniques of walking: a cross between a sneaker and a running shoe. Other footwear in the exhibition had beautiful and complex sole designs that would soon have been obliterated by any strenuous activity. A London critic hammered the 1999 Nike Air Zoom Seismic as a "nylon tube" with a "thick coating of style that will look redundant next month" and is designed to degrade its own intricate sole ornament. (Expendability is not a new principle; some women's dress shoes of the early nineteenth century could be ruined in a single night of dancing.) Leading fashion brands like Tommy Hilfiger, Nautica, and DKNY began to introduce their own lines of designer sneakers through department stores—a trend warily noted by the AFA. Is this the beginning of a new cycle of popularity or a sign of the end of an old one?[40]

There has been still more bad news for the industry, the unwanted celebrity endorsers. The Nike "swoosh" logo began to turn up on capsules containing the illegal hallucinogen/stimulant Ecstasy, sold as "Purple Nike Swirl"; athletic shoes at one time were de rigueur at the rave dances Ecstasy helped animate. In 1997 the members of the Heaven's Gate cult outfitted themselves in new black Nike tennis shoes before their mass suicide, announced on the Internet, as they prepared to rendezvous with extraterrestrials in the wake of the Hale-Bopp comet. Whether it was a sardonic comment on the consumer society they were leaving or an invocation of the manufacturer's advertising slogan of the time, "Just do it," the gesture showed the exposure of megabrands to uses beyond the powers of trademark lawyers. In January 1999, the U.S. retail chain Just for Feet ran an advertisement featuring the capture and drugging of a barefoot Kenyan male marathoner by a largely white paramilitary team who lace running shoes on his feet; it provoked accusations and denials of racism and a round of financial troubles and lawsuits. Not that these necessarily harmed sales. In one of the protests against the low wages paid to workers in the factories owned by the shoe manufacturers' contractors, a rioter at the December 1999 meeting of the World Trade Organization was observed smashing windows of the Seattle Nike Town showroom while wearing Nike shoes.[41]

OLD SHOE

The great surprise for athletic-shoe manufacturers at the turn of the century has not been such incidents. It has, rather, been the graying of the sneaker. The AFA reports that younger people are a static or diminishing market, and middle-aged and older people a growing one. Between 1992 and 1998, teenagers' share of sneaker purchases slumped from 18 percent to 16 percent, college-age buyers' share was almost unchanged, and twenty-five- to thirty-four-year-olds' share dropped from 24 percent to 20 percent. The greatest growth was among forty-five- to fifty-four-year-olds (who went from buying 9 percent of athletic shoes to buying 12 percent) and those aged fifty-five and up (whose share went from 10 percent to 13 percent). Slippage among younger groups concerns the industry because young people are the most enthusiastic consumers of higher-priced models. As a spokesman for the National Sporting Goods Association observed, "If a fifteen-year-old sees his or her parents wearing a pair of Nikes or a pair of athletic shoes, they think, 'That must not be so cool if my parents are doing it.' " The trend also suggests that sneakers may be coming full circle after a generation or two. In 1961, the recently elected California attorney general Stanley Mosk looked into a newly notorious and allegedly sinister conservative organization, the John Birch Society, and dismissed it in a letter to Governor Edmund G. Brown as "wealthy businessmen, retired military officers and little old ladies in tennis shoes," thus launching an immortal cliché. Mosk later recalled attending Birch Society meetings, where those little old ladies "were the first to arrive and the last to leave and they were always making motions—and their feet hurt, so they wore tennis shoes." But then, older people may have always been more enthusiastic buyers than manufacturers liked to acknowledge. During the 1980 New York transit strike, when wearing sneakers to and at work first became fashionable, one specialty store manager was struck by "the old ladies leaving the store in red, white and blue and orange shoes."[42]

The popularity of athletic shoes in an aging society raises the question of whether they are healthier than the footwear they have displaced. Although running-shoe manufacturers retain some of the world's leading biomechanics specialists, a group of researchers at Concordia University in Montreal, including the physician Steven E. Robbins and the engineer Robert Gouw, have challenged assertions that contemporary running shoes are safer than yesterday's sneakers, especially for older wearers. In a variety of laboratory conditions, they have found that older volunteers are more likely to lose their balance in footwear with softer midsoles. They have pointed to other studies suggesting that injuries were 123 percent higher among runners wearing the most expensive shoes than among those equipped with the cheapest models. Softer soles appear to encourage higher impacts during running, as thick, soft mats in gymnastics result in 20 percent to 25 percent greater impact on landing. Advanced midsoles, these researchers argue, promote subtle misjudgments by reducing the sensory feedback (proprioception) that runners and walkers get each time a foot makes contact with a surface—the feeling we get of where we are and what we are doing. Robbins, a barefoot running enthusiast, has cited a number of reports suggesting low rates of impact-related foot injury among barefoot athletes and customarily unshod populations. Like people with neurological disease, runners can damage their feet when feedback is inadequate. Shoes also tend to interfere with the natural extension and contraction of muscles and tendons. Athletes unconsciously tend to land harder to improve stability by compressing the material. Runners and walkers wearing hard-soled shoes, on the other hand, unconsciously adopt a better technique, bending knees and hips more deeply and getting natural shock protection in dipping slightly closer to the ground. Further, Robbins has argued, arch supports interfere with the natural cushioning of foot bones. More recently, he has even suggested that not the intrinsic properties of the material but "deceptive" advertising by manufacturers may be to blame for injuries. Volunteers landed with significantly higher impact on a surface when told it was designed for especially effective absorption of shock. The mystique of protection, Robbins and his coauthor argued, encourages people to take greater risks.[43]

Robbins's ideas are intriguing, but they are disputed by other biomechanics specialists. The one study he cites of actual runners uses data from 1983 and 1984. Laboratory research can suggest human performance in training and competitive conditions, but very little has been published about the everyday safety of sport-shoe designs. Shoes differ not only in cushioning materials and thickness but in outsole and upper materials and even lacing patterns, all of which may affect injuries. Manufacturers point out that Robbins and his colleagues have tested only a small number of midsole designs. Specialists outside the industry say that trial and error is a better shoe-buying strategy than choosing the lightest and cheapest brand. Yet as some of Robbins's critics acknowledge, his results may help develop better shoe designs even if his conclusions are unproved. Indeed, there has long been a small group of serious runners who have their own techniques for slicing out cushioning and supports with razor knives.[44]

Robbins and his collaborators are not so far from their critics' positions as they sometimes seem. They realize that barefoot running is not an option for most people in industrial countries. They also recognize that older people rate thin, hard soles much lower in comfort than younger people do. And in their study of footwear and balance in the elderly, Robbins and his coauthors had to concede that on the balance beam used in the test, walking barefoot resulted in instability 171 percent more often than did walking in shoes with thin, hard soles—and 19 percent more often than walking in shoes with thick, soft soles. Bare feet may be hazardous to the health of the elderly; doctors should suggest to older patients with a history of falls that they "avoid barefoot locomotion completely, and wear footwear with hard soles at all times when upright." If further research confirms their conclusions, sneakers with high-traction rubber soles may replace our grandparents' hazardous smooth-soled slippers. And running-shoe manufacturers, always alert to changing markets, will no doubt create a new generation of dynamic-looking, firmer-soled but comfortable sneakers for aging baby boomers, especially those who join the growing number of senior runners. Just as it helped link inner-city culture with suburban sports fandom, the sneaker may connect the youth movement of the 1960s, the fitness boom of the 1970s, and a new elder culture that is still in its infancy.[45]

And what of the sneaker as a global shoe? A recent *National Geographic* magazine article on Pakistan illustrates a barefoot rural schoolboy, shipbreakers (dismantlers of oceangoing vessels) apparently in sandals, and sneaker-wearing students at an exclusive boys' secondary school in Lahore. But it also quotes a young, American-educated progressive landowner who no longer wears his San Francisco–bought sneakers in the fields but prefers indigenous khusas, slippers of pliant kidskin that, as he relates, let him "feel the soil through the soles of my feet and know, as I walk the land, just what to plant where." Footwear can promote not only biomechanical performance but the body techniques by which we relate to the earth.[46]

CHAPTER FIVE

Sitting Up Straight

Posture Chairs

JUST AS BILL BOWERMAN was testing new soles in the late 1970s, industrial designers and architects were planning a new generation of office seating. High-performance sneakers and chairs—instruments of the active and the sedentary life, respectively—might seem to have little in common. One moves with us, and the other generally stays put. (Wearable seating has won design prizes but rarely appears on the market.) But for the techniques of the body, seating is to the back what shoes are to the feet. Chairs both reflect and shape how we move and how we rest. Like shoes, they can become intimately linked with their users, extensions of the self, molding themselves to our bodies but also reshaping the people they contain. Shod people and chair-sitting people, even in agricultural and artisanal societies, are already transforming themselves technologically, beginning to fuse with the objects around them.

The television commentator Andy Rooney has described this feeling insightfully. For thirty years, he wrote in 1989, he had used a leather living room chair that would have been more than three sizes too large, if seating were sold like shoes—as he believes it should be. Because his legs could not bend if he sat back, he sat "slumped way down in it with my legs resting on a footstool," a position so relaxing it induced sleep. Yet he hesitated to give the chair away, partly because it had become so shabby, but also because it had taken his shape "the way a pair of shoes finally conforms to the shape of your feet just before you have to throw them away." Discarding it would be like putting an ailing but beloved dog to sleep. Having made chairs in his home workshop, Rooney observed that no matter how much measurement, study, and thought one puts into a chair, it can be tested only by actually building it. By then it is too late to modify the shape.[1]

Shoes and chairs have been linked not just in popular culture but on the frontiers of science and design. It was no coincidence that the tighter men's and women's shoe fashions became in the mid- to late nineteenth century, the greater the interest in comfortable furniture. As the English costume reformer Ada S. Ballin wrote in 1885: "It is only natural that when progression is so painful, as I have known it to be in fashionable boots or shoes, people should prefer to remain at home on an easy chair or on a sofa." At the origins of modern ergonomics, the German-born Zurich paleontologist and anatomist Hermann von Meyer (1815–1892) wrote not only the first popular scientific work on shoe fitting, starting the movement for healthier footwear, but also a pioneering article on the mechanics of sitting and its implications for school furniture. In our own time, both athletic footwear and seating have combined traditional and new materials, from canvas and leather to (beginning in the 1960s) synthetic meshes, to simplify production and improve performance. In 1968, the American designer Richard Neagle even invented the first armchair with a vacuum-formed plastic shell—coincidentally, it was called the Nike chair. Its cushions, like the midsoles of many later athletic shoes, were of polyurethane foam. The English designer Jasper Morrison claimed inspiration from the padding of his girlfriend's Prada loafers for the upholstery of his Low Pad armless lounge chair, now in the Tate Modern museum. And as we will see, the premium version of Niels Diffrient's new office chair uses a gel earlier employed in some sneaker midsoles.[2]

AN EXCEPTION BECOMES THE RULE

While most complex societies have developed some kind of footwear, only a minority of world cultures used chairs until a century or two ago. Even in Europe, movable furniture spread only in the Renaissance and early modern period, and chairs were beyond the reach of many well into the eighteenth century.

Before the chair spread around the world, humanity had the greatest repertory of postures of any species. Other animals acquire and transmit tool-using behavior culturally, but their positions and movements have not appeared, so far, to vary from one region to another. Humanity is the only species that can tell its young to sit up straight, or otherwise. We can shape not only our own posture but the behavior of other animals into gaits they may not regularly adopt in nature, down to the high-stepping maneuvers of the Lippizaner horses of Vienna's Spanish Riding School. Human technology also opens new resting possibilities for other animals. I once saw a tame parrot in Chicago's Lincoln Park Zoo that enjoyed lying on its back, feet clasped together, as a docent slid it down a concave polished bronze railing. Cockatoos can even learn to roller-skate.

In a landmark article in 1957, the anthropologist Gordon W. Hewes observed that humans can sit, squat, kneel, and stand steadily in about a thousand ways. These are only the static positions; only limited documentation exists of walking, running, and jumping styles. While many Western travelers have brought seating furniture with them into the field, others have discovered the rationality of some traditional ways. In South America, the U.S. explorer William Beebe learned the Indian technique of squatting on his heels. By varying it—changing foot positions and resting his chin, his armpits, or his elbows on his knees—he discovered that he could stay in one spot for hours while relaxing every muscle. Of human postures, this deep squat may be as widely distributed as chair sitting, being used by about a quarter of humanity when Hewes wrote. It is also closest to the resting stance of the chimpanzee, is comfortable for small children in all societies, and, Hewes speculated, might be a lifelong practice if cultures did not socialize the young into other positions. Among these, chair sitting has been the minority practice, a technique and a technology that originated about five thousand years ago in the Near East. In Egypt and Mesopotamia, chairs were reserved for high personages, and stools, too, were luxuries. The wealthy spent much of their lives at ground level. Workers had no seating. In the remarkable miniatures of the tomb of a senior official of Egypt's Old Kingdom in the Metropolitan Museum of Art, butchers, bakers, weavers, and scribes work either standing or squatting. Even in the New Kingdom, luxurious chairs were often broad and close to the ground. (On the other hand, Egyptians rich and poor could not sleep well without headrests that we would consider the height of discomfort.)[3]

In Egypt and elsewhere in the Middle East, the Arab conquests ended the prestige of the chair for a millennium. Wood was rare and precious in Arabia, but leather and wool were affordable. Elevated sitting was mixed with reclining on a variety of cushions and raised platforms; even the words *divan* and *ottoman* are Middle Eastern in origin. When we find an illustration of a sixteenth-century Ottoman official sitting on a Renaissance X-frame (Savonarola) chair, it turns out to be an English convert to Islam, Hasan Aga, originally Samson Rowlie. An early-nineteenth-century illustration of the Ottoman governor of Egypt, Muhammad Ali Pasha, receiving Western European advisers depicts the governor sitting cross-legged on a raised, cushion-backed, carpeted platform; the Ottoman advisers are standing, and a scribe sits on the floor. The Europeans are sitting with their feet on the floor.[4]

It was not only the Arabs and Ottoman Turks who conquered and governed empires and cultivated the arts and sciences without chairs. The complex cultures of Asia flourished at floor level. In China, mat-level living prevailed in the low-ceilinged buildings from the Shang dynasty (which ended about 1000 B.C.) through the Han dynasty (which ended in A.D. 200). As in ancient Greece, squatting was considered vulgar and disrespectful. Etiquette books taught kneeling and cross-legged sitting until the third century A.D., when good form began to dictate extended legs, leading to the first armrests and cushions. Beginning in the fifth century A.D., political disruptions and the influence of northern nomadic peoples began to change Chinese ways. For reasons that are not completely clear, houses were taller and more spacious. Folding stools, which had long been known from trade with the West, now took their place in Chinese households. Some but not all Buddhist figures in murals of the period sit on stools and beds with legs hanging down. Gradually from the seventh through the tenth centuries, the Chinese started to use folding stools and other furniture. To the art historian Sarah Handler, economic and social development during the Song Dynasty (960–1279) became decisive. The rise of a cash economy spread wealth and education, undermining paternalist hierarchies and leading even ordinary people to aspire to the raised seating that had been introduced at court. Chairs in turn promoted more democratic dining on raised tables, and by eliminating mats encouraged the wearing of shoes indoors. When European furniture was still massive and built in, Chinese artisans were developing elegant sets of movable chairs and tables.[5]

Why did Japan not follow China in elevated living and working? One reason may be that just as Chinese material culture was changing, Japanese society was becoming more independent of China, developing its own scripts and some of its greatest literary works. Buddhism, with its cross-legged and kneeling meditative positions, and the later prestige of the tea ceremony, promoted mat-level living. The traditional Japanese house also remained much more suitable for Japan's climate, warmer and more humid than China's. Folding chairs were occasionally imported and even made locally, but they were difficult to integrate into Japanese dwellings. Their legs could damage the woven rush surface of tatami mats, around which traditional Japanese domestic space is organized. And even if floors could be protected, the view from a chair would compromise the elegant proportions of the house, while the chair itself would interfere with the deployment of the futons that otherwise could turn any room into a bedroom.

Industrialization accompanied chair-sitting but did not cause it. Traditional Middle Eastern, South Asian, and Japanese administrators, scholars, and merchants had no need for Western chairs; less obviously, neither did machine operators. An American ergonomist visiting India recently discovered that an apparently Westernized factory, with machine tools on pedestals and operators sitting on stools, had once functioned with equipment and men at ground level. In the absence of visitors, operators would assume their traditional positions, with their legs folded under them on the stools. Even today, at least one standard ergonomics text reminds engineers to take low, chairless positions into account when designing equipment for non-Western factories. In 1968, the Stanford Research Institute even developed an experimental "yoga workstation" with a floor-level cushion.[6]

The spread of chairs was less functional than social. It began with cross-cultural encounters of diplomats and officials. Even before the European domination of South Asia and Africa, the rulers of non-Western societies were providing chairs for their European and North American guests. But this concession created a dilemma in protocol. Universally, a higher seating position is a sign of dominance, as is sitting in the presence of others who must stand. Some societies, like the Ottoman Empire, had their own raised seating. In others, especially in South and East Asia, elites learned Western seating styles. A first step was the construction of special rooms for these encounters as annexes to traditional public buildings and residences. More and more public business was transacted in offices with stools and desks, while the family and social spheres remained at mat level. In an 1838 engraving of a fashionable Damascus riverside coffeehouse, for example, customers sit cross-legged on carpets or mats.[7]

Africa followed a different pattern. While sub-Saharan Africans lived at mat level, many African societies used low, carved stools, often highly personal objects even considered to embody their owners' souls. The Golden Stool of the Asante in present-day Ghana still is believed to contain the soul of its nation, may not be occupied even by royalty, and is displayed only occasionally. When Portuguese merchants brought the first Western armchairs to West Africa in the sixteenth and seventeenth centuries, chiefs commissioned their own artisans to create new seating, embellishing European forms with indigenous motifs. The high-backed chair became a super-stool, a throne.[8]

Up to a point, mat-level and chair-sitting practices can coexist. While urbanization has been spreading European-style chairs in Africa, traditional African stools and chairs were generally low and used without tables. In other societies, including China, chairs were at first introduced for men alone; women's use of them was a serious breach of etiquette. In colonial South America, even into the nineteenth century in a few areas, men sat in chairs while women occupied low platforms called *estrados*, which were covered with rugs and cushions. During the Moorish occupation of Spain both sexes had lived thus, but in Christian Spain and some of its New World colonies the *estrado* became a female zone. Women dined off low tables, and sometimes even off cloths laid on the carpet, picnic-style. (Late in the colonial era, the Franciscans may have promoted the use of low side chairs by the Zuni of Mew Mexico as a form of cultural control.) And globalization has not completely erased sexism from seating. In a capsule of Year 2000 objects assembled by the *New York Times* is a stool from Zimbabwe called a *zvigaro*. As a young bank clerk explains in a label, it "is meant for the father to sit on. The mother sits on a traditional mat," showing that "in our culture a man is more important, and that we cannot all be equal."[9]

Despite such attitudes, the growth of cities has been drawing mat-living rural people into structures, however simple, designed for Western furniture, however modest or battered. International standardization of machinery and equipment will continue to encourage the spread of Western sitting styles. Even when children are raised practicing both positions, the workplace will accustom them to the task chair and reduce the flexibility needed for traditional sitting. The trend is already evident in Japan, where almost all new houses still are built with a tatami room but many young people find the ancient meditative position of *seiza*—kneeling upright while sitting on one's heels—scarcely bearable. In Japan as in the colonial Americas, upper-class women are expected to be able to sit gracefully at mat level, while many traditional restaurants offer low seats with backrests and leg wells for their male executive clientele. (Japanese doctors still debate whether *seiza* is healthy for the legs, strengthening knees and protecting them against arthritis, or whether it retards growth.) Chairs are creeping into even traditional Japanese life. Some Buddhists now practice Zen meditation sitting in chairs rather than in the lotus position, and even tatami producers are selling chairs covered with the material.[10]

The spread of chair-sitting may reflect world prosperity, but it also may be hazardous to our health. Chair-level societies have higher rates of varicose veins; sitting to defecate seems to promote hemorrhoids. Japanese children who have grown up kneeling on tatami mats and using traditional squat toilets have higher thighbone densities than others, and Japan has half the rate of thighbone fractures seen in Western countries. And although the Japanese have lower calcium intake and lower bone density than Westerners, they experience only 40 percent the rate of American hip fracture. Some Japanese scientists believe that frequent kneeling and rising of the older generation of Japanese, along with exercise of the pelvic muscles in traditional toilets, have developed both strength and agility that persist in old age; yet the chair-sitting Chinese also have low rates of osteoporosis.[11]

Galen Cranz, a professor of architecture, has argued that all chairs are inherently unphysiological, that they deform the body by straining the spine and weakening the muscles of the back. If Cranz is right, the technology of chairs, even those scientifically designed for comfort, promotes a self-sabotaging technique. As sitters become accustomed to the support of a backrest, their back muscles weaken and they must recline even more. The chair is a machine for producing dependency on itself. When Cranz examined photographs of West Africans taken by a friend who had taught English in Upper Volta (Burkina Faso), she learned that the two men who seemed to embody ideal physical development and absence of muscular strain were the ones who had grown up in villages without missionary schools with their tables and chairs. While Cranz never saw these men and the circumstances are only suggestive, it is certain that once people have become accustomed to sitting in chairs, they cannot work without them— yet often feel uneasy in them.[12]

THE WESTERN WAY OF SITTING

The redeeming feature of chairs is their variety, and the ingenuity of Europeans (and, independently, Chinese) in matching them with literary, social, and commercial needs.

Greek and Roman intellectual life were strongly oral. The scribes who copied ancient books did not transcribe them from exemplars but rather wrote them from dictation, if only because they had no writing tables.

Their equipment, like that of other unfree and lower-class workers, hardly concerned their masters. A slave might sit on a stool and a wealthy man on a chair, but each had only his knee as a writing surface.[13]

The first technological breakthroughs in sitting and writing occurred in the Middle Ages. The earliest clear illustration of a scribe using a writing table to prepare a book is in the Lindisfarne Gospels of late-seventh-century England. By the fourteenth century, European scripts had evolved to make writing more comfortable and natural. While movable chairs remained rare in medieval Europe, people did sit on stools (including the portable kind still common today), benches, and chests. Old Mediterranean sitting habits were preserved, although men and women squatted far more often than they do now. Learned men sat on high-backed wooden built-in furniture. And these arrangements have defenders even today. Galen Cranz believes that flat, unupholstered seats give our sit bones healthy support and that plain wooden or stone backs encourage free movement. One of the most recent ideas in seating, the stool with a weighted base that can tilt to support workers in a half-standing, half-sitting position also recommended by Cranz, derives from the seating of medieval choir stalls. To use their full lung capacity, choristers had to stand for prolonged services, and small projections on the undersides of their seats supported their sit bones while they stood. Medieval technology could be both ingenious and humane.[14]

It could also be more usable than modern furniture. The slanted surfaces of medieval writing desks positioned books at an optimal angle for sitters. As the Swiss architectural historian Siegfried Giedion suggested, our horizontal reading tables and desks date from the late eighteenth century, when English libraries were refurbished to accommodate large-format engraved books. Today, mail-order houses specializing in equipment for readers and writers sell tilted reading stands, but medieval and early modern Europeans had rotating stands and desks that could support three, four, or more volumes, pioneering now familiar multitasking unknown in antiquity.[15]

The Middle Ages, then, had remarkably sophisticated reading equipment, but no chair designed especially for business or reading. The word *bank* is derived from bench, the same substantial seating that men of authority might use. (In those insecure times, furnishings were either virtually built in or light and transportable.) And the seventeenth century was not very different. While the poor continued to use stools or to improvise, people of authority sat in high-backed armchairs built for ruling, not relaxing. The characteristic seating of the rulers of the seventeenth century's greatest empire, Spain, were the *sillones fraileros* (friars' chairs), square armchairs of mortise-and-tenon construction with leather seats and backrests. Whether in Spain, the Netherlands, France, or England, the immediate physical well-being of the occupant was unimportant to the makers and users of this furniture. Just as vital as the decisions he made was his presentation of himself to his peers and subordinates. Early modern etiquette books agreed on the need for an upright, steady, and dignified appearance. Fidgeting—now praised as healthy by some medical experts—and crossed legs were violations of what one contemporary writer called "measure and consonance." Today we expect our mechanically adjustable chairs to support the person; once it was the person who conformed to the chair. Chairs were favorite objects in formal paintings of nobles and cardinals because they framed the august sitters as well as supporting them.[16]

Eighteenth-century etiquette, elaborate as it remained, shifted its attention from formal authority to ease and grace. Chairs were now built with more sinuous lines, encouraging new and varied body positions, as we will see in the next chapter, on reclining. In fact, chair making as we know it was born in the early eighteenth century, as a new trade combined what had been four separate skills: joining, turning, carving, and upholstering. Chairs grew lower and wider. Arms curved to envelop the sitter. This union of techniques in turn made possible more sophisticated and specialized forms. Among these were the earliest library and desk chairs. Some, also called cockfighting chairs, had narrow, upholstered backs and curved arms at shoulder height. (Male) readers could straddle the chairs backward and read books on stands mounted above the backs. Other armchairs for reading and writing had square seats oriented so that a corner pointed forward, toward the table. Their French equivalents, with rounded edges, were called *fauteuils de cabinet* and later *fauteuils de bureau*—apparently the first office furniture named as such. Practical Americans grafted writing arms and drawers onto the comfortable and economical Windsor chairs that had originated early in the century as garden furniture.[17]

The most revolutionary design of the eighteenth century was a "French chair for a writing table" illustrated in a book of new designs by a Nuremberg furniture maker named Jacob Schuebler. He described it as a "French comfortable chair in which the backrest is padded to the hollow of a man's back and is provided with a spring so that the backrest will flex backwards, bending but not breaking." Since even the prodigiously learned Giedion found no French predecessor, Schuebler was probably exploiting the prestige of French craftsmanship. I have found no record of an actual chair built to Schuebler's design; he seems to have been an intuitive harbinger of an age of massive business organization that did not develop until the later nineteenth century, and of a science (ergonomics) not named until the mid-twentieth. He also invented a merchant's desk with an enormous built-in rotary file for books and papers.[18]

Even Schuebler's reading chair was not office furniture in our modern sense. It was still a mainly domestic object for reading and writing. The specialized library as a feature of the aristocratic household dates only from the late eighteenth and early nineteenth centuries. In merchants' offices of that time, the clerks still worked on stools while the employer sat in an upholstered armchair. By legal wording he (or, occasionally, she) was a master, and even respectable clerks were servants. As late as 1882, a popular etiquette manual described the proper form for standing while writing, illustrated with a man working, with no stool visible, at an ornate slant-top desk. But even some rulers valued dignity or simplicity above comfort. The austere Emperor Francis Joseph II of Austria is shown at a writing desk in a print of around 1900, surrounded by the sumptuous trappings of the Hofburg palace but leaning forward awkwardly at the edge of a simple wooden armchair as he composes a letter.[19]

A CENTURY OF POSTURE

Long before either royalty or plutocracy began to adopt work chairs for their own health, nineteenth-century opinion turned to children's posture. Mass education was bringing ever larger numbers of pupils into elementary schools. Beginning in the 1850s, writers in Europe and America expressed alarm at the ills brought on by uniform desks and chairs, too big for some pupils and too small for others. Children were twisting and slumping, and one medical authority in 1880 claimed that from 83 to 92 percent of schoolchildren had misaligned spinal cords. The catalogue of troubles also included myopia (as we will see in Chapter Nine), stooped shoulders, and respiratory problems. Two physicians, a Swiss, Hans Konrad Fahrner, and a Swiss-Russian, Fedor F. Erisman (born Friedrich Erismann), developed a system of eight graded sizes of desks and chairs. According to a Russian-born civil engineer and inventor, Gabriel Bobrick, Russia was the first country to establish health standards for school furniture. Bobrick proposed his own designs, featuring screw adjustments for seat and backrest height as well as for distance between desk and chair. While Bobrick viewed student comfort sternly—settings were adjustable only with tools—the seats he proposed were among the earliest with any kind of movable lumbar support. His chairs would have been costly to produce, and they probably never went into production. Yet the reform movement did encourage school furniture with sloping writing surfaces, back support, graded sizes, and other features friendlier to growing bodies.[20]

Furniture reformers also turned their attention to women workers, especially in the United States. As early as the 1850s, American writers began to investigate the need for lumbar support, but the most notable innovations were proposed for railroad seating—for an already industrial space—rather than for the home or even the office. In 1871, a U.S. patent was granted for a chair "[c]onstructed on scientific principles to prevent many of the diseases to those operating sewing machines." Sloping gently forward, it had spindles forming "a deep recess at the lower part of the back," by which "the muscles of the thighs are relieved from pressure whilst the back just below the shoulders receives a suitable support." It was followed twenty-five years later by a patented "typewriter's chair," with a curved rest for the lower back connected to the bottom of an adjustable swivel seat by three curved, flexible metal rods. Shortly thereafter, other inventors began to link the back support of secretarial chairs to spring mechanisms beneath the seats, creating what were described as the first posture chairs.[21]

Chairs for female office and factory workers underscored not just the role of machinery in changing furniture design, but the importance of gender in techniques of sitting. The cultural historian Kenneth L. Ames has noted that American Victorian parlor furniture offered a template for privilege: larger "gent's chairs" for men, smaller "sewing chairs" and "lady's chairs" for women, even if families did not consistently observe this division in everyday life. The rocking chair, seating developed by household improvisers in the eighteenth century who attached rockers to ordinary Windsor chairs, became an American specialty in the early nineteenth. Women favored small, low-seated models called nursing chairs or sewing chairs. Generations of Americans grew up associating gentle oscillations with the tenderness of childhood. American men had more massive versions of their own. As Giedion summarized international differences in body techniques in the mid-twentieth century: "The American farmer, at the end of the day, will instinctively move to the rocker on his porch. The European peasant sits immovable through the twilight as if nailed to the bench before his cottage." American males did not need special seating to kick back, though. They were renowned for their habit of tilting on the rear legs of their chairs, sometimes resting the seat backs against the wall, sometimes keeping their feet atop a railing or even on their desk or table—a sign of insolence or of manly self-assurance, depending on the observer. No wonder American sitting techniques have fascinated Europeans.[22]

American male tilting had a technological counterpart in the tilt-and-swivel chair that spread in the later nineteenth century. Americans, less concerned about maintaining the formality of upright positions, were developing specialized chairs for travel, haircutting, and surgery. In 1853, an American named Peter Ten Eyck worked out a system of springs that would let the sitter move back and forth as though in a rocking chair, though his chair rested on a four-legged pedestal with casters rather than on rockers. This was the beginning of a new technique of sitting, forgoing the dignity of the chair as static frame and embracing regular changes of position. Introduced in domestic furniture, it gradually spread to office chairs. We are used to this style of sitting now and would find baroque ideals of decorum intolerable, but it was once a new technique that had to be learned. The tilting office chair remained a male object, and not just because most white-collar men in precomputer days still used pen and ink rather than machinery that required a semi-fixed position. Women's boned corsets, prevailing in polite society until World War I, limited flexing and reclining. In fact, corsets were longest and most restrictive from 1900 to 1914, just as overstuffed parlor furniture made sitting a special challenge.[23]

In the late nineteenth and early twentieth centuries, manufacturers introduced springs mounted under the seat with knobs for adjusting tension to suit the weight and work habits of the occupant. This is still a common arrangement even in premium office seating. With medium tension, the tilt-and-swivel desk chair afforded not only the notorious feet-on-desk posture but more genteel rocking. As growing numbers of women entered business as secretaries and typists, two sets of body techniques prevailed: the uprightness of the female machine operator and the flexibility of the male manager. The male clerk, on his feet or perched on his stool, was becoming an anachronism.

Before World War I, not all female office workers sat in secretarial chairs. Many women must have continued to work in straight-backed side chairs like that of the secretary in the anteroom of a prominent Madison,

Wisconsin, lawyer and judge, Albert G. Zimmerman, recorded in a photographic self-portrait in 1898. Zimmerman is visible through a doorway. He is reading papers while tilting back in a wooden armchair pivoting on a four-legged base, as though his secretary's corset-enforced uprightness accentuated his ease. (He evidently set up the shot for exposure by a third party.) But even Zimmerman's chair appears to have provided no support for the lower back. Nearly all seating specialists today recommend support for the lumbar region to restore the natural S-curve of the spine, which tends otherwise to have an outward bow (kyphosis) when the pelvis rotates backward during sitting. Early secretarial chairs not only had lumbar support, though not always adjustable; they also had the gap between seat and back that today's specialists recommend. Continuous backs, present even in some costly executive chairs today, tend to push the pelvis and sacrum forward and produce kyphosis.[24]

BATTLING FATIGUE

It was World War I that brought concern about chairs to public attention—once more, not the furniture of managers and executives but the seating of workers, especially female workers. The early industrial engineers paid surprisingly little attention to furniture, probably because so many of their clients and employers were in heavy industry, where workers customarily stood. Until the late nineteenth century a factory worker was lucky to have an upturned box or a stool. Characteristically, Frederick Winslow Taylor, who gave his name to the efficiency movement, illustrated his speeches with a story of a laborer and his shovel. Even Taylor's almost equally celebrated disciples, Frank and Lillian Gilbreth, introduced only one significant innovation, a chair elevated on a movable platform at a high desk to allow alternating sitting and standing positions. Illustrated in textbooks, this chair was never, apparently, produced. No wonder Frank Gilbreth said that the only comfortable chair he had found was a church pew.[25]

The search for scientifically designed chairs began not in offices but in factories. Throughout Europe and North America, industrialists, managers, and scientists were discovering that prolonged factory work could actually retard production. Doctors and physiologists developed a theory of fatigue that regarded the body as a "living motor" (in the phrase of one expert, Jules Amar) subject to the conservation of energy. The laws of thermodynamics replaced the precepts of religion in efforts to increase industrial production. Some manufacturers began to support regular rest periods and limits on hours in the interest of higher profits and lower accident claims. During World War I, when the military realized that industrial output could determine victory and hours were increased sharply, governments began to sponsor studies of work physiology, hygiene, and safety—research that extended, for the first time, to seating.[26]

Some cultural historians have seen a sinister side to the movement for employee welfare. Even campaigns against industrial accidents can appear to be pretexts for extending paternalist surveillance over workers' bodies. And it is true that, like earlier campaigns against fatigue, efforts for seating reform appealed strongly to employer self-interest and national power politics. In 1916 the Police, Factories, etc. (Miscellaneous Provisions Act) empowered the British Home Office to regulate industrial seating. In 1920, a new Home Office report called the strain of prolonged standing "of a most exhausting description—and so absorbs energy which should, by right, be at the disposal of the employer and the employee in the work under discharge." But the alternative was equally exploitive and far more unpleasant. Many employers, according to the London *Times* medical correspondent, still believed that sitting "encourages slackness and that men and women work harder when they are forced to stand." In 1922, the U.S. journal *Industrial Management* still had to tell its readers: "Back-rests do not mean lazy workers, as is asserted by some of the old school manufacturers," though it acknowledged that bad backrests could cause fatigue that was misinterpreted as indolence.[27]

In England, tens of thousands of women worked in munitions and heavy industry. In 1914 there were only ten women in the Royal Factories, by 1918 over 24,000. Their work exposed them to explosions, fumes from TNT and other toxic substances, and asbestos. Their equipment reflected the crude standard of the day. In a photograph of munitions workers filling machine-gun ammunition belts in London, the women work four abreast on backless benches facing a table with compartments. Some factories, especially in newer industries, appointed "welfare superintendents" who arranged for canteens with comfortable seats and even "rest rooms"—an early instance of the present euphemism—away from shop-floor noise and equipped with lounge chairs for naps, either during scheduled breaks or on referral by a nurse or forewoman.[28]

Amid concern about declining productivity and increasing illness among women workers, one manufacturer began to sell the first industrial chair with a health claim. The maker, the Tan-Sad Works of Birmingham, is now familiar only to a handful of design historians. But while its products never shared the high modernist aura of the Bauhaus, they were the closest ancestors of today's ergonomic seating. An American executive later heard that the chair was a direct response to a drop in productivity among women workers.[29]

The Tan-Sad chair had four tubular steel legs supporting a shallow (apparently about 1.5 feet) slightly concave seatpan that centered the worker. A wing nut on each leg of the chair allowed adjustment of a telescoping steel tube within the leg to raise or lower the height of the chair for each worker. A curved backrest fit in the worker's lumbar region and swiveled up and down with the position of the back. A bracket running along each side of the chair had a series of holes for curved metal rods supporting the backrest. By moving the end of the rod forward or back, and by adjusting the position of another rod anchored at the back of the seat, the backrest could be positioned higher or lower, forward or back for each worker. Four steel stretchers, attached with nuts that could be tightened as needed, connected the legs and stabilized the chair, equipped with cast-on steel feet claimed to rest securely even on uneven surfaces.[30]

After the war, Tan-Sad advertised heavily. The company's claims for its product were unprecedented in the furniture business. It was said to hold a "Gold Medal awarded for 'The Only Chair Designed for the Worker,' " to promote health by eliminating fatigue, to increase output 25 percent, and to pay for itself in twelve working days. Lewis's of Liverpool was said to have ordered two chairs one week, thirty chairs the next, and four hundred chairs the following week in response to improved productivity. Tan-Sad sold two models, the original factory product with seat and back in three-ply veneer wood for twenty-eight shillings and an office model with a seat upholstered with "Rexine" for thirty-three shillings. By December 1921 it appears to have achieved national distribution, an impressive record since business magazines at the time advertised Thonet-style bentwood chairs for eight shillings ninepence. In 1923 Tan-Sad listed the Great Western Railway Company, Lever Brothers, and the governments of South Africa and New South Wales among its customers.[31]

Meanwhile, Tan-Sad attracted some of the first professional and press testimonials recorded for a work chair. When the firm sent a review copy to the London *Times,* the medical correspondent approved its light weight and "firm, strong support" for the back, noting that "the body is kept at a correct degree of tension," reducing the chances of injury. A *Times* employee who had been using the chair praised its comfort. The Tan-Sad chair appears among a number of other models illustrating good features. By 1925 it was featured as an "anti-fatigue chair" in an article on productivity and equipment in the London edition of *Success*.[32]

Other manufacturers and businesses were discovering posture seating, especially in the high technology of the time—telephone exchanges and electrical and electronics factories—with hundreds of women workers. As a new generation of managers sought to import the techniques of industrial rationalization into commercial organization and equipment, seating ideas conceived for the factory migrated to the office. The American Posture League, uniting orthopedists with specialists in physical education and industrial efficiency beginning in 1914, evangelized college deans and manufacturing executives alike. North American and European inventors responded to posture consciousness with a new wave of office chairs. As usual, outsiders brought innovation. A French engineer named Henri Liber established a company called Flambo in 1919 to market a secretarial chair with a backrest adjustable up and down along a U-shaped metal track. Meanwhile, postwar Germany turned its drive for standardization and public health to seating. In the later 1920s, the office supply manufacturer Fortschritt of Freiburg im Breisgau, best known for state-of-the-art card file systems, marketed a Fortschritt-Stuhl ("Progress Chair") with a spring-loaded back support system—unlike the rigid Tan-Sad and Flambo—and a lever under the seat for adjusting the height of the chair without rising, fifty years before this feature became common. An advertisement noted proudly that the Prussian Ministry of Welfare (Volkswohlfahrtsministerium) had awarded the chair a winning ninety-point score in an evaluation of seating that must have been one of the first conducted by any government.

While Fortschritt hired the rising young photographer and graphic designer Anton Stankowski to produce superlative advertising in the modernist style, the architects of the Bauhaus made surprisingly few technical contributions to office seating. Marcel Breuer's pivoting metal secretarial chair of 1926 had a planar seat and fixed straight back, both features discouraged by posture experts of the day. Sitters could not have been enthusiastic, either, but modernist functionalism no less than baroque decorum demanded sacrifices.[33]

THE AMERICANIZATION OF POSTURE

Curiously, European chair makers of the 1920s were slow to develop posture chairs for managers and executives. The executive office was still more a throne room than a work space, a theater of prestigious self-representation. Specialized furniture would have broken the spell. Thus it was an American entrepreneur who took the next step in ergonomic seating. In the early 1920s William J. Ferris, an Elkhart, Indiana, manufacturer of metal baby carriages, was looking for an adult chair design to produce. According to later company lore, he saw a drawing that appealed to him in the portfolio of another visitor in the U.S. Patent Office waiting room. That visitor was one of two young American veterans who had worked with Tan-Sad during the war, probably as military coordinators, and he put Ferris in touch with the company. No records of the companies' dealings or of their patent numbers survive; Tan-Sad was liquidated in 1975. (It had long dropped its seating business and concentrated on prams, which parents were spurning in favor of push strollers.) But the relationship must have been cordial, because by 1925, Tan-Sad was using the trademark of Ferris's company: the Domore Chair.[34]

Do/More—the original U.S. spelling—offered an industrial line adding more graceful swivel models with radial bases and casters. It appealed to executives and managers with armrests, leather upholstery, and optional simulated-wood paints on metal parts. It sold health benefits aggressively. Slouching in conventional chairs, Do/More's literature warned, concentrates blood in the abdomen and overtaxes the heart. The Do/More Chair, now Do/More Health Seating Equipment, could help prevent "hemorrhoids—kidney trouble—constipation—a slowing down of the digestive tract—chronic dyspepsia—prostatic trouble and hardening of the blood vessels." Do/More salesmen privately called this the "blood and guts posture story." And they also knew how to sell from below. Riding streetcars with sample chairs in hand, they would carry them up to offices, approach managers, and leave one for a single worker's trial, adjusting it carefully for her. Whether because of the chair's innate features or the attention, employees usually wanted to keep the chair, and jealous colleagues would want their own.[35]

In the early 1930s, Do/More appears to have appealed increasingly to comfort as well to health. It established a Posture Research Corporation (later Institute), under the nominal direction of a consulting physician, that developed fitting procedures and even calisthenics that could be performed while sitting in Do/More chairs. The Woodfield Executive Chair had "special back construction [that] enables a user to take exercises in his office that will strengthen the abdominal muscles and help pull down any waistline bulge." The spring-loaded reclining function in its executive models was not new, but Posture Research literature glorified it with a silhouette of a male executive in upright and reclining positions against the background of a muscular man operating a rowing machine. The arms and seat were more substantially padded. For the rank and file, Do/More offered Air-Duct Chairs with ventilated seats and backs. Thanks to this range, Do/More became a supplier to the U.S. government and the Bell Telephone System. Trained "posture specialists" working for the company's distributors fitted each chair to the user and instructed employees on proper sitting; Do/More literature insisted that individuals, who often had adapted their bodies to bad chairs, could not be trusted to follow their immediate sense of comfort. In fact, the ventilated Postur-Matic model was supposed to nudge occupants into uprightness. If the sitter slouched forward, his or her sit bones would encounter the discomfort of a recessed duct running across the seat and the correct position would be restored. (Likewise, Frank Lloyd Wright defended the disconcerting three-legged chair he designed for the Johnson Wax headquarters in 1939 with the argument that its instability forced good posture. Wright eventually relented and helped add front legs. Workers in his Larkin Building of 1904 had called the seat's predecessor the suicide chair.) Do/More initially encouraged distributors to establish paid service agreements for cleaning and lubricating chairs twice a year. The New York distributor had, in addition to twelve salesmen, six full-time field mechanics.[36]

High-handed and even authoritarian as the system seems now, its appeal to scientific expertise sold chairs. The line was hierarchical, ranging from "factory" to "clerical," "junior executive," "executive," and ultimately "mediator," a judge's chair for top executives, later used by Harry Truman in the Oval Office. And the company gathered testimonials. In the New York City garment district, manufacturers reported reductions of back problems and unauthorized breaks. The secretary of State Farm Mutual Insurance Company wrote in 1939 to attribute its over 99 percent employee attendance rate to the company's 739 Do/More Chairs.[37]

Of course, there were other posture-conscious firms. A man named Charles E. Pipp offered a "Pep Chair" in 1921 with adjustments and features resembling the Do/More's. The Gunlocke Company, which specialized in luxurious suites for top executives, introduced its Washington Series in 1923, and Evan S. Harter produced an Executive Posture Chair in 1927, one of the first chairs with controls for synchronizing back and seat reclining. But Do/More, as a newcomer, was most systematic in relating the technology of its chairs to the proper techniques of sitting. Its health campaign helped prepare for the postwar design renaissance.[38]

The metal shortages of World War II interrupted the refinement of office chair mechanisms; Do/More and no doubt other companies were able to continue producing steel executive posture chairs only with medical

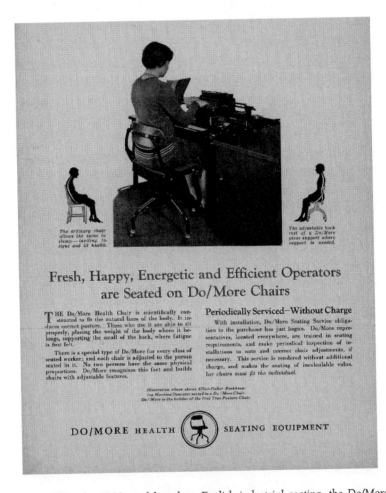

The ordinary chair allows the spine to slump—inviting fatigue and ill health.

The adjustable back rest of a Do/More gives support where support is needed.

Fresh, Happy, Energetic and Efficient Operators are Seated on Do/More Chairs

THE Do/More Health Chair is scientifically constructed to fit the natural lines of the body. It introduces correct posture. Those who use it are able to sit properly, placing the weight of the body where it belongs, supporting the small of the back, where fatigue is first felt.

There is a special type of Do/More for every class of seated worker; and each chair is adjusted to the person seated in it. No two persons have the same physical proportions. Do/More recognizes this fact and builds chairs with adjustable features.

Periodically Serviced—Without Charge

With installation, Do/More Seating Service obligation to the purchaser has just begun. Do/More representatives, located everywhere, are trained in seating requirements, and make periodical inspection of installations to note and correct chair adjustments, if necessary. This service is rendered without additional charge, and makes the seating of incalculable value, for *chairs must fit the individual.*

Illustration above shows Elliot-Fisher Bookkeeping Machine Operator seated in a Do/More Chair. Do/More is the builder of the first True Posture Chair.

DO/MORE HEALTH ⬤ SEATING EQUIPMENT

Introduced in the 1920s and based on English industrial seating, the Do/More Chair was the first to be marketed nationally in the United States for its health and morale benefits. The line was extended upward to include more richly upholstered and imposing executive and managerial models. More recently, the Domore Company produced some of the first chairs especially made for air traffic controllers, emergency service dispatchers, and other intensive users and is still an active supplier. (Courtesy of Lux Steel Inc.)

prescriptions. But indirectly the war had an immense influence. Steel was a mature technology, but defense work had helped industry metallurgists develop improved techniques of tubing, stamping, bending, and welding. And the war forced a new round of attention to human-machine interaction, from aviation seating to the design of controls. Military sponsorship of human factors studies helped create a new cadre of professionals.[39]

During the 1950s and 1960s, work chair design was demedicalized. Do/More continued its system, but it had new competitors with less physiological and more aesthetic goals. After a fifteen-year break in office construction, a building boom began, but the large corporate architectural firms and their clients shunned the shop-floor culture of exposed metal of the 1920s and 1930s in favor of sculptural forms using not just metals but plastics developed during and after the war. Knoll International's classics swept architects and clients off their feet: Eero Saarinen's tulip chairs, Charles Pollock's swivel armchair with its one-piece plastic shell, and above all Charles and Ray Eames's fiberglass series and their aluminum group became icons of the postwar years. As popular culture, medical thought, and etiquette books alike moved away from the full upright position of the interwar years, seating emphasized ease and flexibility, not support. Posture-enforcing technology was becoming passé. Niels Diffrient, who worked for Saarinen in the early 1950s and knew most of the other star designers, recalls that none of them expressed interest in human factors. In a Knoll company history of the 1980s, Pollock described his executive chair of 1965 as built around its extruded-aluminum rim, observing that it has "no inner spine. . . . It doesn't rust, it doesn't tarnish, it doesn't fade." Even Do/More (now Domore) hired the influential designer Raymond Loewy to rejuvenate the lines of its seating, though not to change the mechanism.[40]

FROM POSTURE TO ERGONOMICS

Most of the celebrated 1950s and 1960s chairs departed from the old posture standards. The Pollock armchair lacked a lower-back support, and its shell enforced a rigid angle between back and seat. Gradually a new posture movement emerged, first identified as "human factors" and later as "ergonomics." In 1967, air traffic controllers sat in rickety tubular-frame swivel chairs and leaned over tables to use horizontally mounted instruments. As part of a reform of their work conditions, the Federal Aviation Administration (FAA) tested dozens of chairs and issued a set of guidelines

The Aeron chair, the Leap chair, and the Freedom chair offer complementary features. The first (above) is celebrated for its cooling mesh and a deep, rocking recline motion. The second (facing page, top) has a seat pan that slides forward, separate upper- and lower-back controls, and a lumbar support that curves snugly. The third (facing page, bottom) needs minimal user adjustments and provides an especially deep reclining position, and (as options) a synchronized headrest and an ottoman. The shared ideal is no single posture but promotion of healthy variation—especially reclining. (Courtesy of Herman Miller [photographer: Nick Merrick of Hedrich Blessing], Steelcase, Inc., and Niels Diffrient)

for controller seating, governing everything from the backrest action to the shape and adjustment of the seat. Domore won the competition to produce a new chair for every air traffic controller. These Domore chairs may have been the first major postwar line available in A, B, and C sizes for a single style. For Domore, the contract opened an important niche, round-the-clock seating. (The FAA still uses 16,000 Domore chairs.) And it also helped turn the attention of architects and their clients back to the health side of seating. By the mid-1970s, the two largest office seating companies had posture chairs of their own: the Steelcase 430 Series (1974), derived

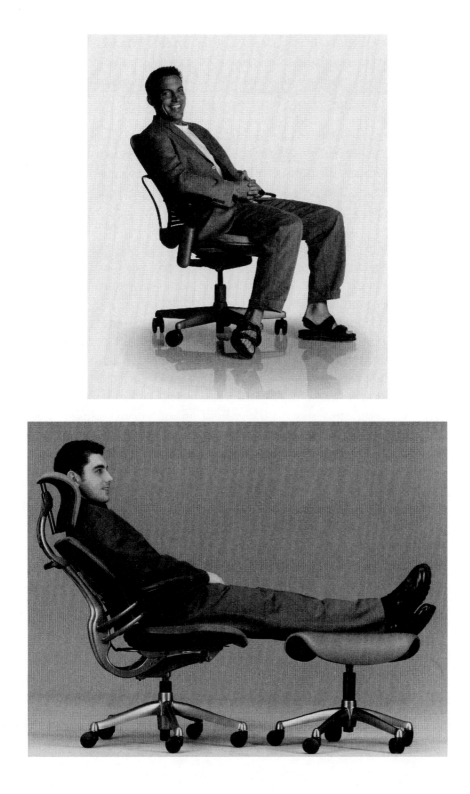

from aerospace human factors research, and the Herman Miller Ergon chair, based on time-lapse photographic studies by the designer, William Stumpf, at the University of Wisconsin. With its high back, body-fitting curves, comfortable edges, and ingeniously placed armrests, the original Ergon helped make posture chic again—under the new aegis of ergonomics. It was the first chair to take full advantage of new foams and plastics to create a distinctive shape without compromising comfort or motion.[41]

Stumpf's ideas on posture reflected a restless, experimental decade. The Tan-Sad and Domore generations had sold efficiency and productivity. Stumpf doubted openly that the ergonomic design could boost office output by more than 1 percent at best. As *Progressive Architecture* put it in 1980, Stumpf recognized that "people sit right-side-up, sideways, and upside down as they please, lumbar support or no." His was the first U.S. chair to use a gas cylinder—a German invention—to regulate seat height and help cushion the shock of sitting down.[42]

Independently, other designers were pursuing similar goals. Niels Diffrient of Henry Dreyfuss Associates, who had worked with the early medical chair expert Dr. Janet Travell on airline and industrial seating, patented the first knee-tilt mechanism that rotated the sitter's body back and downward instead of tilting it where the center column met the seatpan. Feet could now stay on the ground, making deep reclining more comfortable. After studying orthopedic and vascular health, the Argentineborn American polymath Emilio Ambasz and his Italian colleague Giancarlo Piretti introduced the Vertebra Seating System in 1978, a series of chairs with ingeniously hinged backrests and spring-loaded sliding seats that encouraged relaxation. Their patented joints covered with butyl rubber bellows helped define a new high-technology look in chairs.[43]

Even as these chairs came to market, offices were changing. The rise of computing was putting terminals and keyboards on the desks of some managers and even executives. But pens, papers, and telephones weren't going away. The ideal chair would help the body through this growing range of activities, by positively encouraging changes of motion. The Cyborg chair, developed in Denmark by the designer Jacob Jensen and the ergonomist Vibeke Leschly, had pneumatic cylinders for adjusting seat height, seat angle, and seat depth. But its greatest innovation was the seat angle cylinder, which automatically and gradually changed the seat back position in response to variations in weight. Users were induced unconsciously to change the position of their spines through about four degrees at the rate of about a degree a minute, so loads on muscles and vertebrae varied. By changing points of contact between chair and body, the Cyborg also promoted circulation and prevented muscle tissue malnutrition, according to Rudd International, its U.S. manufacturer. Lifting the body's weight, even by reaching for a pencil, reset the cylinder for a new cycle. The Cyborg chair went out of production quietly in the early 1990s, no doubt held back by its price of over $1,300. But it showed how much (and how little) attitudes toward body techniques had changed. Like the Domore Postur-Matic, it induced motion gently. But while Domore had discouraged user adjustments and was designed to nudge the occupant back to a single optimal attitude, the Cyborg intentionally and subtly destabilized the sitter, as though symbolizing a shift from high modernism to postmodernity.[44]

All these chairs of the late 1970s had one important feature in common. They were advertised for a range of positions and activities; reclining appeared almost as important as upright sitting. And as women sought greater equality in the workplace, secretarial chairs gained more features and adjustments for greater comfort. In fact, Emilio Ambasz told me in the early 1980s that he considered the headrest of his executive model to be a functionless "nimbus" and personally used the secretarial model of the Vertebra chair.

THE ERGONOMIC EXPLOSION

From these beginnings the ergonomic seating movement flowered in the 1980s. Once more, the military and the schools led the way. In 1978, the National Aeronautics and Space Administration (NASA) published the *Anthropometric Source Book,* derived from research on weightlessness in the space program. It determined, among many other dimensions, a natural 128-degree angle between upper body and thighs and between the thighs and the lower legs. NASA's was not the first measurement of weightless posture; an American surgeon named J. J. Keegan had established in 1953 that the natural resting postion of people asleep on their sides with relaxed muscles was a 135-degree angle, and we will see in Chapter Six that European researchers had reached similar conclusions in the 1930s. (Even before them, in 1878, *Nature* had identified the bamboo veranda chair of colonial India, with its **W**-shaped profile, as physiologically optimal seating.) But the *Anthropometric Source Book* was the first work of its kind widely available to engineers and designers. At the same time, a Danish surgeon, A. C. Mandal, alarmed by the rate of increase of back problems in an increasingly sedentary society and aware of Keegan's work, pointed to the deformation of young backbones by conventional school furniture. Backward-sloping seats with lower-back support were deforming the lumbar spines of children, who spent their time not sitting up like typists but leaning over desks to read, write, and reach for objects, distorting the lumbar spine. Mandal's solution was also to open up the thigh-trunk angle with a seat that inclined forward 15 degrees and a desk slanted toward the student at 10 degrees. Indeed, he observed that children, and adults with back problems, often tilted chairs forward for relief.[45]

While forward-sloping seats had been recommended as early as 1884, the 1980s were their great age. Schools in Scandinavia and France were refurbished with chairs and desks built on Mandal's principles. And the angle of adult seating was influenced, too. In America, a Texas professor of mechanical engineering and former aviator, Jerome Congleton, designed a new series of chairs now best known under the Bodybilt trademark, with highly contoured seatpans to hold the sitter in place in the angle recommended by NASA and Dr. Mandal. In the late 1970s the Norwegian company Håg introduced the Balans chair, a kneeling stool with a forward-inclined seat and a shin rest that kept the sitter from sliding forward. The Balans, still in production and widely imitated, shows how technology and body techniques can interact. The rise of the personal computer multiplied the number of professionals who sat at keyboards and monitors all day; it also coincided with a rise in back problems caused by poor posture. But like many of the other technologies we have seen, the Balans had to be learned. According to Håg, users need one to four weeks to activate muscles that have not been used in conventional seating. There is also an art to sitting down and getting up. For many people, the Balans chair demands a new set of body skills. Satisfied users found relief from back pain and considered the effort worthwhile; others complained of sore shins and circulation problems. For the first time, abstract ideals of design confronted not only the differences in human dimensions but also those in body habits.[46]

The athletic-shoe industry, selling replaceable products to individuals, solved a similar problem by categorizing foot types (low and high arches) and running styles (pronators and supinators) and offering a variety of shoes selling directly to the consumer for $50 to $150. But the ergonomic chair industry was selling $500 to $1,500 durables largely to employers through architects and interior designers with aesthetic as well as physiological goals. When employees have a chance to comment on sample chairs, consensus is rare, if only because (as studies have shown) most users do not study the instructions for adjusting their permanent chairs, let alone each of a number of floor samples. For most athletic shoes, lacing is the only user adjustment; for office chairs of the 1980s, there may be ten or more, including three adjustments of armrests alone. One survey of office workers with adjustable chairs showed that more than half did not know how to adjust tilt tension or back height, and one in five did not even realize the seat height could be raised and lowered.[47]

Meanwhile, the spread of computers was changing the body techniques of the office. Until the 1980s, most office workers used computers either almost continuously or not at all. Executives shunned them. As their attitudes changed—and we will see the triumph of the keyboard in Chapter Eight—their body techniques shifted, and they alternated between keyboarding, reading, reaching for the telephone and fax, and talking face-to-face. In fashionable open-plan offices, the totem pole of executive, managerial, and task furniture seemed anachronistic. Like Ambasz, many executives discovered that smaller chairs fitted the new office society. In 1989, the *New York Times Magazine* featured the Texas billionaire Ross Perot in a cloth-upholstered "manager" version of Klaus Franck and Werner Sauer's FS chair, designed for Wilkhahn, one of a new generation of seats adjusting automatically to the sitter's weight and motions. Even Thomas S. Monaghan, the chairman and founder of Domino's Pizza, was revealed to work not in his $5,000 leather throne but in a "high-quality" armless secretarial chair kept in a tiny private room within his two-story, 3,500-square-foot office suite.[48]

While many corporations were shedding employees, and income differentials were starting to grow in the United States, designers campaigned successfully for more equal access to seating comfort with minimal adjustments. The chair was supposed to respond to the sitter; active people needed "passive" seating. Herman Miller worked with William Stumpf and Donald Chadwick on the Equa chair (1984), using a one-piece glass-reinforced thermoplastic polyester shell cut out like an **H** and mounted on a knee-tilt mechanism. When the sitter leaned back, the shell flexed automatically to support both the upper and lower back during reclining. Herman Miller's neighbor and archrival, Steelcase, introduced the Sensor chair, licensed from the German designer Wolfgang Muller-Deisig in 1986, with sculpted, contoured foam support and a polypropylene shell that flexed with the user's back. As Muller-Deisig put it, the chair "moves when the body moves." While available with multiple back heights and fabric grades, the Sensor and the Equa were among the first chairs marketed up and down the corporate ladder, and the first chairs to successfully substitute responsive materials for mechanical linkages. Seating hierarchy became cosmetic; power might bring leather upholstery and a chromed base, not a better mechanism.[49]

FLEXIBLE SEATING

As is often the case in design, especially design for the body, the simple and elegant solutions of the Sensor and the Equa had taken years of painstaking studies, trials, and development work. By the late 1980s, a competitive chair needed not only to be ergonomically sound but also to be comfortable on a first trial and (with a few exceptions like the Balans) to look comfortable. It took the Italian architect Mario Bellini six years to develop his own body-conforming chair, the Persona, introduced in 1985 by the Swiss firm Vitra and still considered a style leader in its category by many architects.[50]

In the dawning twenty-first century, it is no longer new ergonomic theories that produce new seating. Regular changes of position—what the ergonomist Karl Kroemer has called free posturing—have largely replaced the fixation of the older posture movement on uprightness. But there is still a search for new styles of body motion. The most successful and controversial innovation has been the Aeron chair (1994). Designed, like the Equa, by Stumpf and Chadwick, it is also made by Herman Miller. The Aeron took the idea of a full range of body motion farther than any other chair produced by major furniture companies. At the onset of the Internet era, it offered the most boldly technological look in mainstream office seating, with a polymer mesh called a pellicle stretched across a curving molded plastic frame that swelled across the back and shoulders. (This was not as radical an idea as it seemed; some American swivel armchairs even before 1890 had backs and seats of woven cane, fashionable for executive seating.) Even the original Equa had two back heights; the Aeron has large and small variants but they share the proportions of the standard model. The key to the Aeron is an usually deep tilt, up to about 30 degrees, from the ankles rather than from the knees. The sitter's body rotates away from the desk and down. It takes a powerful spring to balance the chair; the tension knob moves through one hundred revolutions, or up to three hundred turns of the hand. Some ergonomists frown on this, and on the Aeron's continuous back, which pushes the sitter's buttocks forward. The pellicle lets air circulate around the body but does not allow full cushioning of the seat edge and needs a movable block for lumbar support.[51]

The Aeron chair has still been an overwhelming market success, not just because of its edgy but elegant appearance, but also in response to what some designers call its ride. Every chair interacts differently with the body. Once the tension is adjusted properly, the Aeron swings smoothly and can take the sitter from a tedious keyboard task to an easy, cushioned recline. Herman Miller's brilliant campaign for the Aeron featured young men and women sprawling joyously. Locked upright, the Aeron can still be a formidable throne for an authority figure. Released, it is even more of a crypto-rocker than other spring-loaded seats. Miller spokespeople have called it a "cross-performance" chair, as though it were the seating counterpart of the cross-training athletic shoes and the industrial skills cross-training that were becoming so popular in the early 1990s. Rival manufacturers claim—and Miller officials deny—that the reclining Aeron sitter tends to sink too low to allow the sitter to use a keyboard or monitor efficiently.

Steelcase had an equally remarkable advanced project, the Leap chair (1999), with a radically different style of motion. Steelcase's consultants underscored the need for continuous and separate support for the lower and the upper back. The Leap chair is the first to have separate controls for the firmness of each. It has a ride as distinctive as the Aeron's. The lower part of the backrest conforms unobtrusively to the spine, and the force on the upper back can vary with the sitter's weight. After comparing Aeron and Leap in the same showroom, I left wishing the designs could be crossbred. The Aeron had not just its rocking action but a wonderful upper-back cradling missing in the Leap. But the Leap combination of upper- and lower-back adjustments, trademarked as the Live Back, was the most sophisticated and comfortable I have found, once a user is familiar with it. When I press back to recline, I notice complete support without the usual gap between my lumbar region and the seat back. In a phrase coined by a patient of Janet Travell, the Live Back had raised my standard of comfort.[52]

The Aeron and Leap chairs also show how corporate spirit affects design. Miller's Calvinist culture has always been disciplined yet often surprisingly playful, hence the rocking motion. Steelcase, still renowned for its massive filing cabinets, built lots of metal into its latest seating. Even the casters of the sixty-pound Leap chair are plastic-coated steel. While the Aeron swings away from work for breaks, the Leap seat remains level and slides forward as the back reclines, permitting continued work in what Steelcase literature calls "the vision and reach zone"—and also helping cost-conscious facilities planners by saving some of the space other chairs need for reclining.[53]

If there is a single word that captures both the supple back of the Leap and the bounce of the Aeron, it is *flexibility.* The anthropologist Emily Martin has called attention to the similarity between the metaphors of the healthy immune system that spread during the AIDS crisis and emerging notions of business organization. Shedding middle management layers, corporate executives want their employees to be risk takers, ready to drop old projects and form new teams in the interest of global competitiveness. *Agility* rather than *efficiency* or *output* is the watchword. Chair design reflects this thinking. For both Aeron and Leap seating, stability is less important than a new openness to surprises and changes of direction. The adolescent sprawl of college students is no longer a bad habit to be purged with proper seating, but a potential corporate asset that can be nurtured with the right technology, just as the student's sneaker is the model for a more flexible style of footwear, shaping itself to conform to the wearer's motions.[54]

The new flexibility is not yet universal. Many financial companies and law firms believe their clients expect traditional decor. They are still a strong market for the Gunlocke Washington series chair, the posture seating of several U.S. presidents. And the straight-up sitting style, like the Balans chair, has a strong following among independent programmers. Dennis Zacharkow, a Minnesota physical therapist and posture theorist, has for over ten years marketed a work chair with no recline mechanism, the Zackback chair, based on his research into a zone of optimal performance. Zacharkow believes that conventional lumbar support promotes unhealthy slumping. Like the original Domore, the Zackback needs adjustment by a second person while the user sits in it. Instead of a conventional backrest, it has a heavy steel tubular frame on which are mounted supports above the hips (sacral region) and below the shoulder blades (lower thoracic spine)—points that sitters can't easily locate alone. According to Zacharkow, who exhibits widely at computer shows, over 95 percent of Zackback buyers are computer-intensive workers.[55]

For all the persistence of the full upright position, the future seems to be on the side of the opposite, reclining with flexed knees, as in the veranda chair praised by *Nature* as the ideal antifatigue furniture: the "chair in the shape of a straggling **W,** which the languor consequent upon a relaxing climate has taught the natives of India to make, and which is known all over the world." It is to the parallel story of reclining seating in the West that we now must turn.[56]

CHAPTER SIX

Laid Back

Reclining Chairs

I F TODAY'S advanced desk chair faces a challenge, it is the appetite of sitters for reclining. The more horizontal we wish to be at work, the dearer the seating. Niels Diffrient's Jefferson chair, introduced in 1984, was an executive chaise with a built-in headrest that adjusted automatically; upholstered in leather over a steel frame, it cost $7,500. The desk chair, even one with an exceptionally deep reclining position like Diffrient's Freedom chair, is still a compromise. We would really rather have a bed. And reclining has not only been natural; at times, it has been prestigious. The story of the reclining chair is one of the richest in the history of the body's interaction with technology. It starts with the wealthy of the ancient world. It unites French ancien régime gentlewomen, Victorian bibliophiles, and twentieth-century German invalids. The recliner's greatest modern inventors were not chair makers by training, but a history teacher, a woodworker, and a farmer. And it became a cherished if ambiguous emblem of mass prosperity.

BEDS AS DESKS

Sleeping, like nursing, walking, and sitting, is universal and natural. But, also like them, it is a technique, and thus cultural. In the absence of artificial light, people seem to sleep in two phases separated by a quiet wakeful interval of an hour or two—a pattern that prevailed even in early modern Europe. In the United States and most other industrial countries, children grow up sleeping alone and without the sounds and smells associated with communal sleeping in other cultures, so the children alternate between sensory overstimulation and deprivation. The anthropologist Carol M. Worthman believes that early sleeping habits may even shape how people respond to stress later in life. Rural Balinese children, carried constantly, learn to fall asleep amid loud sounds and confusion; as adults, they may react to threatening situations by rapidly falling into what they call fear sleep. Body techniques of sleeping, like those of working, continue to evolve in industrial societies, often in response to medical authorities. In the 1920s, American children, who had once slept in cradles with soft linings near their parents or in beds with siblings, were isolated in cribs with firm mattresses. Until recently, 75 percent of American infants were put to bed in a prone position, but since the early 1990s pediatricians have urged supine sleeping to reduce the probability of sudden infant death syndrome (SIDS)—with both benefits and unintended consequences, as we will see in Chapter Ten. The sleep techniques of adults, too, change. For adults, extra-firm mattresses go into and out of favor, perhaps following economic cycles. We actually know surprisingly little about the effects of bed technologies on the techniques of sleep and on health.[1]

Sleeping and resting have a material culture, too. The Japanese futon was part of a complex of objects that included zori and tatami mats, just as beds are part of a system that includes closed shoes and raised furniture. The ancient Greeks introduced not only chairs to the West but also beds. Unlike massive modern bedsteads and heavy innerspring mattresses, the beds of the Greeks and Romans, whether of wood or metal, were portable. Today we recognize a reclining meal, nibbling from suspended clusters of grapes, as one of the decadent scenes immortalized by nineteenth-century academic painting.

While Western culture generally regards working in bed as a suspect activity for a healthy person, it is striking how many great authors wrote while reclining. Lawrence Wright, in *Warm and Snug,* lists Cicero, Horace, the Plinys, Milton, Swift, Rousseau, Voltaire, Gray, Pope, Trollope, Mark Twain, Robert Louis Stevenson, Proust, Winston Churchill, and Edith Sitwell. Fantin-Latour drew in bed, and Glinka and Rossini composed.[2]

In the 1960s, psychologists at the University of California–Davis even confirmed the suitability of reclining for serious work. They noted the advice of many student handbooks to choose simple, straight-backed chairs rather than comfortable ones and to avoid beds, sofas, and lounge chairs; relaxation was said to impede concentration. Probably reflecting this theory as well as college budgets, dormitory study chairs had no padding. And the university even had a "study table" rule for first-semester students.

SARGENT'S "MONARCH" RECLINING CHAIR.

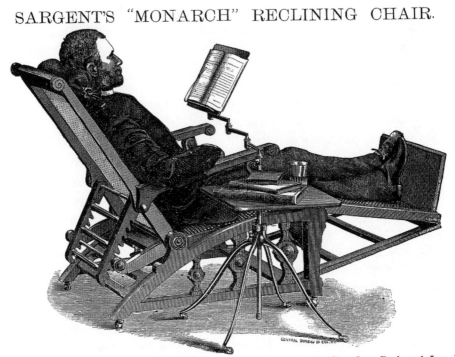

The above illustration represents the "MONARCH" Reclining Chair with Cane Seat, Back and Leg Rest.

This late-nineteenth-century reclining chair (above), often equipped with acces-
sories for holding books and papers, was a familiar type sold equally for invalids
and able-bodied bibliophiles. The American designer Niels Diffrient created the
upholstered Jefferson Chair (facing page) in 1986, as a premium-priced, leather-
upholstered working lounge chair for top executives. Its manufacturer unfortu-
nately did not survive the financial crisis of 1987, and it is out of production.
(Courtesy of Warshaw Collection of Business Americana, Archives Center, National
Museum of American History, Smithsonian Institution, and Niels Diffrient)

Dormitory advisers monitored freshman women, who were required to
spend stated times on weekday evenings seated at their desks. (Why there
was no such rule for male freshmen in dormitories is not explained.) After
surveying 331 students, 171 of whom studied at desks and 160 on beds,
they found no difference between the grade-point averages of the groups.
One reason for the popularity of beds (besides the hardness of the chairs),
it turned out, was that many assignments required more space than the

standard desks provided. No wonder Niels Diffrient once declared that the best chair is a bed.[3]

Reclining was once a much more serious activity than it is now. The wealthy men and women of antiquity memorialized themselves more often on couches than sitting or standing. The ancient body technique of reclining was confined to the wealthier classes. Beginning in the eighth century B.C., the Greeks, and after them the Etruscans and Romans, emulated the rulers of Assyria and Phoenicia, who dined on couches. The Hebrews adopted the custom for banquets, too, and the prophet Amos denounced those "that lie upon beds of ivory, and stretch themselves upon their couches" (Amos 6:4). Lounging looked indolent but, like other aristocratic body techniques, was a high skill. As the classicist and cultural historian Margaret Visser has observed, it took years to learn how to rest gracefully on the left elbow and eat with the right hand without showing fatigue. The original Greek symposia were drinking parties of such reclining gentlemen. Romans maintained the custom but preferred to share their couches with two or three other men, sometimes even in a continuous semicircle. Except among the Etruscans, women had to sit in chairs when they were allowed at all. For a young Greek or Roman man, admission to the conversational world of reclining parties was a great transition in life.[4]

The reclining banquet lasted in aristocratic villas until the very end of the Roman empire in the West, but the wealthy also began to entertain guests seated at tables. As the privileged life of late antiquity disappeared, so did both the furniture and the social and body skills of the reclining banquet. Reclining was no longer a custom of gentleman equals; it was the occasional prerogative of royalty in certain court and legal ceremonials. But the Roman couch did not die completely. It was preserved in the visual record of antiquity. It represented a style of reclining that might be called Convivial, facilitating friendship and conversation.

RECLINING FOR HEALTH

Monarchs not only continued the social use of beds and couches; they also developed the first seating furniture with adjustable backrests and leg rests. The furniture historian Clive Edwards has traced the earliest mechanical seating to a "stool" made for Elizabeth I of England (r. 1558–1603), with cushions upholstered in cloth of gold with silk and gold fringes, and a cushioned back raised and lowered "with staies, springs and staples of iron." The chair has not survived, and it is impossible to say how often it was used and whether it was designed to relieve a medical condition. Much more is known about an invalid armchair constructed by Pierre Lhermite, a Flemish noble, for Elizabeth's great adversary, Philip II of Spain (r. 1556–1598). Philip would be a familiar type of early-twenty-first-century executive, overwhelmed and bewildered by information from his global enterprises, sleep-deprived, and under constant self-imposed stress as he attempts to scrutinize every detail of his realm. Lhermite's design had a back with quilted padding, a footrest, and two curved ratchet bars with teeth that could be used to lock the sitter's position from upright to fully reclining. Lhermite boasted in his memoirs: "Though it was but of wood, leather and ordinary iron [it] was worth ten times its weight in gold and silver for his Majesty's comfort." According to the historian Pamela Tudor-Craig, this was the first time the word *comfort* was applied to physical and secular well-being, as opposed to the spiritual consolation signified in the phrase *comfortable words* in the Book of Common Prayer.

Mechanical furniture and micromanagement thus share sixteenth-century roots.[5]

In the seventeenth century, reclining furniture spread in royal and noble circles. "Sleeping chairs" were owned by Charles I of England and Charles X of Sweden, who died in one. The writer John Evelyn described one in Rome in 1644 in his diary, and fully thirty are recorded in the royal French court by 1687. By the late eighteenth century, upholsterers were fashioning "metamorphic" chairs with hidden functions. One of these, a wing chair made in Denmark and upholstered in soft brown gold-tooled leather and attributed to the Danish royal court architect, C. F. Harsdorff, was auctioned at Sotheby's in April 2000. Metal bars with hooks engage brackets to let the back recline, and the arms and board beneath the T-shaped seat cushion constitute a platform that can be rotated out to form a footrest. This sumptuous if well-worn object was the preferred working chair of its last owner, a Danish antique dealer in New York. The search for healing comfort and the ability to recline, far from being an American populist innovation, had impeccably upper-class origins.[6]

Even as these luxurious chairs were produced, though, health furniture that reclined was already spreading to the less wealthy. By the end of the seventeenth century, some English furniture makers appear to have specialized in these mechanisms. Sleeping chairs took on the now familiar wings for protection against both falls and drafts. Some were designed for "lying-in" by mothers of newborns. Caned reclining chairs were relatively affordable; one used cords running through the arms to synchronize the lowering of the back and the elevation of the footrest. In 1766, a pair of London cabinetmakers patented a medical bedstead with a winch that could adjust the elevation of the back and turn the piece into a settee as the patient recovered. In the early to middle nineteenth century, both cabinet-makers and surgeons received further patents on new designs for easy and reclining chairs to meet the needs of soldiers disabled in the Napoleonic and Crimean Wars. There was also an increased interest in relieving chronic illness. Beside these convalescent chairs appeared a few intended more for the library than the sickroom. One of these, designed by a William Pocock, was celebrated in the early nineteenth century for its mixture of practicality and fantasy. It had an adjustable back and a long, slender footrest that extended from beneath the chair when the back reclined. Attached to the frame with what appears to be a carving of a coiled snake was a slanted bookstand with a lamp; the front legs supported winged lions. Pocock's chairs could be called the beginning of the Cogitative style of reclining furniture.[7]

THE ETIQUETTE OF REPOSE

It took more than inventiveness and medical concern to revive the ancient custom of reclining. Until the eighteenth century, only the sick, convalescent, or elderly were entitled to lean back. The kings of France would impose their will on the aristocratic Parlements while reclining in a formal ceremony called the Lit de Justice, but the point of the monarch's ease was to dramatize his power over the assembled sitting, standing, and even kneeling subjects. A portrait of Mary Tudor, the wife of Philip II, depicts her sitting stiffly on the edge of an upholstered armchair, no doubt partly because it was too large for her, but also because royal status demanded this bearing. (Philip was shown in his reclining chair only in the inventor's long unpublished sketch.) It was thus a major change when healthy people experimented with new techniques of sitting that ultimately changed both the furniture and the social life of the West.[8]

We have seen that in antiquity reclining was a male banqueting custom imported from the Near East. In early modern Europe it was a mainly female social innovation, and also of exotic origin. With the end of the civil wars of the sixteenth century, there could be new attention to luxurious interiors and the arts of living and conversation. Male aristocrats still preferred the grandeur of large, high-backed armchairs designed to set off their splendid wigs. An emerging group of cultivated women had other ideas. The earliest of these, the Marquise de Rambouillet (1588–1665), supervised the design of a new palace that included alcoves, niches in the walls influenced by Spanish and ultimately by Muslim practice. In one of these, a small chamber annexed to her bedroom, she received the leading literary men and women of her day, establishing what later became known as a salon. In supposedly delicate health, she saw visitors while reclining in a daybed set up in the alcove, reestablishing the ancient connection between physical ease and cultivated conversation. (The daybed or *lit de repos* was invented around 1625 or 1630.) It was only a beginning—guests sat in chairs—but the Château de Rambouillet was still a turning point. If Philip II legitimized comfort for monarchs, Mme de Rambouillet helped extend it to aristocratic women; later in the century, they took to reclining on a sofa—a word derived from the richly covered, raised sitting platforms of the Ottoman and Arab worlds.[9]

Despite the existence of sofas and movable-back chairs, the court of Louis XIV, which inspired both the furnishings and the body techniques of much of Europe, had not been an encouraging place for reclining. Seating was graded from square cushions for the lowest ranks of courtiers to stools, side chairs, and armchairs. Daybeds and sofas were foreign to the court's protocol. The reduced importance of courtly life in the eighteenth century gave new popularity to furniture that encouraged leaning. Curves were now modeled after the body, and cabinetmakers produced an array of seductive new types: *chaises longues, duchesses,* and *veilleuses.* New postures prevailed in private gatherings. A painting by Jean-François de Troy, *The Reading of Molière* (ca. 1728) depicts women in a luxurious parlor, leaning far back in low armchairs as their gowns flow elegantly over the seats and arms. The century's characteristic chair was probably the *bergère,* a well-padded armchair with a gently reclining back that the designer Karl Lagerfeld considers the perfect form of seating and that the *New York Times* has described as "ergonomic in its user-friendliness." These chairs acquired a voluptuary aura. One English version of the *bergère,* a half-couch called a birjair, had an adjustable back that, according to a contemporary reference book, "is made to fall down at pleasure." A daughter of Louis XV, when asked whether she planned to enter a convent like her sister, replied that she loved the amenities of life too much, pointed to her *bergère,* and said, "That chair is my undoing." (In fact, at least in England, one form of reclining couch was even known as a *péché mortel,* a mortal sin.) The eighteenth century had its own word for these informal postures, lolling.[10]

By the early nineteenth century, then, reclining needed no medical reason. The corsetless, high-waisted styles of the early century removed the structural obstacles women had faced. In 1842, an English etiquette writer warned ladies who received their guests while extended on "Grecian sofas" (Jacques-Louis David's portrait of Mme Récamier, another great hostess in the Rambouillet tradition, comes to mind) that their "vile and distorted positions" were unsuitable for good company, male and especially female. The Victorian age had begun, and with it momentous changes in both the technology and techniques of reclining furniture.[11]

MECHANIZATION OF RECLINING

While men and women had been leaning back for centuries, reclining furniture as we know it, like infant feeders and tennis shoes, was born in the later nineteenth century. It emerged in a middle-class society that (as the historian Katherine C. Grier has shown brilliantly) had two not quite compatible goals, propriety and public display on one side, comfort and domestic intimacy on the other. As working-class living standards rose and society became more fluid, especially in the United States, middle- and upper-class people used their homes as theaters to display manners and cultivation. The casual grace prized by the eighteenth-century upper classes yielded to a literal and figurative uprightness reflected in the parlor furniture of the later nineteenth century. The early typists' chairs that we noted in the last chapter encouraged the straight posture that was thought most appropriate for respectable women. The corsets worn by middle- as well as upper-class women enforced it by making slouching uncomfortable.[12]

Not that the Victorians were indifferent to comfort. It fascinated them. Americans were especially excited about the coiled metal springs that began to appear in the 1830s and 1840s, even if their chairs still seem stiff to us. What appealed to them was not only the more elastic feeling of the chair but the sense of participating in the stunning technological progress of their age, for springs were turning up everywhere, from railroad carriages to beds, promising to take some of the shock out of jolting change. Later in the nineteenth century, the lounge—a sofa with one raised and one level end—became a popular item of parlor furniture. But the softened appearance of late-nineteenth-century furniture can be misleading. Middle-class people, and especially women, were expected to sit up straight in it, not to lean back. These lounges, were, like running shoes, examples of what might be called potential consumption, suggesting activities that might be, but usually were not, attempted.[13]

Outside the parlor and its strict etiquette, new generations of reclining chairs were emerging. Some were luxurious library chairs, continuing the line begun by Pocock. Others appealed to the sick and frail and their caregivers. As early as 1830, the London upholsterers and cabinetmakers George and John Minter patented one of the earliest automatic recliners that moved with the sitter's weight; by 1850 George Minter had sold more than two thousand of these "self-acting" chairs. In the Crystal Palace exhibition of 1851 he introduced an "Archimedean screw" mechanism for adjusting the reclining angle. Minter also seems to have begun the industry tradition of patent litigation, filing three lawsuits for infringement. Minter's advertising of 1850 still referred to the product as an "invalid chair"; in the next half-century it would be American entrepreneurs who began to market recliners to the able-bodied, and especially to men.[14]

Furniture makers in New York and Philadelphia were imitating English invalid and library chairs as early as the 1830s. But American inventors and manufacturers soon developed their own recliner styles. Unlike the English predecessors, who worked with fine woods, Americans boldly introduced metal furniture into their sitting rooms and even into parlors. The relatively long distances of U.S. rail transportation stimulated new approaches to comfort. J. T. Hammitt's reclining chair, patented in 1852, may be the first design of its kind with a built-in footrest and levers for adjusting both its position and the seat angle. Americans' enthusiasm for mechanical furniture seemed unbounded. At its peak, between 1879 and 1900, about twelve hundred residents of Chicago alone received patents for furniture and its accessories, designed for people like themselves: middle-class city dwellers with limited space and an appreciation for technical ingenuity. America took the worldwide lead in producing sturdy metal seating. In 1870 and 1871 a Chicago inventor, George Wilson, patented a metal chair frame. By 1876, in the *Pictorial Album of American Industry*, the Wilson Adjustable Chair Manufacturing Company of New York presented its product made of the "best wrought iron and rivets" and adjustable with knobs and levers to assume twelve illustrated positions, including "parlor chair," "easy chair," "half-reclined" and "fully reclined lounger," "invalid position" with knees raised, "bed," and "reading position." It even folded down compactly for shipping. The makers promised eighteen other positions, thanks to an ingenious system of pivots and ratchets. This variety was intended especially for invalids, but it also was suitable for the larger population "of sedentary habits" affected by the lower-back pain prevalent in the nineteenth century—an appeal echoed in catalogue and Internet advertising for back-relief chairs 125 years later. In the 1880s another Chicago inventor, George F. Child, made a chair that could rock, recline partly or fully, and raise the sitter's feet. The Marks Adjustable Chair Company in New York brought a new standard of flamboyant advertising to the industry, and its product became a transatlantic hit in the 1880s and 1890s. The bulky, opulent look in recliners can be traced to this era.[15]

This first wave of American fascination with the variety of body positions lasted only a generation. The World's Columbian Exposition that opened in Chicago in 1893, as Siegfried Giedion observed, promoted classical ideals against obtrusive metal appliances. But at that very time another trend in reclining furniture was gathering force.[16]

THE MORRIS CHAIR ERA

The turn of the century's rival of the mechanical recliner was the Morris chair. The designer and writer William Morris (1834–1896) did not originate it; his associate, the architect Philip Webb, adapted it in the 1860s from a traditional design he had found in a Sussex workshop. Instead of the iron framework of American mechanical chairs, it had a solid wooden frame and a movable rod for adjusting the rake of the back. Morris's goals were not always compatible: a return to high standards of preindustrial craftsmanship and the uplifting of working-class life. A properly hand-made original Morris recliner upholstered with vegetable-dyed fabrics could never be popularly priced. Yet the design was sturdy and straightforward, and the idea of reclining continued to appeal to people of all classes. The chair was an international success, and American manufacturers were perhaps the most enthusiastic.

Around the turn of the century Gustav Stickley adapted the designs of Morris and Company for high-quality machine production. The simple and direct lines of these chairs mark them as precursors of twentieth-century modernism, but they also helped set the pattern of male reclining. They were made with heavy oak frames; walnut, previously favored, was disappearing from North American forests. Stickley designed them to be "massive" furniture, especially intended for reading, that would not be moved often within the household. Elbert Hubbard, a more flamboyant and self-serving American apostle of Morris's arts and crafts movement— he had made his fortune selling soap with a nineteenth-century version of multilevel marketing—produced his own handmade style of the chair. But Stickley and even Hubbard were too devoted to the craftsman ideal to enter the mass market. (Stickley called his product a reclining chair rather than a Morris chair, as though it were crass to exploit the master's name.) Mass sales were left to another tier of manufacturers, who used the latest machinery to simulate the features that idealists had prized as "structural ornament"—hallmarks of hand construction, especially the pegs that secured (or appeared to secure) joints. Sears, Roebuck sold thousands in the heartland, declaring in its 1902 catalogue that "no household is complete without one of these chairs." By 1908, the Sears catalogue was advertising "Morris" chairs with extravagantly carved front posts, upholstered in black imitation leather for as little as $3.65, and had naturalized William Morris posthumously and brazenly as "a New England Yankee."[17]

The Morris chair, no less than the Wilson and Marks chair, soon reached the end of a generational cycle. The vogue for its signature material, dark oak, passed with the mission style around 1916. Production shortcuts and dubious ornament dissolved what remained of the Morris aura. In the 1921 *Ziegfeld Follies,* Irving Berlin placed the lonely protagonist of the song "All By Myself" dealing himself cards in his "cozy Morris chair"; in a 1950s cover of the song, the Morris reference was dropped, apparently having become unrecognizable. In James Agee's novel *A Death in the Family,* published in 1957, a deceased father's "morsechair" (as the young son calls it) has taken on the imprint of his body and his entire personality. It is a relic of a lost age, as well as of a departed person, but shows the emotional bond between man and chair that had been established by the early twentieth century. Other chairs might have been recognized as seats of authority, and wealthy individuals may have bought earlier patent reading chairs, but the Morris chair had become an object as personal as a pair of shoes or a hat.[18]

The Morris chair boom had another lasting influence. It showed furniture manufacturers that consumers had a passion for recliners as long as they looked domestic rather than mechanical like the patent chairs of the mid-nineteenth century. The industry began to offer reclining club chairs with concealed devices, with and without the Morris name. In 1908 Sears sold "Davis Automatic Morris Chairs" with footrest attachments. The chairs featured "high carbon Bessemer steel coil springs" and were claimed to adjust to the desired back angle automatically in response to the sitter's pressure, with no need to rise to change back settings. Other makers sold Morris chairs that replaced open with concealed rods and ultimately with more sophisticated ratcheting mechanisms. In the late teens, one Michigan firm introduced a patented Royal Easy Chair, with a spring-loaded push button at the sitter's fingertips for setting the back angle, a step toward later railroad and airline seating.[19]

In the early twentieth century, the reclining chair remained furniture for spaces usually considered masculine, domestic studies and libraries. Advertising models were male. Women did have mechanical reclining furniture of their own, though, despite the corset's persistence. In the 1908 Sears catalogue is a "Combination Roman Divan, Sofa, Davenport, and Couch," with high arms at each side that could be lowered by lifting them upward, releasing a concealed ratchet mechanism. In the illustration a lady in a flowing gown is lounging against one half-raised arm; the other arm is completely lowered.[20]

RECLINING, EUROPEAN AND AMERICAN STYLE

With the decline of the Morris chair, designers and manufacturers reversed national styles. While Americans were burying hardware in upholstery, Europeans were building reclining seating that openly announced its origins in the realms of medicine, science, and technology, as American metal chairs had once done. Neither group had great commercial success before World War II, yet both helped to prepare for the explosion of reclining in the 1950s and 1960s.

The most important early U.S. company was a Cincinnati firm, C. F. Streit, founded just after the Civil War, a large maker of diverse institutional and residential furniture. Streit was an early producer of Morris chairs but also had brought out by 1908 a fully upholstered armchair with matching ottoman called the Streit Slumber Chair. Unlike mass-produced Morris chairs, the Streit was upholstered as upper-middle-class living room furniture. While few examples appear to survive, and Streit's advertising and catalogues never explained the mechanisms, the chair had a fixed angle between seat and back, which tilted as a unit so that the sitter could choose from three positions between mild and deep reclining. Since no lever is apparent in Streit advertising, the most likely mechanism is a spring counterbalance adjusted by a catch beneath one of the side panels. The Slumber Chair may have been the first widely produced form of seating to tilt at the sitter's knees rather than at the center of the seat. (We have already seen that this feature did not appear in desk chairs until the 1970s.)

Streit survived the Depression and continued advertising Slumber Chairs at least through the early 1950s, but its furniture never entered popular culture. It was another Midwestern company that brought reclining to the masses. The young founders were cousins in Monroe, Michigan: a woodworker named Edward M. Knabusch and a farmer named Edwin J. Shoemaker. Their first products, as the Floral City Furniture Company, were homely novelties like those they had been making for friends, including doll furniture and a telephone seat called the Gossiper. One of these was a folding outdoor chair of wooden slats. When a retailer suggested producing an upholstered version, the cousins quickly designed a new mechanism, applying for a patent in early 1929. The chair used a now familiar form of construction—a parallelogram of steel bars on each side, linked with the back—but its movements would probably surprise most of today's recliner users. The sitter's pressure on the backrest not only tilted it but raised the seat. The patent application claimed that this arrangement helped balance the chair and made it responsive to the user's pressure.[21]

The first Floral City chair, solid but not stylish, became so successful in 1929 that Floral City began licensing its production to other companies in return for royalties. By 1931, a Milwaukee company was making twelve hundred a month. The chair acquired the immortally folksy name La-Z-Boy, chosen in an employee contest over such other suggestions as Slack-Back and Sit-N-Snooze. (The corporation did not take the chair's name until 1941.) Sales dropped by 1933 as the Depression struck the furnishings industry, but Knabusch and Shoemaker learned from their hardships. They diversified into retailing, opening a large showroom that taught them the industry from the retailer's point of view. The store profited from Monroe's convenience to Detroit and Toledo by the expanding road network. Meanwhile, the cousins were planning a new, state-of-the-art factory. Their backgrounds had prepared them well. Knabusch was a connoisseur of wood and later often went to the forest to select trees personally. Shoemaker was part of a generation of master rural artisans of the Model T age and had a superb intuitive grasp of mechanisms. Following the example of Detroit, they planned a production line that would move frames past a series of upholsterers. The success of the retail store even in the Depression suggested that there was a strong latent market for reclining seating despite squeaky mechanisms and substantial prices ($49.50 and up; in 1939, Sears, Roebuck was selling an easy chair with ottoman for as little as $18.98, and a seventy-five-pound "Sears Ease" tilt-back reclining chair from as little as $19.95 to $26.95). The owners even took produce and livestock in exchange for their chairs. They were tireless inventors who continued to develop new mechanisms designed for efficient mass production.[22]

European reclining was more elitist. Massive wood-and-upholstery library chairs were advertised by a London company called Foot. And there were more graceful innovations. Already in 1883 the bentwood giant Thonet was selling an anonymously designed caned rocking sofa with gracefully bent supports; its sinuous lines, elevated head support, and adjustable back angle make it an ergonomic as well as aesthetic classic. In 1904 the Viennese architect Josef Hofmann introduced a beechwood armchair with an adjustable back and extending footrest for the Purkersdorf psychiatric sanatorium near Vienna; he later adapted it for country house living. It is to Hofmann that we owe the word and concept of the chair as a device for sitting, literally *Sitzmaschine*. In 1922, a Paris physician named

Pascaud contrasted the poor lower-back support of a conventional chaise longue with his recliner design called the Surrepos ("Superrest") that featured elevation of the knees and an open angle between legs and trunk—a profile close to the relaxed W of the veranda chair.[23]

European designers were developing theoretical approaches to the body techniques of sitting. The Danish designer, architect, and teacher Kaare Klint began to study human proportions and to construct chairs and other furniture as rational systems for living. In 1928 in France, the team of Le Corbusier, Charlotte Perriand, and Pierre Jeanneret introduced a chaise longue inspired by illustrations of Pascaud's Surrepos. It had a fixed S curve with an elevated headrest. Made by Thonet in Paris in 1930, it was supported by tubular runners that could change position while resting in a base; removed from the base, the chair became a reclining rocker like Thonet's 1883 sofa. Following Hofmann's lead in developing a modernist furniture doctrine, Le Corbusier claimed a rationalist clarity for the group's work. A chair was not just a "machine for living" but a "human limb object," a supporting prosthetic device. The Thonet chaise was also a brilliant synthesis of the masculine and feminine styles of reclining. For the prototype Le Corbusier used pony skin and declared he had been inspired by the image of a pipe-smoking cowboy of the old West, tilting back at ease with his boots on the mantelpiece. Yet the early advertisements for the chair featured a languorous Charlotte Perriand, like a latter-day Mme Récamier, but with her face turned enigmatically from the camera.[24]

European recliners of the 1930s were luxury goods. Le Corbusier might have been inspired by the mass production methods of automobile and aircraft factories, but his Thonet chaise-longue à réglage continu (continuously adjustable) required even more hand craftsmanship than the company's already high-priced rocking sofa. Polished metal exposes every minor defect mercilessly. Like Hofmann's Sitzmaschine, Le Corbusier's chair commands exceptional prices in today's market not just for its exceptional beauty but for its rarity. Few were ever sold. In the search for comfort, Americans were producing homely, conventional seating in growing numbers; Europeans were announcing bold modernist experiments that reached but a small number of wealthy connoisseurs when they did not remain at the prototype stage, like the varnished-steel and canvas "Grand Repos" (1928) of the designer Jean Prouvé, which tilted backward on ball bearings, counterbalanced by springs. The modernist dream of machine-age living proved a mirage, and not just because of its price. People who worked in factories, as critics observed, shunned the industrial aesthetic in their homes.

BETWEEN TWO WORLDS: ANTON LORENZ

There was one man who realized that European modernism could be fused with American mass production: the Hungarian-born inventor and entrepreneur Anton Lorenz (1891–1964), a link between the high modernism of the Bauhaus and the pragmatism of the U.S. furniture trade.[25]

Legendary in the furniture industry of the 1950s and 1960s, Lorenz never had his own manufacturing company in the United States. Even as an inventor, he remained in the background after the early 1950s. He had no children and no close friends or confidants; the associates he financed were independent inventors and designers. His extended family remained in Hungary, and his personal correspondence with them has been lost. What remains of his business correspondence is dispersed and uncatalogued. As one of his attorneys puts it, "he lived for his wife and his chairs." Yet Lorenz's intensity affected not only the companies with which he worked but their tenacious competitors. For all his thick Hungarian accent and Old World manners, he accelerated a change in American popular culture.[26]

Lorenz was born in Budapest in 1891, and in 1913 he began to teach history and geography, probably in a secondary school; details of his education, military service, and employment are sketchy. He married an opera singer and moved to Germany when she accepted a position in Leipzig. Lorenz somehow entered the lock manufacturing business in the early 1920s and was successful enough to abandon his teaching career and relocate to Berlin. There he met the Hungarian architects Marcel Breuer and Kalman Lengyel, affiliated with the Bauhaus in Dessau, and managed Lengyel's company, which manufactured tubular steel chairs of their design. Through aggressive management of the patents of others, especially the Dutch designer Mart Stam, he dominated the growing tubular steel furniture industry.[27]

Lorenz must have been aware of contemporary interest in reclining furniture as equipment for optimal rest. One of his associates, the Bauhaus architect Hans Luckhardt, began to design "movement chairs" in the early 1930s, including a slatted wooden chaise longue with a neck roll and linkages that extended a footrest as the sitter reclined, cradling the back and thighs and supporting the lower legs. A knob mounted at the side edge of the seat could be screwed down against a slotted wooden link to permit continuous adjustment between slight and full reclining. Lorenz helped develop it for Thonet, which called this wooden chair the Siesta Medizinal and still produces it today. Unlike previous convalescent chairs, built by cabinetmakers or metalworking firms, the Siesta had a theoretical agenda: allowing the greatest possible relaxation of the sitter's muscles. Luckhardt had been studying physiology since 1934, and Lorenz also began to devour medical texts. Just as coaches and architects helped make the modern running shoe, nonscientists were among the founders of ergonomic seating. Lorenz financed research at the Kaiser Wilhelm (now Max Planck) Institute for Industrial Physiology in Dortmund to validate the chair's design. Subjects were lightly supported in tanks of salt water, then photographed to determine the angles that trunk, thighs, and lower legs assumed in near weightlessness. A scientist and later director of the institute, Gunther Lehmann, wrote that this experiment was the first attempt to determine the true resting position of the limbs, although as we have seen, the relaxing position had been known from the 1870s.[28]

The Siesta chair appears to have been successful in a mainly institutional market. Air France was testing an upholstered version before the war broke out. Even the Nazi taboo on Bauhaus design was inconsistent; Anton Lorenz saved in his files an undated photograph of Adolf Hitler himself sitting stiffly in one of Lorenz's tubular-steel lounge chairs. By 1940, German military hospitals were using a tubular-steel wheelchair version of the Siesta. Lorenz, who happened to be in the United States on business when the war broke out, remained, escaping the destruction of his Berlin apartment and office.[29]

THE ROAD TO BUFFALO

Lorenz first settled in Chicago, possibly because of his association with Ludwig Mies van der Rohe. Though they were once legal adversaries in Germany, they respected each other and at least one U.S. patent bears both men's names. Cut off from his European businesses, Lorenz enrolled in a two-year course in human physiology and claimed to have studied two thousand books and articles in that field. Meanwhile, he was introduced to Nelson Graves, president of the Barcalo Manufacturing Company of Buffalo, New York, at the 1940 Chicago Furniture Show. Lorenz offered him an exclusive license on the reclining chairs he had been developing.[30]

Located near Buffalo's steel mills, Barcalo had been best known for metal beds, porch furniture, and hand tools. But its management appreciated Lorenz's passion for human factors. After brass and steel bedroom furniture went out of fashion, Barcalo had turned to making hospital beds. Some of these had cranks to raise the back and knees, achieving an optimal relaxing position, the lazy **W,** for recovery from surgery.

Lorenz was soon on the Barcalo payroll. In 1942, the company—like Thonet in Germany—produced reclining wheelchairs based on his patents. Immediately after the war's end in 1945, Barcalo began to advertise a high-back version of this chair to furniture retailers as a rolling recliner ("more than a wheel chair—it's luxurious comfort for the thousands of invalids and convalescents in your market area!"). A popular reclining lawn chair called the BarcaLoafer appeared in 1946.[31]

It took several years for Barcalo to begin producing upholstered chairs. Its license included all embodiments of Lorenz's "floating in water" position, but it had no facilities for making living room furniture. Through Nathan Ancell of Baumritter & Company, a dynamic manufacturing and marketing organization that later became Ethan Allen, Barcalo looked for sublicensees. In January 1946, a maker of commercial and industrial seating, Ernest F. Becher, saw a German outdoor invalid chair—probably the Siesta—at a Chicago show. Becher's company, Chandler Industries, was located across town in Buffalo from Barcalo. Becher, familiar with the principles of hospital beds, shared Ancell's enthusiasm for the design, telling him that there would be a large market for a version suitable for the home. Ancell eagerly accepted the idea and arranged a license from Barcalo, which built the mechanism.

Becher may have done more than any other manufacturing executive to make possible the reclining-furniture boom of the postwar years. His specialty was automotive seating. In the 1930s, car makers were among the first businesses to appreciate a reclining angle; automobiles previously had straight backs that echoed the design of carriages. As a serious student of posture, Becher saw the opportunities of the principle developed by Luckhardt and Lorenz, and was so enthusiastic that he agreed to merge his larger company into Barcalo in 1947, becoming executive vice president for manufacturing as well as the largest shareholder. After the merger, the company introduced a series of reclining upholstered chairs called the BarcaLounger (the internal capital was later dropped) in autumn 1947, conservatively styled with an attached pillow. Unlike the BarcaLoafer, the BarcaLounger concealed its medical heritage. But advertising for both the Loafer and the Lounger proudly referred to the "floating in water" position.

Lorenz was not the only designer of the BarcaLounger. Graves and Becher also retained another Buffalo furniture man, Waldemar Koehn. Koehn had been president of the Sikes Company, makers of premium leather executive chairs. Traveling to Washington, D.C., Becher and Graves obtained government specifications for a high-back chair with a head roll, a variation of what decorators call a Lawson armchair. This design, developed by Koehn, became a series of Barcaloungers upholstered in full-grain leather and in plastics; the federal government and commercial furniture dealers bought it enthusiastically, encouraged by veterans of the Sikes sales force. Soon other reclining-chair makers, including La-Z-Boy and Berkline, also adopted the design. Symbolically, it evoked male authority figures like judges and cabinet secretaries; ergonomically, it provided welcome support for the sitter's head in the full reclining position. By the early 1950s, the classic image of the recliner was fully established: a white male executive, back from a hard day at the office, kicking back in his suit or his shirtsleeves, puffing at his pipe. Other makers had used similar themes, but Barcalounger advertised more broadly and consistently and created a new mix of the Convivial, Convalescent, and Cogitative. It promoted the chair as a Father's Day gift. The appeal succeeded. Between 1946 and 1955, according to Barcalo Company estimates, an average of 30,000 Barcaloungers were sold each year.

FATIGUE AND RELAXATION

Advertising was not the only reason for the success of this heavy, expensive furniture. Part of its appeal was technical. The Lorenz design was the first to offer a built-in ottoman and a balanced, neutral position activated by the sitter's motion rather than by knobs or buttons. Even more important was the growth of popular interest in relaxation. Health claims for chairs appeared as early as a 1927–28 La-Z-Boy brochure for the "Recline-Relax-Recuperate chair," a well-padded, fully upholstered armchair promising a zone of blissful, invigorating repose: "the most soothing, *healthful* softness you have ever felt." Addressing the Depression-era middle class, the best-selling popular psychologist of the day, Walter B. Pitkin, called for an "Easy Way of Life." World War II reinforced the search for relaxation. As in the first war, the demands of both combat and civilian production pushed men and women to their limits, and the military sponsored crucial research in human comfort and fatigue to maintain morale and accommodate injuries. In England, Spitfire pilots returning from missions could lean back in Morris chairs. In America, one of the first ergonomic postwar recliners was designed by Marie LeDoux, the wife of an injured tank corps officer, with the help of a St. Louis upholsterer. It was introduced in 1947, the year of the first BarcaLounger, and was selling a thousand copies a month at $195 to $300 by 1949. According to *The New Yorker,* customers included Charles Boyer, Betty Grable, Ida Lupino, James Mason, and Eleanor Roosevelt. (Like Anton Lorenz, Marie Le Doux had an unorthodox background in physiology—in her case, limited to a course in a chiropractic hospital in Los Angeles. In civilian life, her French-born husband was a professional mind-reader. The chair was later produced under the Craftmatic brand until the 1990s.)[32]

Just as the pace of industrial work in the early twentieth century inspired desk chairs that promoted an optimal upright posture, the beginnings of an information economy made people more conscious of their leisure. Barcalo successfully courted physicians, who recommended and even prescribed its chairs. A 1951 article, "Learn to Relax," in *Today's Health,* recommended stretching out on a couch or bed with the head supported by a pillow: "You relax by letting yourself go limp. If you shift or fidget, speak unnecessarily or lie stiff and uncomfortable, you are not relaxing." Two years later, another writer in the same magazine recommended "muscular ease," recommending that readers emulate "a youngster lying on his back, gazing pleasantly into the sky, a blade of grass between his teeth," keeping limbs "as limp and soft as possible," and banishing all thoughts from the mind. (Yoga, zazen, Transcendental Meditation, and other Asian mind-body techniques reached the Western mainstream only later.)[33]

Hygienic relaxation was not, of course, the only influence on postwar reclining-chair design. Suburban living was equally powerful. While most of the detached houses in new developments were modest by present U.S. standards, they gave former apartment dwellers additional space for furniture. Dens and family rooms were gaining in appeal. With motorization and air conditioning, front porches disappeared, but there was ample garden space for outdoor reclining seating. The outstanding outdoor-indoor chair of the postwar years was the Barwa, designed and at first made by Edgar Bartolucci and Jack Waldheim in Chicago. It consisted of a cloth cover stretched on an aluminum frame; the user could either sit upright or recline with feet above head by shifting his or her weight. (The chair could rest stably in two different modes, thanks to the ingenious geometry of the frame.) Like the even more popular but far less comfortable Hardoy (butterfly) chair, the Barwa came to represent a new informal spirit in living and entertaining. This attitude affected high design as well; Charles and Ray Eames studied seating preferences and produced a lounge chair for Herman Miller with a rosewood shell and leather cushions filled with feathers and down. Charles Eames promised "the warm, receptive look of a well-worn first baseman's mitt," and generations of owners have paid substantial prices for that appearance and feeling.[34]

In the late 1940s and early 1950s, seating equipment and sitting habits reinforced each other, just as sneakers and the fitness movement were to do in the 1970s and 1980s. But while medical doctrine and popular culture alike had long recognized the value of exercise, the middle years of the century were especially keen on therapeutic lounging. By 1955, a *Life* magazine article on leisure gave as its first answer to the question "How Does the American Relax?": "He collapses." Lying down, the editors continued, once was restricted to the outdoors and the bedroom. Now "the growing informality—and fatigue—of modern life" had made it ubiquitous. With family members and cocktail guests alike, Americans were lifting their feet above their heads "or lolling in an elongated basket like an oyster on the half shell." Even alone, Americans had embraced horizontal listening and reading. To clinch its point, the editors illustrated thirteen common types of lounging furniture. There were BarcaLoafer-style lounge chairs for small children and even a "dog couch." In 1964 a satirical writer in the *New York Times Magazine* foresaw "the end of the chair as we know it," as Americans (unlike Europeans) sought ever more horizontal positions and new domestic arrangements like conversation pits. Except at crowded gatherings, Americans were putting their feet up and letting their spines slide down.[35]

THE RISE OF AN INDUSTRY, THE FALL OF AN IDEAL

Recliners still faced challenges. Like the original La-Z-Boy, the Barcalounger was expensive, even covered in Naugahyde. A skilled workforce assembled it as upholsterers had worked for centuries. Inserting the mechanism complicated the job—and taught the industry to work with more precision, because the tolerances of the metal parts could be highly sensitive—but in the end, a skilled workman upholstered one chair at a time.

Expense was not the only problem. Americans loved to recline, but ever since the waning of patented steel furniture in the 1890s they had resisted having machinery in their homes. And early Barcaloungers were big. Lorenz himself was over six feet tall and developed his mechanisms and prototypes accordingly, though Barcalo eventually offered smaller models. Early mechanisms were bulky, too, and needed ample space within the chair's frame. Add this to the patriarchal image of the high-back chair that inspired many early designs, and women's hostility to postwar recliners becomes understandable. To many, it seemed an intrusion and an aesthetic blot in the living room. Even later variants that could be placed within a few inches of a wall took up six feet of space when extended. Because reclining furniture gets up to six times as much wear as conventional chairs and its construction complicates reupholstery, 1950s and 1960s models often used tough vinyl fabrics, many of which nevertheless discolored. Men loved recliners as soon as they tried them, but women controlled the selection of decor, so manufacturers did their best to assure them that they, too, would love sitting in the chairs.

As recliner sales grew, manufacturers looked for ways to increase production and lower prices. La-Z-Boy's 1941 plant, leased for aircraft parts manufacture, resumed assembly-line production, but it still could not match Barcalo's output. Morris Futorian, a Russian-born Chicago furniture maker, met Lorenz and licensed his reclining chair ideas. Lorenz had sold exclusive rights to Barcalounger, but controlled other patents that had been seized by the Alien Property Office during the war because of his joint ownership with the architect and German national Hans Luckhardt. He licensed these to Futorian. Barcalo executives felt betrayed but took no legal action.

Meanwhile, Lorenz was developing a web of hundreds of patents and an intricate global licensing system. He continued to help drive the industry. Furniture people sometimes asked Futorian, known as a strong-willed, cost-conscious businessman, why he did not use an alternative or imitation mechanism. He replied that he was buying not only the patents but Lorenz's advice—a tribute indeed, because Futorian was famous for making intuitive changes, as small as a quarter inch in a single dimension of a chair, that multiplied sales. Futorian was also the first to see the potential of northern Mississippi, with its extensive timber and low-wage labor, as a major furniture manufacturing center. Workers who were otherwise unskilled were trained to perform a single operation, such as upholstering a left arm. Barcalo specialized in the upper middle class and La-Z-Boy in the middle class; Futorian saw that in a sprawling nation, the masses also wanted to recline. He had been one of them; he knew their tastes. He would change the prototype of a budget chair if it did not look cheap enough; the people who bought those models, he explained to his associates, were suspicious of furniture that seemed to have too much padding for its price. Following the Sears, Roebuck principle of Good-Better-Best, Futorian also produced excellent higher-priced chairs. In 1958 his Stratolounger division offered chairs in fourteen styles at retail prices from $59.50 to $359.50, enough to stock a full department. By 1959, after less than a decade, Futorian could boast in a *Home Furnishings Daily* advertisement that the Stratford Company had made a million reclining chairs under its Stratolounger brand; by 1963, the firm had made 1.6 million.[36]

La-Z-Boy, still a relatively small company, rose to the challenge. In 1952 it produced its first chair with an integrated footrest rather than a detached ottoman, continuing to improve it and adding low-back models (the sitter had to raise the back for reclining support). Knabusch and Shoemaker developed their own system of patents. In response, Lorenz designed a new generation of recliners that were to offer a third position between uprightness and horizontality, with the ottoman board fully extended but the back only partially reclined: the Television Chair. Peter Fletcher, a young engineering graduate and U.S. Army veteran whose English parents had been Lorenz's neighbors in Buffalo, was set up as an independent collaborator in a Florida workshop to help Lorenz turn this and other intuitions into workable mechanisms that could be manufactured economically and withstand years of operation. Fletcher had (and has) a brilliant understanding of the geometry of linkages—kinematics—and spent several years developing a system that enabled the sitter to cycle through all three positions through shifts of body weight, without using external levers. Lorenz himself watched little or no television. He remained a reader and music lover, and Fletcher remembers joining Mr. and Mrs. Lorenz in their oceanside winter mansion in Boynton Beach, Florida, as they sat in matching Barcaloungers and listened to radio broadcasts of the Metropolitan Opera on a state-of-the-art high-fidelity system.[37]

When the TV chair appeared at the Chicago market in June 1959, a *Home Furnishings Daily* correspondent could not contain her enthusiasm. The new television and reading position, she predicted, would soon "obsolete all others" and spread to hotels and offices. Already Lorenz had licensed the design to four manufacturers. Most of the chairs followed the high-back judicial head-roll design that Waldemar Koehn had introduced, though Barcalounger also offered a German-designed high-leg model with

Scandinavian lines. Demand was overwhelming. And henceforth reclining and sports broadcasting would be linked in popular culture. Athletes and other celebrity endorsers were soon dramatizing the features of the new recliners on television.[38]

By the early 1960s, recliners were no longer tied to medical ideas of rest. They were becoming a staple of furniture retailing, and thanks to assembly-line production were available for as little as $59. A Topeka dealer who displayed them prominently reported that they had become "popular with the local farming and worker populations." The mechanical recliner had followed, in other words, the same trajectory as the Morris chair, beginning as an elite handmade product reflecting a cultural movement and turning into a mass success compromising quality for price. Manufacturers and dealers not only dropped references to healthful living but were happy to promote sedentary viewing. What was originally promoted as a "heart-saver" for convalescents and a source of wholesome relaxation for tired men and women became in some circles a symbol of passivity and obesity. According to Clark Rogers, a veteran independent chair designer, Europeans do not mind getting out of a chair to extend a footrest; the U.S. market "caters to couch potatoes."[39]

What Lorenz felt about all this is unknown. He was passionate about the health benefits of regular reclining, but he loved money, too, and he was becoming one of America's richest independent inventors.

THE DEATH OF LORENZ AND THE MATURITY OF THE RECLINER

In the new frenzy, many manufacturers copied Lorenz's designs without a license. He ran a full-page advertisement in July 1960 in which he threatened not only makers but dealers, whom he urged to insist on tags bearing Lorenz patent numbers. The courts supported the alleged infringers. In a key case, *Lorenz v. F. W. Woolworth* (1962), the federal appellate court in the Second Circuit (which includes Manhattan) ruled that Lorenz's associate's design had been obvious. But Judge Harold Medina objected in his dissent that a dozen or more other inventors had failed to make a workable version of this "obvious" idea. Lorenz still had a vast income from his licensees, but his hold on the market was broken. Peter Fletcher and others believe that Lorenz's complex patent system may have distracted the judges—none of whom had a technical background—from the originality of Lorenz's basic ideas. Lorenz died of liver cancer two years after the *Woolworth* decision.

In the early 1960s the recliner industry consolidated and moved south. Morris Futorian acquired Barcalo in 1961 after labor disputes, moving all production to a plant in North Carolina; soon thereafter, he sold his interests to the carpet manufacturer Mohasco. Many of his employees went on to start or run reclining-chair companies of their own in Mississippi and elsewhere, proudly calling "Futorian University" their alma mater. Major firms like Lane built flourishing recliner divisions. But the ultimate design and marketing coup was La-Z-Boy's Reclina-Rocker, introduced in 1961, combining the most popular forms of motion with a mechanism covered by forty separate patents and offered in fifty thousand combinations of style and fabric. The Reclina-Rocker may be the most profitable single piece of furniture ever patented; according to a company document, it increased the company's sales from $1.1 million in 1961 to $52.7 million by 1971. No analysis of its success has ever been published, but La-Z-Boy advertisements from the 1960s give a clue: they feature men reclining and women rocking. Dr. Janet Travell, an authority on healthy seating as well as rehabilitative medicine, had prescribed a rocking chair for President John F. Kennedy's back pain; now rocking could be masculine and fashionable. And many customers of both sexes must have liked being able to alternate between the two sitting techniques for which Americans had already been famous in the previous century. At least a few foreign markets were growing. By 1966, La-Z-Boy chairs were being made in England, West Germany, South Africa, Australia, and Mexico, as well as the United States and Canada, although industrious German customers reportedly disapproved of the name.[40]

The domestic recliner industry has been expanding since 1964. Wall-hugging mechanisms led to sofas with reclining seats; families could lean back together. And comfort stayed affordable. In a seating contest sponsored by New York magazine in 1974, a $99 black vinyl recliner from Macy's tied for second place with an Eames leather-covered aluminum recliner by Herman Miller selling for $1,070 with ottoman. (The winner was a $1,535 custom-order Finnish reclining chair.)[41]

Most other features are refinements of earlier technology. Models are ingeniously designed to resemble conventional traditional high-legged furniture: two-part ottoman mechanisms fold out from concealment under the seat. Upper-back supports and headrests either pop up or swing up to avoid the telltale high-back look. Some chairs have fabric panels that fill in the gap between footrest and seat, giving the chair a chaise silhouette. Newer mechanisms are safer for curious children, and also for kittens tempted by the potentially fatal warmth of extended footrests with moving metal parts, though recliners remain a mortal hazard to domestic ferrets. Experience and competition have also made mechanisms smoother, quieter, simpler to manufacture, and easier to use with small as well as large chairs.

There is less work for independent inventors today; the big manufacturers prefer to hire in-house staffs. There are also few name designers. Raymond Loewy developed a Barcalounger in the 1960s, but today's best known signature models are those that La-Z-Boy offers inspired by the patriotic, nostalgic paintings of Thomas Kinkade.[42]

The health appeal of the recliner has not disappeared, but it has been redefined. In postindustrial society, popular concern has turned from the heart to the back, and specialty catalogue and Internet outlets sell dozens of chairs with fixed curves similar to the Lorenz "floating-in-water" position. The most popular premium health feature is massage, but the chairs are a far cry from 1950s vibrating models with big motors. At least a dozen suppliers offer a more compact system of tiny motors (vibrotactile massagers) of a type originally built into pilots' vests by the U.S. Air Force to signal the direction of incoming missiles. Manufacturers have been using them enthusiastically. Other chairs return to the medical market of Lorenz and Luckhardt's original Siesta, offering sleep settings and power-assisted rising. Some Japanese companies have revived the turn-of-the-century idea of resting rooms with napping chairs—recliners with built-in lights and fans to limit sleep to the twenty- to thirty-minute intervals that refresh without producing grogginess. (The new equipment refines an old insight. In the sixteenth century, Emperor Charles V of Spain, who napped on his throne, clutched a heavy key as he dozed off; before he could sleep more than twenty minutes it would drop, waking him. Thomas Edison is said to have used iron balls similarly.)[43]

These models have helped reclining chairs sustain the growth that began in the 1950s. In 1998, they accounted for almost a quarter of the $8.16 billion U.S. upholstered furniture market, in California for over 35 percent. In the last published survey of recliner market share, in 1997, La-Z-Boy remained the largest with an estimated $386 million in sales, followed by Action Lane with $265 million. (Barcalounger is still a vigorous and growing brand, but ranked only seventh, at $43 million.) Nearly seventy-five years after the first La-Z-Boy and more than fifty years after the first Barcalounger, recliners are said to be in 25 percent of U.S. households. Many are hard to distinguish from other cloth-upholstered living room furniture, or from elegant leather club chairs. Yet the core of the U.S. market remains what the furniture journalist Susan M. Andrews calls the Bubba chair. Despite the elegant new models, "Bubba will always find his chair." Embarrassing as these customers are to an industry now targeting aging, affluent baby boomers, the Bubbas testify to the entrepreneurship of people as different as Knabusch, Shoemaker, Lorenz, and Futorian, who brought a regal style of sitting to the masses.[44]

Will recliners ever be found worldwide, like the other body technologies described in this book? We have seen that Indians developed the veranda chair by the nineteenth century; they even built in drink holders. The weight and size of today's upholstered recliners works against them. But who can rule out new materials and forms that might do for the reclining chair what the modern athletic shoe did for footwear?[45]

CHAPTER SEVEN

Mechanical Arts

Musical Keyboards

W HAT THE CHAIR is to the back, the keyboard is to the fingers: an innovation in body habits that emerged in the ancient Mediterranean and has made its way around the world. Sandals and shoes help shape how we move, and chairs how we work and rest; keyboards affect not only our physical positions but our mental performance. They influence both musical performance and composition. As producers of text, keyboards transmit intimate messages once reserved for voice or pen. Like the sandal and the reclining chair, the keyboard shows staying power as a body interface. The layout of both musical and writing keyboards has barely changed in the last hundred years, for all of the upheavals in twentieth-century culture. Philistine as it sounds to compare the output of the piano and organ keyboard with the clacking (and more recently clicking) of text production, playing and typing both demand complex techniques that psychologists are still not completely able to explain. The mind turns out to have prodigious powers to create patterns from a series of discrete strokes. In fact, those abilities have produced a paradox. Musical and text keyboards have resisted all reform movements. They are like the Staunton chess set, introduced in 1851 and used continuously without a serious rival ever since: not ideal, but good enough so that most skilled professionals have shunned alternatives. And as change grows more rapid, our attachment to these familiar, imperfect devices seems to become even stronger.

FROM THE ORGAN TO THE PIANOFORTE

Unlike many other important technologies, keyboard equipment, musical as well as typographical, is distinctively Western. In the mid-third century B.C., Ctesibius of Alexandria (according to most scholars) used water pressure to build a giant mechanical flute called the hydraulis, the ancestor of all organs. The ancient organ, far from a sacred instrument, was a crowd-pleaser at athletic events and celebrations, like the electronic organs of today's baseball parks. It was played with open hands or fists, not fingers. The early church fathers disapproved. Even after the first organ reached the Kingdom of the Franks as a gift from the Byzantine emperor in A.D. 757, over a hundred years passed before organs were used in worship. The organ was one of the most complex medieval machines; the instrument (constructed in 1361) of Halberstadt Cathedral was powered by ten men pumping twenty bellows with their feet.[1]

Still more notable was the layout of the Halberstadt organ's keyboard. The user interface, in today's term, had changed in response to the new musical style, polyphony, with its two melodic lines. Earlier organs had used sliding horizontal bars with handles. Holes in the bars opened and closed the passage of air to the pipes. The levers were later spring-loaded for easier playing. The next stage was the key, a counterbalanced wood lever, which permitted even briefer notes. As the diatonic scale, our present white keys, became inadequate for new musical styles, keys were added to complete a chromatic scale of twelve tones. Their positions followed the pattern that nearly every twenty-first-century piano and electronic synthesizer has maintained: accidentals (sharps) higher and shorter than naturals. The spade-shaped keys of the Halberstadt instrument were still played with hands, though. According to one early source, each was the equivalent of about 8 centimeters wide. By the fifteenth century, keys became narrower and rectangular; finger operation was beginning. In the sixteenth and seventeenth centuries an octave span measured about 16.7 centimeters, a standard that persists in our contemporary span of 16.5 centimeters. Modern piano keys are significantly deeper, color schemes have varied, and the precise placement of accidentals has shifted slightly over the centuries. But the keyboard as we know it has existed for five hundred years.[2]

Organ building was competitive high technology. Max Weber underscored the mechanical complexity of the instrument, and noted that early organists were also organ builders. The science writer Thomas Levenson, in his book *Measure for Measure,* has compared Byzantine and Carolingian organ development to the post-*Sputnik* space race. Like space satellites, organs were indexes of the advance of knowledge and civilization. Throughout the Middle Ages and the early modern period, pipe organs were among the most complex mechanisms in use, and the most impressive that most people could experience. The organ's sound overwhelmed the peoples of northern Europe. Yet the keyboard also acquired an intimate, popular side. It helped make possible the small-scale portative and fixed organs ubiquitous in medieval illustrations; these were so heavily used that few have survived. The keyboard could even be adapted to bowed instruments: the hurdy-gurdy is a variation of today's violin family in which a crank bows the strings with a pearwood wheel covered with resin (and produces a droning underlying tone) and keys, originally T-shaped, depress the strings at fixed points to carry tunes. The hurdy-gurdy began as a monastic instrument (organistrum) for teaching music, among other uses, but spread across secular society, from peasants, beggars, and wandering musicians to courtiers.[3]

Despite medieval Europe's debts to the more advanced scientific culture of the Islamic world, it did have a distinctive gift: breaking previously seamless phenomena into smaller, discrete parts for easier teaching and use. Latin, for example, had been written in an unbroken stream of letters—*scripta continua*—in antiquity. Words and sentences were run on. Pupils learned where to place the breaks by reciting aloud or at least moving their lips. In the seventh and eighth centuries, Irish and English monastic teachers and scribes began to teach their pupils to recognize whole words by their images instead of proceeding letter by letter. (Compare this with the challenge of learning the flowing script of Arabic.) The same clerics also invented alphabetical glossaries and dictionaries, unknown in Roman antiquity but now desirable because their pupils were learning foreign languages in preparation for the priesthood. Later medieval authors like Gratian and Peter Lombard wrote books explicitly designed for ready reference; book indexes, tables of contents, and running heads were medieval innovations. Medieval textbooks abound in ingenious diagrams and memorization systems, and it was medieval pedagogues who introduced letter grades. The musical staff, attributed to the eleventh-century monk Guido of Arezzo, mapped the scale to a series of lines and spaces; the historian Alfred W. Crosby has called this the first graph. A keyboard in turn assigns each note to a finger. Like other such strategies—such as the medieval counting board and the Asian abacus—it simplifies relationships for the beginner, yet allows masters to develop highly skilled techniques, just as word separation promoted new scholarly studies. The computer scientist David Gelernter has described the musical keyboard aptly as our own millennium's "ergonomic masterpiece."[4]

The transition from the horn springs of early medieval organs to the diverse actions of keyboard instruments is still imperfectly known. But historians have been able to explore their milieu: a burst of mechanical ingenuity most spectacularly expressed in the astronomical clocks of the day, the first in the world to convert the force of a falling weight into uniform oscillating motion. Cathedrals and royal palaces soon boasted wonders of precision, with animated figures and sound effects, centers of pride and amusement and some of the first objects of mass technophilia.[5]

The same people who were building the new clocks probably played a large part in keyboard design. They were familiar with the latest in metallurgy and with the design of complex linkages like the mechanical crowing cock of the Strasbourg cathedral—the beginnings of the kinematics still used in constructing the reclining chairs we considered in the last chapter. They also were learned in astronomy, mathematics, and music theory. The author of the most important surviving fifteenth-century treatise on musical instruments, Henri Arnaut de Zwolle (d. 1466), was a physician, astronomer, scientist, and builder of clockwork instruments for the Duke of Burgundy. Arnaut's manuscript has precise diagrams and explanations of four distinct ways of using keys to make strings vibrate. He and his colleagues were like the restless aerospace engineers of the twentieth century.[6]

(Of course, didactic accessibility had, and has, a price. The twelve-tone octave of the standard Western keyboard excludes consistently accurate tuning. C-sharp and D-flat should really be two distinct notes, and on some experimental keyboards the black keys are split horizontally for this purpose. A number of tuning systems have attempted to avoid dissonance and preserve a smooth sound, but they all have one thing in common: at least some notes must be altered for the sake of the whole. In the equal temperament promoted by Johann Sebastian Bach in his *Well-Tempered Clavier* and almost universally used today, every note is slightly mistuned. Our ears have become so accustomed to this scale that it is easy to forget that it is a compromise forced by the limits of technology. The twelve-tone keyboard also complicates the tuning of instruments. As clavichords, harpsichords, and pianos, known collectively as claviers, spread in the eighteenth century, more and more musicians sought professional tuners. The construction of nineteenth- and twentieth-century pianos has made them indispensable. Like other simplified technologies, the keyboard needs formidable complexity behind its facade.)[7]

In the early modern period, the instruments described by Arnaut took two main forms. In the clavichord, the keys actuated flattened brass surfaces called tangents that made contact with metallic strings. Because the tangent continued to vibrate the string, the player could control the tone, even swelling it. The sound of a clavichord was clearly audible only within a ten-foot radius, yet in its domestic setting it could produce complex and graceful music. In the harpsichord, the key actuated a wooden bar (jack) holding a quill (pick) that plucked the string. The sound was precise and bright and could fill a room, but it could not be as expressive as that of the clavichord. There was no direct way to control the instrument's volume or shape individual notes, and the quills wore out.

Three hundred years ago, instrument builders began to develop a technology that would open new styles of music to the keyboard. Bartolomeo Cristofori, a harpsichord builder working in Florence, probably was following a suggestion by his patron, Prince Ferdinando de' Medici, for an instrument that could reflect the expression of the human voice. The pianoforte was not just a clever idea waiting for musical material to take advantage of it; it was a response to a desire for new expressive possibilities. Cristofori abandoned the tangent and the pick for a more powerful way of vibrating the string, a hammer made of strips of parchment covered with leather. The hammer assembly was isolated from the rest of the instrument. The key did not activate the hammer directly as it did in the clavichord; it triggered a lever that in turn set the hammer in motion but broke off all contact with it. Thanks to Cristofori's escapement mechanism, the hammer bounced off the string as though off a trampoline, awaiting a new stroke, even if the key was still depressed. The strings themselves were thicker and wound to higher tension than those of the clavichord, producing a stronger sound and also permitting more rapid repetition. A system of levers increased the force of the hammer eightfold. Even so, the sound was softer than that of the harpsichord, and softer than that of a modern, iron-framed piano. Linked to the keys were dampers— felt-covered blocks—that rose when keys were struck and made contact again to arrest the vibration of strings after the keys were released. Later in the eighteenth century, pedals were introduced to keep all the dampers lifted and prolong the sound.[8]

To keep the key from unwanted rebounds, Cristofori included a silk check-cradle to hold it back after it had fallen, without obstructing the next stroke. He changed the internal architecture of the harpsichord, moving the strings closer to the action. He developed most of the central concepts of the modern piano. Yet despite its possibilities and some interest from Johann Sebastian Bach, Cristofori's invention remained mostly a curiosity for the rest of the eighteenth century, especially in his native Italy. Some of his pianofortes were converted back to harpsichords, and in 1774 Voltaire deemed the pianoforte an instrument for a tinker (*chaudronnier*) "in comparison to the magnificent harpsichord."[9]

What Voltaire missed, in his scorn, was the miraculous capacity of the pianoforte—then barely explored, it is true—to produce a range of shading and timbre even though the musician has no opportunity to change the color of a note after the hammer contacts the string, except by using the pedals. With control only over the time and strike speed of any given note, a pianist can appear to defy the laws of physics. Generations of psychophysicists and acousticians have wondered how the instrument can work as it does; Sir James Jeans even remarked that it did not matter whether a key was struck with a finger or an umbrella handle. Brent Gillespie, a mechanical engineer and musician, has argued that musicians produce "impossible" shadings by the overlap of notes in phrasing. He gives the example of an arpeggio played with slurred notes, which has a timbre distinct from one with articulated tones.[10]

The most important eighteenth-century innovation in keyboard technique was independent of Cristofori's invention: the development of a new fingering system by Carl Philipp Emanuel Bach (1714–1788) and his *Essay on the True Art of Playing Keyboard Instruments* (1753–62). C. P. E. Bach developed a keyboard technique so valuable to his contemporaries that Haydn, Mozart, and Beethoven all considered it essential. His best-known innovation was a new emphasis on the thumb and especially its use as a pivot; most music before his time was played largely with four fingers. His father, Johann Sebastian Bach, had been the first to advise turning the thumb under the other fingers. Both technique and technology were ready for a new phase in the history of the keyboard: the piano's emergence as the central instrument of Western music, and its global diffusion.[11]

THE MODERN PIANO

Practitioners did as much as inventor-craftsmen to bring about this transformation. Surgeons have long commissioned new instruments not just to improve existing procedures but to make new ones possible. Artists working with their suppliers turned acrylic paints from a World War II military expedient to the medium for achieving the flat surfaces of the New York School. For the spread of the keyboard, few individuals were as important as Ludwig van Beethoven.

Beethoven was the greatest representative of a generation of composers and performers who sought new volume, range, and dynamic shadings in their music. They deliberately exceeded the limits of the instrument with the assumption that if the pianoforte could not produce the sound they desired, its design rather than their music would have to change. Even when performing Mozart's music at the Viennese court in his youth, Beethoven played with such expressiveness that his page turner later recalled spending all his time removing broken strings and freeing up hammers. Until Beethoven's time, Viennese pianos had been known for a softer, more "singing" sound than their English counterparts. Beethoven's hearing loss, which began in 1802 when he was still in his early thirties, led him to demand even louder adjustment, which he achieved by working with local manufacturers, and to play with even more force. (Musicians favoring a more delicate touch worked with other makers. Chopin was known for his affinity with the Paris firm of Pleyel.) As heavier frames and hammers and higher string tension prevailed, sounding the same note quickly in succession became more difficult. Another famous maker, Erard, responded with a repetition mechanism that kept the hammer poised to strike the same string again.[12]

To these improvements in technology were added electrifying innovations in technique. As usual, the migration of skills from one domain to another was crucial. The twenty-year-old Franz Liszt, after hearing the astounding (and in popular lore diabolically inspired) effects of Paganini at a Paris concert in 1831, resolved to bring the same style to the piano, and spent several years developing a mixture of technical virtuosity and emotional intensity that made him the most celebrated musician of his time. Liszt's widely imitated technique in turn helped force changes in technology. In the early 1840s, pianos went out of tune and strings snapped when he performed. His demands helped drive the instrument's next great change.[13]

Nineteenth-century pianists, teachers, and manufacturers formed an exceptional community in which the techniques and criticisms of outstanding users were constantly employed to refine and improve products. The ultimate response to the perceived need for a bigger sound became the foundation of the modern piano, the cast-iron frame. Early in the century, makers were bracing wood frames with metal to support higher string tension of up to ten tons and prevent warping with seasonal changes. But cast iron, not reinforced wood, proved the ideal material for piano frames because of its exceptional compressive strength. Today's frames remain stable under as much as twenty-seven tons of tension in the strings of some grand pianos. Several European and American makers introduced these frames in the 1820s, but the most influential variation was developed by Jonas Chickering of Boston in 1840; by the 1870s, Steinway and Sons of New York had devised the frame and stringing used for grand pianos ever since. Steinway boasted that its design could resist seventy thousand pounds of tension.[14]

Manufacturers maintained concert halls in major cities as showcases for their instruments and affiliated artists. Long before the golf equipment and athletic shoe industries, they were pioneers in engaging star performers to help market high-technology products. Late-nineteenth-century instruments overwhelmingly incorporated the cast-iron frame and Steinway's cross-stringing. Each maker had, and their successors often retain, distinctive innovations and features, but by the 1890s the grand piano was a fully mature technology.

The cast-iron frame fostered careers by making possible solo performances in ever-larger auditoriums and, for the few, princely incomes even in the decades before recording royalties. For amateur musicianship, the great innovations were industrial production and commercial promotion. The instruments of the early century, regardless of frames or mechanisms, were high-priced artisanal products for the upper middle class and upper class. The piano became part of mass culture beginning in the 1850s. Not only was it the most practical way to enjoy music in the household before—and even after—the appearance of the first recordings; it was also a pedagogical tool, thought to build character through exercises and forming an essential part of middle-class education, especially for women. By the turn of the century, economies of scale, including the proliferation of firms supplying actions and other standardized parts for manufacturers, had brought pianos within the reach of the better-off working class. Uprights could rival small grands in tone and occupied far less space than the discredited square piano. In England, pianos were available for as little as £15 at a time when artisans made forty to fifty shillings a week. High-quality pianos were perhaps the greatest bargain; a superb small Bechstein, the equivalent of an instrument costing thousands of pounds today, sold for only about £50. In the United States, Sears, Roebuck advertised a model in the 1890s for $98.50. American piano production reached a peak of 356,000 uprights and 10,000 grands in 1909. Seldom had mechanization so transformed and diffused what remained a complex product of craftsmanship. By 1920, the United States had 7 million pianos for 105 million persons or about one for every four households. Low-priced pianos may have been shameless knockoffs of the great brands, yet their sound satisfied most untrained ears. The keyboard had at last reached the people; for those who could not afford even a cheap piano, there was the accordion.[15]

SUFFERING FOR BEAUTY

The piano was not just a half-ton of machinery in a wooden cabinet. It had become the centerpiece of a middle-class way of life, in which young women were destined to sacrifice their leisure. For girls of Northern Europe and North America, the skill of playing was an "accomplishment," a character-building domestic counterpart of the tireless work of the male head of household. The keyboard was a technology that enforced a discipline of practice; it became a kind of Prussian parade ground for young fingers. Most American teachers had learned pedagogy from followers of Siegmund Lebert and Ludwig Stark, authors of a classic treatise on method that went through seventeen editions between 1858 and 1884. All students were expected to follow drills and exercises for hours on end to strengthen the fingers for their "conquest" of the piano. In the 1930s, a contemporary recalled the "torture" of his female contemporaries, comparing practice to "the binding of the feet of the Chinese female child, and for the same purpose—to increase her social prestige when she grew up."[16]

Nothing reveals the inexorable demands of nineteenth-century practice as much as its technological fringe, the apparatus sold to develop skills. For building strong fingers, there were spring-loaded devices with names like La Mère de l'Élégance and the Dactylion, and practice pianos with variable resistance. The Technicon and the Manumoneon were multipurpose finger gymnasiums. Gustav Becker, inventor of the second, revealed the mechanical, and dictatorial, side of piano pedagogy in the copy he wrote promoting his invention: "The fingers of the performer are compelled to make the desired motion in a perfect manner, and thus by attentive and continued practice, as per special direction, the student cannot help learning soon to make the movement of his own volition." Simpler pocket exercisers used rubber bands, and manuals of hand gymnastics recommended corks and napkins. A number of surgeons had thriving practices "liberating the ring finger" by severing the tendons between the fourth and fifth fingers to increase the player's span. The Ontario engineer J. Brotherhood, not satisfied with the Technicon he had invented, had his right hand modified by a surgeon and was evidently so pleased (and confident) that he later cut the tendons of his own left hand. The leading piano journal, *Étude*, even endorsed the procedure.[17]

MECHANIZATION TAKES THE BATON

The triumph of mass manufacturing brought into the skilled worker's parlor an instrument he and his family no doubt considered superior to the harpsichords of the eighteenth-century aristocracy. But it had a more ambiguous side, and not only for its part in the slaughter of elephants for ivory and the destruction of tropical forests for veneers. As rationalized production and aggressive marketing spread the piano beyond the leisured classes, first-generation piano students discovered that mastering their instrument took more time and sometimes painful practice than salesmen were willing to acknowledge. Indeed, even early in the century some of the aristocratic London clients of the virtuoso teacher Ignaz Moscheles were requesting "brilliant but not difficult" pieces so that their daughters could impress listeners with minimal effort. The search for superficial éclat was the other side of the dictatorial pedagogy the instrument also inspired. And it proved the more durable. We have seen the variety of chairs produced in the late nineteenth century using mechanical systems to relax the upright sitting that had been a Victorian mark of good breeding. It was only logical that the same ingenuity would be applied to tame the imperious piano.[18]

Mechanical music had a long but restricted history before the late nineteenth century. As far back as 1430, wheels with pins were built into stringed instruments, and a related playable instrument, a barrel organ, survives from 1502. Nineteenth-century technology gave new vitality to this old idea. The first systems were not great improvements on their medieval predecessors. A British patent was issued for a crude barrel piano, activated by pins on a large cylinder, in 1829. Most of these devices, though, were intended for taverns, dance halls, or street performances rather than bourgeois homes.[19]

Nineteenth-century automation promoted new devices and ambitions.

Beginning in 1815, the silk mills of Lyons and other cities were using strings of cards with punched holes to reproduce designs on special looms developed by the manufacturer Joseph Marie Jacquard. This was one of the earliest forms of automatic process control in industry. Pianos, organs, and other keyboard instruments were obvious candidates for more advanced systems for reproducing recorded notes. The first automatic pianos were rudimentary instruments for places of cheap popular entertainment. The 1880s and 1890s saw a wave of more sensitive players, beginning with organs, that used air pressure controlled by holes in rolls of paper. In 1896 the most influential of the automatic pianos appeared, the Pianola, an initially bulky apparatus that fit over part of a standard piano keyboard and struck the notes in an approximation of a human musician. The pneumatic-activated paper piano roll was some of the first music software: instructions encoded in a flexible medium that could be reproduced and transferred from one playback device to another. In little more than a decade, the flourishing industry was able to build the roll reader and keyboard control systems into the piano itself.

The player piano substituted technology for skill with unexampled boldness and ubiquity. Automata built in the late eighteenth century had executed drawings, and had even appeared to play chess, but they had been precious playthings. Player pianos were affordable and widely available. Small motors replaced foot pedals as sources of air pressure. By 1910, a second generation of "reproducing" pianos appeared, using a system developed by the German firm Welte to record the subtleties of performance. Nearly every celebrated pianist of the time was recorded with this system. By 1926, one maker was even able to capture the speed with which hammers hit the strings, making such accurate and complete data recordings that—when converted for playback on today's CD-controlled pianos—they produce uncannily vivid musical performances with no trace of the original roll's honky-tonk effect. Pianists were even able to edit master recordings to correct their false notes.[20]

High-quality recorded music led to bitter controversies over copyright laws, much like present-day litigation over shared MP3 files, but disputes went beyond intellectual property issues. Composers, performers, and critics wrangled over the consequences of the new devices for the study and use of the piano. Did technology threaten musicianship or promise to enhance it? Foreshadowing some of the points later made by Walter Benjamin and other members of the Frankfurt school, the bandleader and composer John Philip Sousa wrote a widely reprinted article, "The Menace of Mechanical Music." Player pianos and phonographs, he warned, were about to "reduce the expression of music to a mathematical system of megaphones, wheels, cogs, disks, cylinders, and all manner of revolving things," diminishing music by soulless uniformity. Cultural technophiles among music critics and teachers countered that what detractors scorned as "canned music" was welcome competition for inferior live performances. Not only would musical reproduction elevate tastes, they believed, but the player piano would aid teaching. Musical notation could be printed on the rolls themselves, though it seldom was in practice. Speed could be adjusted to help students learn complex pieces gradually; between the wars some celebrated performers, including Fats Waller, learned this way.[21]

The player piano was ultimately undone by its own logic. Early phonographs, primitive by present standards, sounded surprisingly real to people overwhelmed by the medium's novelty. A good phonograph was only a quarter of the price of a satisfactory piano. Radio broadcasting offered classical as well as popular music. Automobiles and cinema absorbed increasing middle-class time and money. Piano lessons were no longer universal. By the mid-1920s, critics were predicting that pianos would soon become anachronisms like horse-drawn carriages, or antiquarian curiosities like harpsichords and viols. Their pessimism was unwarranted, for the industry revived after the Depression and World War II. By 1980, the United States was again making 248,000 pianos annually and Japan was producing 392,000. The piano had lost its privileged place in middle-class life but had confounded the pessimists who feared the disappearance of musical skills. Recorded and broadcast music did not help the keyboard, but neither did they kill it. For better or worse, they changed it.[22]

While piano lessons are no longer a middle-class rite of passage, keyboard skills have flourished. Today's young pianists have the most proficient technique ever, in part as an unintended consequence of the rise of high-fidelity recording. Sound engineers splice retakes of flawed passages into an artist's work undetectably, much as producers and artists doctored the old player rolls. The bright sound of CDs and most high-fidelity equipment, many believe, has also conditioned both audiences and performers. Today's listeners expect perfect play, and students demand more of themselves. André Borocz, founder of the Menton Music Festival in southern France and a half-century veteran of classical performance, told the writer Rudolf Chelminski that between the nineteenth and twentieth centuries, "technique has been projected to a previously unheard-of perfection." Yet many critics also sense that something has been lost; Borocz continued that he "would pardon any number of bad notes to hear a little emotion." The near flawlessness of recorded music has worked against the emergence of idiosyncratic but passionate figures like Glenn Gould, whose imperfections were part of his fascination. Servitude to the mechanics of the piano keyboard is now a voluntary discipline, but it can be demanding beyond the dreams of the nineteenth-century drillmasters.[23]

TOWARD A NEW KEYBOARD?

Between the extremes of relentless practice and passive consumption, a handful of visionary musician-inventors sought a third course. Why not take the protests of young piano students, and the increased medical complaints of professionals, more seriously? The word *ergonomics* was not coined until after 1945, but these critics acknowledged the deficiencies of the familiar interface. It was generally impractical to build smaller pianos for instructing children; but full-size keys and octave spans could be a painful stretch for little fingers. Even mature, expert performers found the black keys awkward when music was written in keys using many of them. The skips, arpeggios, and glissandos of nineteenth-century music could also require challenging manual gymnastics. Transposition into a different key could be a nightmare. The compass of the piano had grown to eighty-eight keys, a challenging stretch. Since the piano was a machine, why not make it more comfortable and convenient to play?

The mildest change was an adjustment for transposition. In early-nineteenth-century instruments in which strings were still parallel to the keyboard, keys, hammers, and all could be moved away from or back toward the artist, so that the same performance struck a different set of strings. Or the keyboard could shift laterally. The popular composer Irving Berlin used such a transposing piano from Weser Brothers, now in the Smithsonian Institution. Less elegantly, a false keyboard could be mounted above the real one, so that each key mapped to another. Other innovators as early as 1780 devised organlike keyboards that curved to better fit the musician's reach. A handful of production instruments were built in mid-century by firms in Vienna and Paris. Around 1870, the illustrious house of Bechstein sold a keyboard with a curved arc for each arm; it was surprisingly awkward to play, and it failed. Finally, an Australian-born inventor, Ferdinand Clutsam, apparently not a pianist, patented the first widely admired concave keyboard in 1907. It was taken seriously enough to be fitted to a Bösendorfer grand, one of the finest pianos. In 1908, Ibach, a German firm, invited artists to play and comment on a prototype in its Berlin showroom. Two future stars were initially impressed: amid lively applause Ernst von Dohnányi told his friend Rudolph Ganz that he had played the first Chopin étude faster and better than he ever could on a conventional keyboard. Dohnányi and Ganz both used such instruments for concerts the following year.[24]

The problems of the Clutsam keyboard were social, not mechanical or aesthetic, and they reveal much about technology and technique. The Clutsam was, Ganz recalled "agreeable and easy," yet on a concert tour even a major company like Ibach could not guarantee it would always be available. Switching keyboard styles from one concert or practice session to the next was disorienting. There was also a legal problem: contracts between inventors and manufacturers, and between the makers and leading artists, were a quagmire and international licensing efforts fell apart, Ganz recalled later.[25]

The technology of the piano was bound up with the skills of elite artists. Even when, as individuals, they endorsed change, it was impossible to distribute an innovation rapidly to every concert hall—and gradual acceptance proved worse than none at all. But what about radical new keyboard configurations appealing to the amateur: not just a new curvature but a different size and arrangement of keys? The Jankó keyboard, a second fin-de-siècle device with an even more impressive lineage, met the same fate.

Paul von Jankó (1856–1919) was a member of the minor Hungarian nobility with impeccable academic credentials including engineering, mathematics, and music courses in Vienna and Berlin. His teachers included Anton Bruckner and Hermann von Helmholtz, and his theoretical work on temperaments of more than twelve tones is still highly regarded today. He was not the first to address the problems of raised keys; previous experimental keyboards had lowered the black keys almost to the level of the white, or equalized all the keys. Jankó's innovation in 1882 was not only to use uniform keys narrower and shallower than conventional white keys and with rounded profiles, but to arrange them in two staggered tiers of whole tones:

        C# D# F   G   A   B
                                    etc.
      C   D   E   F# G# A#   C

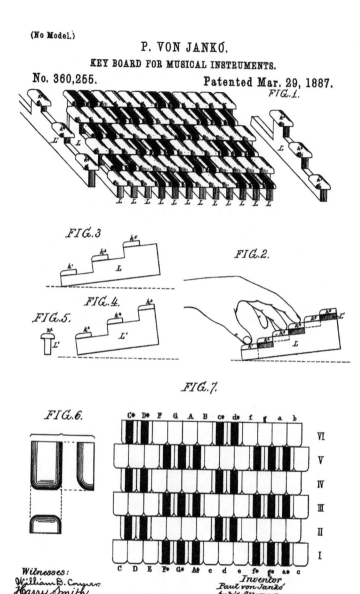

(No Model.)

P. VON JANKÓ.

KEY BOARD FOR MUSICAL INSTRUMENTS.

No. 360,255.   Patented Mar. 29, 1887.

Paul von Jankó, a mathematically and scientifically trained Hungarian musician, believed his keyboard would benefit pianists from beginner to virtuoso. Some leading manufacturers offered it, a few artists expressed interest, and a school devoted to his invention opened in New York. But most pianists and critics prevailed with their belief that it took the edge off performance by making challenging passages too easy.

The Jankó keyboard looked forbiddingly complex but actually simplified playing. The octave was six keys wide, relatively easy for a child or any person with a small hand; people with normal and large hands have a correspondingly greater reach with the Jankó arrangement. It was also easier to play arpeggios, glissandos, and other complex passages, especially with the addition of banks of keys identical to the home rows above and below them, which the piano historian Edwin M. Good has compared to seats in a balcony. Three touch pieces (gently rounded playing surfaces) were mounted in different locations on the same key.[26]

The greatest advantage of the Jankó arrangement was that any composition in a major key could be played in another major key merely by shifting the hands, without the mechanics of other transposing keyboards. There was also the option of reaching either upward or downward to play the same adjacent note. Many experts extolled the keyboard. It was not only easier for beginners and amateurs but permitted new music with otherwise unplayable chords and arpeggios. Its only real problem for the performer, common to other designs with uniform key sizes, was tactile and visual confusion. The hands can navigate a conventional keyboard just by following the familiar layout of black and white keys. The Jankó keyboard needs more complex cues.[27]

At first, Jankó made striking progress. Leading newspapers reviewed his London and New York demonstration recitals respectfully. Piano companies in Europe and the United States began to sell production models in the early 1890s. A conservatory in Berlin gave courses in the Jankó piano, and the inventor's teacher, Hans Schmitt, published études especially for it. Several prominent European pianists adopted it, and Liszt expressed interest. Enthusiasts formed a Jankó Association in Berlin. In 1891 a leading industry journal, the *Musical Courier,* serialized a long technical exposition of the keyboard's advantages. Few inventions in the arts have been so acclaimed shortly after introduction.[28]

Slowly Jankó's celebrity faded. There were technical problems. While his original design had a stiff action, especially when keys in the upper banks, with less leverage, were struck, another inventor improved it. Evaluating the modified action in 1911, the inventor, manufacturer, and historian Alfred Dolge called it "epoch-making." But Dolge had to acknowledge the resistance of most pianists, teachers, and music publishers. Jankó's innovation posed the same difficulty as the Clutsam keyboard: few performers could take their pianos on tour, and many provincial concert halls had no access to innovative instruments. Jankó's real predicament was deeper. He had invented an interface that, according to many musical authorities, worked too well.[29]

In sports, as we saw in Chapter One, sheer performance usually wins disputes over style, as the crawl, the safety bicycle, and the reactive resin bowling ball illustrate. But music is not just a proficiency contest. Turn-of-the-century music critics, like many today, deplored the rise of purely technical skill at the expense of musical intelligence and expressiveness. As one skeptic, Constantin Sternberg, wrote in the *Musical Courier* in 1891, " *'No! no more technique; sense, meaning, feeling!'* " was "the cry of the masses, of the music lovers, and even of those pianists who happen to be at the same time musicians." The Jankó keyboard seemed to foster the professionals' empty showiness and the amateurs' false brilliance. More recently, Edwin M. Good has noted a more subtle flaw in the design. Nineteenth-century composers had written for the standard keyboard. Their challenging passages, demanding manual acrobatics, intimidated piano students. But there was a positive side to their difficulties. The tension and struggle of the pianist to control the instrument and hit all the right notes contributed to the excitement of concert and serious amateur performance. The music was not only hard, it was *supposed* to be hard. To remove the tension by making playing easier and more natural was to break the music's spell. Perhaps a great composer could have shown the hidden expressive possibilities of the keyboard without falling into the flashy virtuosity critics were condemning. None appears to have tried.[30]

The Jankó movement thus lost almost all its cadre of manufacturers, publishers, performers, and teachers by the 1920s, leaving only a few instruments in the leading museum collections of Europe and North America. Dolge reproduces a carte de visite photograph of Paul von Jankó, by 1911 a section chief in the Turkish state tobacco monopoly in Constantinople—no doubt a comfortable position but hardly the dream career of the young polymath he had been in the 1880s. He wears a round astrakhan cap and a coat with an astrakhan collar, matched in texture by an expansive mustache and neatly trimmed beard. His sad eyes look away into the distance. His disappointment underscores the power of established skills in the face of innovation. His nemesis probably was not just the resistance of professional musicians to new fingerings. It was the foundation of the Western piano repertory in the very shortcomings of the traditional keyboard. And if Jankó's admirers protested that exceptionally challenging, unheard-of new music could now be written for the new keyboard, the rejoinder was obvious. The amateurs would have as much agony with the new post-Jankó material as they did with the standard keyboard and the old repertory. To beginner or professional, a technology is meaningful only in its effect on a body of techniques.[31]

No keyboard innovation since Jankó's has come as close to commercial success. Hopeful amateurs still have not given up. One of England's leading orthopedic surgeons, Graham Apley, proposed a new arrangement in 1991, in which black and white keys alternated and each major scale began either with one or with the other. The next year, Henri Carcelle won France's leading prize of its kind, the Lepine Inventions Contest, with a similar keyboard alternating short and long keys, linked to a new notation system that presented scores vertically rather than horizontally. Yet these inventors faced the same roadblock as Jankó; the doubt that any advantages of innovation could justify the agony of reeducation.[32]

TECHNOLOGY PRESERVES THE KEYBOARD

Curiously, new music has not seriously challenged the conventional keyboard; if anything, it has unexpectedly extended its life. Just as the acoustic piano was going through its mid-century troubles, new generations of electric instruments were emerging, some emulating pianos and organs, others expressing new sounds. If there was ever a time for a radically new interface, for a physical device that would make a clean break from the past, it was the dawn of the electric age, the late nineteenth century. Succeeding decades saw the prime of artistic iconoclasm, the ferment of movements from futurism to surrealism. Yet none of these changes seriously challenged the familiar arrangement of keys.

The piano keyboard dominated the first years of electronic music. When the American inventor Elisha Gray, Alexander Graham Bell's unsuccessful rival for priority in the invention of the telephone, introduced a "musical telegraph" in 1876, he activated his row of oscillators (really buzzers) with piano-style controls. The English physicist William Du Bois Duddell discovered how to make the carbon-arc lamps of the day produce tones—controlled by the familiar device. The most majestic electronic instrument ever, Thomas Cahill's Telharmonium, had 145 customized dynamos generating currents of different audio frequencies that were picked up by acoustic horns attached to telephone receivers. It was sixty feet long, weighed two hundred tons, and needed to be housed on the ground floor of its own building. And it used keyboards, too, though these were specially constructed with thirty-six rather than twelve notes per octave and not many musicians cared to retrain for it. (Lee de Forest, inventor of the triode tube and of the method of amplification used even in today's radio and television, built an Audion piano of his own with a conventional keyboard but was not a serious musical experimenter.)

In the early days of electronic music there were great hopes for new styles of performance that could bring out the striking sounds of the new instruments instead of emulating acoustic pianos and organs. The most famous (and mysterious) pioneer of this trend was Leon Theremin, the first inventor to observe that the earliest triode tubes—the basis of modern electronics and broadcasting—could be controlled by the human body's storage of electrical charges. Working for the Russian military during World War I, he found that the natural capacitance of a person approaching an electrical circuit could affect it and emit a signal, making it a sentinel or, as he called it, "radio watchman." The musician's body became part of the machine, absorbing and releasing energy wirelessly. High-frequency oscillators using de Forest's tube design turned out to be highly sensitive to capacitances of the body. By changing the position of the hand with respect to the antenna, the operator could produce a continuously rising or falling tone. This was the first instrument played without physical contact—or rather, almost without it. Initially, a foot pedal controlled volume and a switch in the left hand helped separate notes. To make operation more dramatically touch-free, Theremin added a second antenna, controlled by motions of the left hand.[33]

Theremin was a gifted amateur, and a cellist rather than a pianist. His Etherphone had a cellolike range of three or four octaves and a tone ranging from cello to violin. He also adapted his invention as an electronic cello without strings, played by a fingerboard. Yet when RCA executives tried to commercialize Theremin's invention as an electrical rival to the piano, they rediscovered what had made keyboards so popular for centuries. The Etherphone, renamed the Theremin, demanded exceptional musical skills. Empty space had no visual cues for the hands; even singers use the feedback of their own vocal cords. The theremin player needed precise musical intuition and physical orientation to produce acceptable pitch. RCA was able to build and sell fewer than five hundred, and cut its losses. Theremin finally had to devise keyboard models (as well as fingerboard models, with strings for sliding fingers) to sustain public interest, though even these failed. Advanced technology turned out to demand such extremely refined technique that even after Robert Moog (an engineer and nonmusician) revived the production of antenna theremins in the 1950s, there have been few professional players. The best-known of these remained Theremin's protégée Clara Rockmore, a violinist already acclaimed as a prodigy, who had not only absolute pitch but masterly control of her gestures and superb musicianship. A theremin with a keyboard was an eerie-sounding organ, accessible but musically uninteresting. With antennas it was a sonic laboratory for a small circle of enthusiasts. Rockmore and the handful of other theremin virtuosi were unhappy that the instrument's leading use was in science-fiction films and the music of groups like the Beach Boys.[34]

Theremin might have interested more composers and made more sales had he developed his keyboard version further, or at least used an easier-to-play fingerboard like that of Friedrich Trautwein's Trautonium (1932). A fellow cellist-inventor, the Frenchman Maurice Martenot, showed the path not taken. His Ondes (Waves) Martenot (1928) was based on Theremin's tone-generating circuitry but controlled in early versions by a wire attached to a ring on the player's finger, which could be moved up and down a simple piano keyboard layout, solving most of the problems of pitch that challenged Theremin users. Martenot's instrument became the first electronic device to be welcomed into the classical repertory. For the Jankó piano, with its innovative arrangement, only a few exercises had been published; for the Ondes Martenot, there were works by Arthur Honegger, Olivier Messiaen, Darius Milhaud, and Edgard Varèse. The Paris Conservatory appointed Martenot to a chair of his own instrument after the war—an unusual academic recognition for a musical inventor. Martenot was able to combine the familiar interface with control of timbres in a way that Theremin was not.[35]

MUSICAL TECHNOLOGY AND THE FUTURE OF THE KEYBOARD

As the Ondes Martenot was arriving, the most versatile and ubiquitous of keyboard instruments, the piano, was technologically stagnating, with no significant improvements after the 1890s. In 1948, the piano scholar Ernest Hutcheson could preface his influential handbook, *The Literature of the Piano*, with a technological wish list. Most important, he called the key action and sostenuto pedal "somewhat clumsy" and hoped for a new action permitting "greater delicacies of touch and inflection." Yet over fifty years later, the acoustic piano remains almost what it was in the 1890s. Plastics and other new materials have found only a limited—and not always successful—role in mechanical components. The piano, once a hybrid of craft and industrial innovation, is now marketed as a sign of tradition and preservation—a trend perhaps visible even early in the twentieth century, when the ratio of grand to upright pianos sold began to rise. Scarce labor and materials have raised the price of pianos well above the cost of living. Most other complex consumer products, notably automobiles and television sets, need less frequent maintenance than they did two or three generations ago; pianos still require regular service from master technicians. And competently reconditioned and rebuilt older pianos, unlike nearly all vintage electronics, are competitive with current models.[36]

To some, the acoustic piano seems to be a glorious anachronism, a trophy technology like a luxury mechanical watch, always in the shadow of yesterday's masterpieces. Such thinking can easily become self-confirming as designers fear to tamper with proven mechanisms. Fortunately, there are always a few men and women who refuse to be intimidated by stability. They find new and better ways to make wristwatches, draft-animal agricultural implements, and acoustic instruments. Recently, an American firm, Fandrich Piano, has developed a spring-loaded upright piano action with the feel of a grand. Ron Overs, an Australian piano rebuilder and dealer, has used computer-assisted design (CAD) to devise a radically new action claimed to reduce friction by up to 50 percent, improving response, sound output, and durability, and reducing the frequency of tuning: potentially the most fundamental proposed change in the piano since the Jankó keyboard, and the first major response to Hutcheson's call for a more responsive touch. A number of prominent concert pianists have already auditioned Overs's action, built into a concert grand piano, and endorsed it enthusiastically. Overs is planning to produce his own line, and international manufacturers have expressed interest in licensing the action.

The Overs action's success depends not only on its construction but on its effect on the skills that pianists and piano teachers already possess. If it lets them achieve the same effects with less fatigue it may help revive the acoustic piano market; but no matter how well it works, there probably will always be some pianists who will prefer the higher friction of the conventional action, just as Chopin disdained the Erard repetition lever that was so revolutionary in the 1820s.[37]

The greatest impact on the keyboard world has come not from the acoustic piano but from electronic music, with its relatively rapid technical changes. The postwar pioneer of synthesizers was Raymond Scott, a musician and inventor who had made innovative soundtracks for Warner Bros. cartoons. Scott's Clavivox, patented in 1950, used a keyboard to achieve a theremin-like continuous tone, but it never reached the mass audience he had hoped for despite numerous improvements. The electronic synthesizer market was opened to performers by one of Scott's collaborators, Robert Moog. Moog was one of two designers who realized the new possibilities for music synthesizers after the price of silicon junction transistors fell from $1,000 to 25 cents between 1957 and 1964, allowing Moog to build an instrument in the $10,000 range. The keyboard was only one of a number of devices that could be plugged in as modules trigger voltage changes to control the output of these transistors. Trautonium-style sliding strips with variable resistance were equally feasible, and Moog developed a new "ribbon controller," a potentiometer along which a finger could be moved up and down. The sociologists Trevor Pinch and Frank Trocco have discovered that it was Moog's customers—musicians and synthesizer experimenters—who urged him to use the keyboard instead. Donald Buchla, a fellow pioneer but a musician and composer rather than a product developer, rejected the twelve-tone keyboard as a constraint on the range of new music. His original synthesizer used touch-sensitive plates. Buchla proudly describes himself as "an old fashioned builder of instruments" rather than a machine builder. His controllers are boldly innovative, but thus appeal to a smaller number of musicians than those based on the traditional keyboard layout.[38]

Of course, it was precisely mechanical familiarity that shifted synthesizer development to Moog's keyboard model. Early publicity photographs of the Moog put the keyboard in the forefront to reassure prospective buyers that they would be able to play it. And a single LP suggested the wonderful things that could be played with it. Walter (now Wendy) Carlos's *Switched-On Bach*, released by CBS Records in 1968, became one of history's best-selling classical recordings and was praised by Glenn Gould as the best recording of the Brandenburg Concertos he had ever heard. Carlos seemed to show a wealth of expressive possibilities in what had been thought a one-dimensional instrument. The original Buchla remained one of a long line of relatively obscure academic and experimental instruments, favored by composers seeking new kinds of sounds. Thus just before Carlos was using the Moog to pay homage to sources of keyboard music, the composer Morton Subotnick was proclaiming distinctively electronic values with his album *Silver Apples on the Moon*, recorded with the Buchla.[39]

Beginning with the Beatles and the Rolling Stones, generations of rock and pop musicians brought the Moog and its imitators into mainstream music. In the 1980s, Japanese manufacturers helped make the keyboard synthesizer a standard instrument. Pianists could travel with their own keyboards, and some performance places no longer kept a piano. Newer electronic pianos allow a rock musician to hold keyboards like guitars, making possible a full range of body motions. New generations of equipment, especially integrated circuits, have enhanced the electronic keyboard's power to imitate acoustic instruments, making it the piano's successor as the "orchestra in a box." As the musicologist James Parakilas has observed of the rock scene, and the continuing power of the keyboard as musical interface, it now has "such a spectacular menu of sounds that keyboard players could put some of their fellow band members out of work at the same time as they put the instrument they themselves originally played—the piano—out of business."[40]

Even as the manufacture of keyboard instruments has been globalized, so has their use. Some advanced commercial synthesizers are preprogrammed with a variety of non-Western scales available by simple adjustments. In Arab and Arab-American music, for example, where acoustic pianos were never important instruments, the versatile keyboard synthesizer is now ubiquitous as the "Arab org."

THE ROADS AHEAD

The control of musical instruments is diverging. Electronic musicians have come to doubt the power of standard musical keyboards for innovation in sound. Many composers use the computer keyboard that we will be considering in the next chapter. Some of them are working with "microtonal" or "enharmonic" keyboards, which are capable of quarter tones and even finer divisions of the musical scale. John S. Allen, an MIT-trained electrical engineer, has developed a "general keyboard" that superficially looks like a Jankó model (though Allen was unfamiliar with Jankó's patent and drew instead on earlier organ keyboard proposals). But Allen's keyboard can play up to thirty-one tones per octave rather than Jankó's twelve. This and other advanced designs seem inherently limited to a small number of players and listeners with highly developed pitch. Others use variants of the sliding strings, bows, and body position technology we have encountered. The capacitance of the theremin was just a beginning; new technology can translate the motions of a dancer, or eye movements, into sound. Donald Buchla has developed an instrument played with cordless wands. The independent instrument builders, academics, and performers who develop new controllers do not expect to license their innovations to major manufacturers. It is unlikely that a single new instrument will be the subject of a chair in a major conservatory as the Ondes Martenot has been. The innovators form small circles of professionals and hobbyists who give live performances and produce CDs.[41]

The second path of advanced technology is the electronic refinement of the standard keyboard, mainly for performing the existing repertory rather than for creating new music. Some of the best-known devices are advanced revivals of the player piano. In 1978 Marantz introduced the first electronic reproducing piano, using data encoded on cassette tapes. By 1988, the Japanese manufacturer Yamaha had produced its Disklavier, which could both record a pianist's performance electronically and play it back from a computer diskette with more than one hundred levels of audible hammer velocity. A California musician and engineer, Wayne Stahnke, added his own refinements for the Austrian firm Bösendorfer. He used a new generation of sensors to record not the depression of keys but the velocity of hammers alone, and new types of linear motors to duplicate the action of the hammers and pedals during playback without affecting the touch of the instrument when a pianist is using it. Today the Bösendorfer 290 SE and the Yamaha nine-foot Disklavier Grand Pro are our generation's counterparts of the greatest organs of late medieval and early modern Europe, using the virtually unchanged keyboard to produce a stunning range of sounds—the strings of acoustic pianos as well as sampled electronic sounds of as many as five hundred instruments. Fully featured electronic-acoustic concert grands sell for $300,000 or more.[42]

Are these hybrid pianos only a way station toward fully electronic keyboard instruments that make sounds indistinguishable from those of acoustic pianos? Perhaps, but after over three hundred years, the keyboard and the acoustic piano are still revealing new complexities that cannot be duplicated by playing prerecorded samples of individual notes electronically. Strings respond to each other's overtones. Some piano technicians even deliberately leave the three strings that make up a single note slightly out of tune with each other to delay dissipation of the sound. We still have much to learn about soundboards, the loudspeakers of the acoustic piano, and their interaction with the vibrations of the strings. Acoustic piano dealers report that many of their buyers are young owners of electronic keyboards who grow tired of the sound and hope to upgrade.[43]

The greatest challenge of all may be to contrive new versions of the standard keyboard for electronic instruments, especially electronic pianos.

Robert Moog has produced a "multiply-touch-sensitive" (MTS) keyboard controller with key overlays that transmit continuously the horizontal location of the player's finger on the key (X and Y axes) and the vertical motion of the key (Z axis). The output of these sensors is connected to a MIDI (Musical Instrument Digital Interface) circuit. It is possible that this and similar systems will let keyboard players develop new techniques and create musical effects that were not possible with conventional pianos. The inventors' purpose, in fact, is not simply to reproduce music more accurately, but to promote new musical expression.[44]

From the electromechanical side, engineers have been doing their best to duplicate the force feedback of a traditional piano action for use in electronic pianos. Two different designs were announced in 1990. A French musical technology researcher, Claude Cadoz, and his group demonstrated a software-controlled keyboard with sixteen motorized keys; in the same year an American, Richard Baker, patented an "active touch keyboard" in which keys are connected to small motors with keyboards and pulleys with an analog controller. A third inventor, Alisdair Riddell, completed a "meta action" for use with acoustic pianos; it put the hammer at the end of a solenoid responding to contact made by the keys.[45]

None of these duplicated the full cycle of a conventional keyboard from initial finger contact to liftoff. This force feedback helps the pianist produce what we have seen is a theoretically amazing variety of timbres with an instrument that cuts off contact between the finger and the string once the key is depressed. A grand piano action can be modeled as the response of sets of coupled springs and other damping and stiffening parts. These complex interactions are modeled and translated into software that can easily be modified. A virtual keyboard can, for example, be adjusted to feel like any historical or innovative piano keyboard, or like a harpsichord keyboard. If other sensors are included, it should even be possible to achieve effects impossible with acoustic pianos, such as swelling a note. In his doctoral dissertation, Brent Gillespie was able to achieve a good fit between the outputs of his system and the measured responses of acoustic piano actions. He is now developing his system, called the Touchback Keyboard.[46]

Gillespie's work helps illustrate the flexibility of a technology half a millennium years old. The musical keyboard was probably a fortunate by-product of other mechanical innovations. It could control music that could fill a cathedral or a domestic parlor. Its most successful adaptation, the pianoforte, had a surprisingly small initial effect on the musical world.

And the changes that made it the most successful instrument of the nine-teenth century did not come, at first, from the manufacturers but from the great composer-performers. It was their new techniques of playing that pushed piano builders to a century of glorious technological achievement. Because composition and performance diverged as careers in the twentieth century, there was no comparable force for change. Composers including John Cage and George Crumb have "prepared" pianos by prescribing tem-porary modification of strings with small objects like bolts, screws, and rubber bands—but the spirit of hacking has been easier to extend to $2,500 personal computers than to $25,000 Steinway grands. For all the excitement of innovative instruments, new technology has on balance strengthened rather than weakened our attachment to the venerable twelve-tone keyboard. Technique sometimes transforms, but it conserves equally.

CHAPTER EIGHT

Letter Perfect?

Text Keyboards

I T IS SURPRISING that the music-making keyboard should have preceded the text-imprinting keyboard by at least five centuries. Even counting upper and lower cases, there are no more characters on a contemporary computer keyboard than there are keys on a grand piano. A medieval organ was at least as complex as a mechanical typewriter. By the eighteenth century, artisans were building superb mechanisms ranging from marine chronometers to anthropomorphic automata that drew figures and wrote texts. Yet there was no commercially successful writing machine until the 1870s.

It is equally surprising how the text keyboard has spread as a physical interface in less than 125 years. Within a few decades, the typewriter was replacing pens and pencils not only in commerce and government but in academia and literature, despite initially high prices. Nor were keyboards limited to typewriters. They took over typesetting, data entry, and a large part of telegraphy. Equally impressive has been their global reach beyond the lands of Roman and Cyrillic alphabets, especially in the twenty years since the introduction of the microcomputer. And despite massive campaigns for alternatives like speech and handwriting processors, the text keyboard, like its musical predecessor, seems to be increasing its domain. For hundreds of millions, if not billions, of the world's people, keyboarding has become a body technique more natural and intuitive than writing by hand, which in the West at least is increasingly an onerous challenge rather than a graceful art. And while low platforms could make typewriters and computer keyboards perfectly usable in squatting positions, in practice modern equipment around the world is nearly always operated by people sitting in chairs—as, indeed, pianos are built to be played from stools and benches, and as many non-Western instruments are designed for performance by musicians seated or kneeling on mats and cushions. The keyboard is part of a relentlessly expanding set of body technologies and techniques. Yet there was nothing inevitable about it.

WRITING AS A BODY TECHNIQUE

There are few things more necessary, or difficult, for a growing child than writing. In many ways it is more troublesome than computation. Mental shortcuts make it possible to teach children and adults surprisingly fast and accurate arithmetic through paperless techniques developed in Asia, Europe, and North America. Market and securities traders, real estate negotiators, and even fences of stolen goods have developed exceptionally fast and accurate systems. But writing knows no shortcuts, and shorthand is the hardest form of all. In Japan, learning to make the kanji properly can be a lifelong effort. Even in the West, with only twenty-six letters, combining speed, accuracy, and legibility has never been easy.

Medieval scribes faced many physical challenges. Parchment was dear, and paper became widely available only in the fifteenth century. Into the nineteenth century, writers had to sharpen their own quill pens, up to sixty times a day for a scribe. (Oboists still have to make their own reeds.) Parchment also had to be ruled. The surprise was how efficient writing could be given medieval conditions, especially in the late Middle Ages when private scriptoria flourished. Scribes worked with a pen in one hand and a knife in the other for steadying the parchment, sharpening quills, and erasing errors. Their technique of holding a writing instrument was different from ours: the light quill could be held almost effortlessly between the tips of the first two fingers and the thumb and moved by the whole forearm rather than the wrist, with the hand hardly touching the page. The text of a Book of Hours could be completed in a week, and—what will amaze anyone who has ever admired them in an exhibition—two or three of the miniatures could be finished in only a day. One reason for this speed was the ergonomic innovations of the Middle Ages. We have already seen that the Greeks and Romans had no desks or writing tables. Medieval scribes used slanted lecterns, still recommended for reading and writing, and especially convenient for keeping the pen at the optimal right angle to the paper. Their desks often had additional stands for propping open copy, a convenience lacking in many computer workstations today. It should not be surprising that some scriptoria were initially able to compete with early printers, at least in short-run production of high-priced books.[1]

Of course, any attempt to build a writing machine even after Gutenberg would probably have been wildly expensive and produced crude results. But the manuscript was not just a reproduction of an author's text. As the historian Henri-Jean Martin has observed, copying a manuscript gave the scribe "an almost kinetic memory" of its arguments and "an almost physical familiarity" with the writer's argumentation. Bodily engagement produced a mental identification with the author. Errors could be corrected, or introduced. The scribe was an intellectual artisan, a collaborator. Professional copyists offered a variety of styles according to the purpose and formality of the text; a surviving document gives specimens of twelve. Each copyist's hand was so distinctive that today's experts on medieval manuscripts do not consider forgery a serious problem.[2]

Medieval writing had achieved such effectiveness that the rise of printing produced only gradual changes in the appearance of books. Paradoxically, print led not to experiments in the mechanical production of writing but to a new flowering of handwriting, just as railroads (as we have seen in Chapter Four) popularized walking, and motoring and aviation in turn stimulated the railroads to new peaks of technology and service in the mid–twentieth century. The proliferation of printed matter actually increased rather than reduced the need for writing as governmental, religious, and economic activity grew. Printed books still had to begin with manuscripts, and publishers found a ready market for writing guides after technical problems of engraving were overcome. Beginning in the 1520s, Italian writing masters like Ludovico Arrighi, Giovanantonio Tagliente, and Giovanbattista Palatino prepared manuals that went through as many as thirty editions. They spread the hand used by the papal bureaucracy, called Chancery and (in northern countries) Italic, throughout educated circles of Europe. Until the nineteenth century, this hand, based on the everyday writing of medieval people, remained a standard script. It was an intuitive ergonomic solution, like the musical keyboard of the same period.[3]

The writing masters did not stop with Chancery. They were influential in developing Roman typefaces for printing, a number of which are still widely used. In the seventeenth and eighteenth centuries, they also brought the art of calligraphy to new technical heights, but their very artistry helped create a gap between the handwriting of formal documents and the script that people used in everyday correspondence. Among their achievements were the continuous cursive "round hands" that are the foundation of much formal Western handwriting to this day: elegant in expert use but difficult for others to learn properly. Graphic technology also affected formal writing styles: the copperplate engravings that preceded nineteenth-century woodblock needed hand-incised captions. The pen began to imitate the engraver's cutting tool, the burin. With the expansion of commerce in the eighteenth century came new prestige for round and copperplate styles in account books, bills of exchange, insurance contracts, and other commercial documents. Just as today's office employee must be proficient with the complex formatting options of word processors and desktop publishing packages, the eighteenth- and nineteenth-century clerk had to master textual presentation.[4]

By the early industrial age there was so much business to transact, and so many children were enrolled in the developing school systems of Europe and North America, that a market existed for an inexpensive pen that would not need sharpening. Metallic pens had been known for hundreds of years, but they had been luxuries. By the 1830s Birmingham, England, enjoyed access to the metallurgy, markets, and machine-building skills that made metal nibs among the best-selling products of the early Victorian world. Jealously guarded equipment pressed Sheffield steel and split nibs. From a ton of steel, 1.5 million nibs could be made. At the beginning of the 1840s, a single Birmingham manufacturer, Joseph Gillott, was shipping more than 62 million a year. In 1874 the factories of the other leading pioneer, Sir Josiah Mason, were turning out 32,000 gross each week.[5]

Like other Victorian technology, this industrial product soon captured the romantic imagination. George Pratt (1832–1875), Yale class of 1857, wrote:

> *Give me a pen of steel!*
> *Away with the gray goose-quill!*
> *I will grave the thoughts I feel*
> *With a fiery heart and will . . .*

Actually, the early steel pen was, as the melancholy conclusion of Pratt's verse acknowledges, easily "corroded day by day" by the inks of the nineteenth century, and the twentieth-century calligrapher and historian

Donald M. Anderson called the meeting of steel pen and bond paper "[as] icy cold and mechanical as the greeting of an iron lawn dog." But despite these drawbacks and a tendency to splatter the indelible ink of the day, the steel pen did for education what its distant metallic cousin the cast-iron-frame piano with steel strings was doing for music. It provided an instrument of impressive flexibility for the Victorian mass public. Nibs and pianos, kindred iron and steel technologies of the hand and arms, were among the great international metallurgical successes of the nineteenth century—along, as we have seen, with steel-sprung parlor furniture.[6]

Just as the diffusion of the piano produced a wave of pedagogical methods and mechanical aids, so the steel pen helped make possible mass instruction in writing, which would otherwise have exhausted teachers as it tormented geese. Writing became an instrument not just for communication but for physical and mental discipline and character training. As Bernard Cerquiglini, director of the National Institute of the French Language, has remarked, "The blue ink spot on the finger is a badge of French education. It is the mark of the French flag on the body of the French student." Into the twentieth century, little yellow Waterman ink trucks circulated in Paris, replenishing the reservoirs of pupils' steel pens. Yet—especially in France—rigorously drilled youth somehow, or thereby, became individualist adults: in 1885 a single French firm, Blanzy-Pour, offered five hundred different nibs, at prices ranging from 0.23 to 7.80 francs per gross, according to quality.[7]

(The graphite pencil was also prominent in the international rise of mass education in the nineteenth century. America's more forgiving, free-form contribution to writing was the mini-eraser attached by a metal ferrule, introduced a year after George Pratt was graduated from Yale. Many European teachers still cling to the prejudice that attached erasers encourage errors, overlooking the big stand-alone chunks of rubber their pupils use instead. And what is wrong with making and correcting mistakes?)[8]

France retained a distinctive cursive script that probably originated in the late seventeenth century; it is still visible on some restaurant menus. Germany developed its own angular script counterpart of its black-letter Fraktur type. But it was the United States that turned handwriting into a form of drill that was only a short step from the actual mechanization of writing. While the early master Platt Rogers Spencer (whose style survives in the logotypes of Coca-Cola and the Ford Motor Company) was inspired in his flowing script by the curves of nature, instruction was no aesthetic reverie. Writing masters claimed to develop habits of manly self-discipline from which women were excluded—however feminine the graceful letter-forms of Spencer and his competitors appeared.[9]

In the last two decades of the nineteenth century a new technique, still influential today, was developed for the steel pen: the Palmer method. Impatient with Spencerian curlicues, the master penman A. N. Palmer dominated American handwriting instruction with his call for no-nonsense efficiency. Where Spencer had extolled aesthetic contemplation, Palmer taught what we know today as muscular memory: drill and repetition of motions that would make possible rapid and unconscious production of correct letterforms. Perhaps because women were a rapidly growing proportion of the white-collar workforce, Palmer taught writing as a virile, assertive command of the whole forearm rather than the wrist and fingers. In place of meticulous copybook exercises, Palmer's disciples invoked industrial efficiency. In 1904 one manual described the body as "a machine on which writing is done."[10]

The history of script suggests that typewriting was not just the miraculous unfolding of a mechanical marvel but the logical outcome of a social drive to discipline the body. Cultural historians from Michel Foucault onward have seen the rationalization of labor in the eighteenth and nineteenth centuries as baleful, exploitive manipulation. They have often oversimplified, but it is true that to generations of schoolchildren, writing exercises could be physically painful. The steel pen also produced some of the first cases of what are now called cumulative trauma disorders: a report on "scrivener's palsy" by a Dr. Samuel Solly appeared as early as 1862, and "writer's cramp" had been more casually observed even earlier.[11]

FROM THE LITERARY PIANO TO THE TYPE-WRITER

In the realm of the late-nineteenth-century office, the typewriter was not a revolution but a revelation. If the ideal worker was to be a tireless and efficient machine, was not the pen—blotting, needing to be dipped—a weak link? The logical direction of standardizing technique was finding a technology that put information onto paper with the least manual effort. The keyboard began to challenge not only the pen but other information technologies: the telegraph key and the type case.

From at least 1714, inventors in Europe and the United States had proposed dozens of systems for imprinting letters. But until the mid-nineteenth century few of the designs were suitable for replacing handwriting in everyday commerce, literature, or education. Some, like William Austin Burt's "Typographer" or "Family Letter Press" of 1829, produced attractive text but used dials and levers that made the process too cumbersome for longer copy, like today's hand-held plastic tape embossing machines. Others were conceived less as general-purpose machines than as prostheses, extensions of the body for those unable to write efficiently or legibly with a pen. The best known in North America was Charles Thurber's device, patented in 1843, for marking letters on paper mounted on a traveling and rotatable cylinder. The keytops, arranged around a wheel, bore raised letters for the blind. Other American and French inventors introduced systems for embossing letters for the blind; none appears to have been commercially successful.[12]

If the keyboard was not yet viable for the disabled, it appeared far more promising for another market: the growing number of businesses with substantial telegraph traffic. In the nineteenth century, telegraphy was an advanced technology, one that attracted many ambitious young men and a growing number of women. The men, but not the women, were initially encouraged to develop themselves scientifically and technically. They formed a network of avid tinkerers, proud of their skills, simultaneously competitive and cooperative like today's programmers. Thomas Edison's genius emerged in this milieu, and New York City's thriving financial community keenly rewarded innovations offering users a competitive advantage.[13]

While only a few of the operators became inventors, this elite workforce was a challenge for employers. Wages were high and proficient operators in demand. Because messages were generally received by electromagnetic sounders, each telegrapher had a recognizable rhythm, called a fist, and a proficient operator could "rush" a neophyte by sending letters faster than the recipient could transcribe them. At the other end of the marvelous new apparatus there were, after all, just ears, a brain, a hand, and a pen. Employers sought alternatives promising more speed with less skill. The musical keyboard was a familiar interface, and the increasing durability and dynamics of pianos must have suggested that keyboards could control text as well as sound fluently and reliably. In the early 1850s, Sir Charles Wheatstone, a professor of physics at King's College, London, devised a series of typewriters using a keyboard like a piano's, except that black and white keys alternated evenly. Wheatstone's machines were not designed for message transmission, but since they produced letters on tape, they probably were intended for transcription. In 1855 and 1857, respectively, the

Italian Giuseppe Ravizza and the American Dr. William Francis used similar keyboards to imprint paper on a roller, giving their inventions the charming names of Cembalo Scrivano and Literary Piano.[14]

The crucial step probably was not a device but a manifesto. Ten years after the Francis patent, *Scientific American* described yet another "type writing machine" that John Pratt, an Alabaman, had shown in London. What was notable was not the design of Pratt's invention but the sentiment the editors expressed: that the technique of handwriting had become a torture, and that it was time to replace it with a modern technology. The "laborious and unsatisfactory performance of the pen" would be replaced in law offices, newspapers, and the studies of the clergy in a "revolution" comparable to that begun by the printing press. The "weary process" of school penmanship lessons could be limited to "writing one's own signature and playing on the literary piano." Apparently for the first time in a widely circulated periodical, the article also held out the possibility that a machine could write not just more neatly and easily than the pen but also more rapidly.[15]

The first invention inspired by this article continued to take the musical metaphor. The patent demonstration model produced by Christopher Latham Sholes and Carlos Glidden, two Milwaukee amateur tinkerers who haunted a local machine shop, had six white keys alternating piano-style with five black keys. Sholes and Glidden were soon joined by two others: Samuel Soulé, a technical man who had worked with Sholes on other inventions, and James Densmore, a former newspaper publishing colleague. The invention took another turn. Sholes and his collaborators abandoned the piano-style keyboard for an array of circles; this was to remain the definitive arrangement. But the inspiration of musical keyboard instruments was still evident. Each key indirectly activated a hammer that forced the paper against a ribbon, much as a piano key transmits a force through the parts of its own action to make the felt hammer hit the strings.

The technical details of the Sholes typewriter's history interest mainly collectors and other specialists, but the fact that it went through many versions in the 1860s and 1870s was part of the reason for its triumph. The Sholes-Glidden-Soulé design was inelegant. For decades after its introduction, users could not see their work as they typed; the paper had to be removed. Hammers (typebars), clashing frequently, had to be untangled. The machine was limited to a single typeface. While many rival designs were mainly attempts to get around the Sholes patents, others had major advantages. At about the same time, James Bartlett Hammond was developing a writing machine that used a replaceable circular typewheel, permitting multiple fonts. Text produced on the Hammond could be read with the paper still in the machine. Perhaps best of all, the Hammond typewheel not only made jamming impossible but assured an even impression. The typewheel was fixed. Pressing a key triggered a hammer that struck the paper through the ribbon against the typewheel with unvarying force. Hammond was not the only inventor to use such a principle. Thomas Edison, too, after examining and rejecting the Sholes machine for his employers at the Automatic Telegraph Company, invented an alternative machine with a small rotating wheel against which the paper was pressed.[16]

After World War II, the Hammond principle became part of a versatile and successful composition system called the Varitype that remained the most economical way to set mathematical copy before the computer. Beginning in the 1960s, IBM introduced the Selectric typewriter with its tilting and rotating golf-ball style element; a number of manufacturers still make word processors with rotating type wheels of the kind Hammond pioneered. Why, then, did the Sholes typebar triumph? What it lacked in design elegance it more than made up for with one of the best systems for product improvement that the industrial world had yet seen, and the credit goes to Sholes's associate James Densmore, one of America's most underrated entrepreneurs. Densmore not only risked all of his $600 in savings in a quarter-share of the invention, he gradually bought out the other partners. Equally important, he realized the power of steady incremental improvements. Densmore relentlessly pressured Sholes to produce one new model after another. Believing that court reporters would be an important early market, he sent models to James O. Clephane, one of Washington's leading practitioners, for rigorous testing that led to new rounds of changes—as many as thirty models over four years. Densmore also found an ideal manufacturing partner in the Remington Arms Company of Ilion, New York, where expert mechanics had benefited from the advanced machine-shop practice developed at America's national armories and then introduced into private manufacturing.[17]

The Remington Model 1 typewriter, introduced in 1874, employed most of the principles of a twentieth-century mechanical typewriter, from the spring-driven carriage with its rubber platen to the keyboard arrangement and basket of typebars. But although it had been through years of relentless development, the Model 1 was limited even by the day's standards. It was slow and expensive, at $125 about the real price of today's high-end personal computers. Fewer than a thousand a year were sold during the 1870s, even after Remington offered shifting for upper and lower case. The machines were far more popular with court reporters than with businessmen, whose customers often suspected typewritten letters of being printed handbills intended for the semiliterate. The technique of penmanship was still esteemed, and the legal validity of typewritten documents and signatures was debated. Sears, Roebuck & Company sent handwritten correspondence to its rural customers, and the U.S. government did not start to authorize typewriter use until the end of the century. Handwriting was laborious, and many people yearned to do without it, but when the opportunity appeared they hesitated. Palmer's penmanship schools at first spread faster than the machine that was supposed to put them out of business.[18]

What finally overcame obstacles to the typewriter was not any price reduction or further technological breakthrough. It was the changing face of organizations and society. Companies and government bureaus were growing larger, and their internal as well as external communication needs grew even more rapidly. Fine penmanship might still be a welcome courtesy to an individual customer, yet it could not address this volume of business. As the number of public and private clerical employees rose, all the schools' penmanship exercises could not assure the uniformity that new bureaucratic life demanded. Eliminating confusing individual variations in writing, the technology of the typewriter permitted standardized fonts that were as closely tied to the emerging industrial and commercial society as the late medieval *textus quadratus,* Gutenberg's model, had been to the scriptures and devotional books of its own day. Black letter had achieved shortcuts in writing and reading with dozens of ligatures and abbreviations; typing achieved the same through speedier use of a more limited character set. The typewriter fonts familiar well into the twentieth century, pica and elite, were as anonymous in design and as ubiquitous as black letter had been. Although a variety of typefaces was available from the earliest days of typing, there seemed to be an overpowering unconscious will to suppress variation in the interest of interchangeability. It did not matter that each manufacturer used slightly different matrices or that each typewriter produced unique details of letterforms and alignment. Often ornate printers' fonts, especially for advertising, were proliferating in the nineteenth and early twentieth centuries, but the typewriter retained its chaste uniformity.[19]

By 1886, Remington and other makers were selling fifty thousand typewriters a year, and organizations were embracing textual uniformity. *The Writer*, a leading journal for authors, repeatedly reminded subscribers of magazine publishers' preference for typewritten copy. The advantages of standardized text for compositors working with hundreds of manuscripts each year were obvious; book typesetters could afford to be more indulgent, and only after World War II did nearly all of them require typewritten copy. Even so, as early as 1889 the city editor of the *Boston Globe* could declare that a typewritten manuscript improved chances of acceptance by 10 percent.[20]

The uniformity of type had a social advantage, too, at least for employers. Carbon paper banished cumbersome methods of duplicating handwritten documents and made it possible to generate a copy of every outgoing letter automatically. Even more important, the keyboard responded to an emerging management trend: turning the previously flexible definition of clerical work into a hierarchy with more limited opportunities for advancement. As part of the trend, stenography and typing were feminized once the New York Young Women's Christian Association began to offer courses. Women, who had become so fluent at the piano, were now thought to have a special affinity for keyboards. Best of all for the initially skeptical employers, women could be hired for wages 25 percent lower than those men earned for comparable work. In Frank Lloyd Wright's state-of-the-art Larkin Building (1904) in Buffalo, where we encountered the "suicide chair" in Chapter Five, messengers took (male) correspondents' "graphophone" cylinder customer-service responses to the (female) Typewriter Operators' Department, a novelty with a balcony for visitors. Economic rationalization and gender stereotyping went together.[21]

The keyboard turned out to be not only a tool but a management weapon. After an operators' strike in 1907, the Western Union Telegraph Company, which had spurned earlier automatic transmission systems, began to replace its unionized, mostly male Morse operators with women using automatic transmission equipment with typewriter-style keyboards. After sixty years of prototypes and experiments, anonymous keyboarding replaced the distinctive techniques and knowledge of a formerly proud craft. The new generation of female telegraphers would never have the chance to produce its own Edison. Bookkeeping, too, was shifted to machines that combined alphanumeric keyboards with mechanical calculators. The keyboard was dividing the clerical workforce into semiskilled and professional castes. Only the mechanical compositors were able to

continue corporate traditions with the new technology. Ottmar Mergenthaler's Linotype, which prevailed in newspaper work, had a distinctive
keyboard pattern needed to accommodate stacks of matrices. The keys
were arranged by declining frequency of use in English. Because no typewriter used this arrangement, it took decades before typists could be
enlisted in periodical typesetting. Male hand compositors, unlike their
telegrapher brethren, welcomed automatic technology. Women had been
gaining ground as manual typesetters. While the Linotype Company
insisted they could also be excellent operators of its machines, the men
were able to exaggerate the strenuousness and danger of working with
heavy equipment and reservoirs of molten lead. Even after punch tape
operation removed this pretext, hot metal composition remained masculine keyboarding.[22]

A TECHNOLOGY IN SEARCH OF A TECHNIQUE

What did the men and women of the 1880s and 1890s actually do with
their expensive new machines? Ottmar Mergenthaler's company trained
Linotype operators, and Thomas Edison insisted on sending a representative to demonstrate his new phonograph, but typists and their employers
were on their own. The keyboard came into the world with no recommended technique. Many typewriter inventors like Christopher Sholes and
enthusiasts like Mark Twain were former newspapermen who thought of
setting text as the rapid motion of thumb, forefinger, and middle finger. At
first neither they nor their customers considered the fourth and fifth fingers
strong enough to be used regularly—even though pianists had been applying them to fortissimo passages for years. Early typing manuals were cur-
sory, and many were based on the alternative keyboards of other
manufacturers. The Caligraph and other machines had dedicated keys for
capitals, like the Linotype, above or to the side of the lowercase keyboard.
Operators worked out their own systems, nearly all searching for each letter. Typists working from stenographic or plaintext copy would look at a
sentence or phrase, then tap it out. Four-finger typing could be fast enough.

Not inventors but users were the first to see that typing was a new skill
that, like musical keyboard fingering before it, could benefit from systematic analysis and practice. The pioneers formed a competitive and cooperative community of skills, like the late-nineteenth-century bicyclists and
swimmers we have discussed, or the surfers and skateboarders of the
twentieth century. Many early typists probably memorized the keyboard

and used all ten fingers and thumbs, but one stood out. Frank R. McGurrin, a young law clerk in Grand Rapids, Michigan, in 1878 competed with his employer on the office's Model 1 Remington typewriter. The idea of touch typing was originally a joke by his employer, who claimed that a local court stenographer's typist was able to work rapidly from dictation while looking out the window. As McGurrin later recalled, "I made up my mind that whatever a girl could do I could do," and he started to use all his fingers instead of two or three on each hand. By the end of the year he was able to type ninety words per minute—still considered an excellent speed now—from new copy without looking at the keyboard. Only two years later did he meet the stenographer's typist and learn that she had never even tried the new method. His employer's teasing provocation had spurred him to discover the obvious: a musical keyboard style could be adapted for text.[23]

By the 1880s, typewriter manufacturers were sponsoring speed contests, and newer machines needed less force, making higher speeds possible. McGurrin became known as the fastest Remington operator. In an 1888 contest in Cincinnati, he confronted Louis Traub, an agent for the Caligraph, possibly the best selling of the two-keyboard machines. McGurrin won decisively, achieving ninety-five words per minute from dictation and ninety-eight from copying against Traub's eighty-three and seventy-one. Traub's lower score on copy helped doom the Caligraph. Because its keys were too widely dispersed to permit the touch method, he had to take his eye repeatedly from the keyboard. Traub soon replaced all the Caligraph machines in his typing school with Remingtons and began instruction in touch typing, signaling the end of the double keyboard. McGurrin continued on the exhibition circuit, promoting Remington as he achieved up to 125 words per minute.[24]

For all his prowess, McGurrin had no students and published no books. Mrs. M. V. Longley of Cincinnati, whose husband operated a "Shorthand and Typewriter Institute," developed her own system of touch typing in the early 1880s after seeing a Remington and reading in its brief instructions the manufacturer's advice to use only one finger of each hand, at most two fingers of the right hand. She taught classes with the all-finger method she developed, and published the first textbook for its use in 1882. Citing piano and organ fingering, she observed: "If the hands are held over the keyboard the fingers will reach the extreme right and left, and each be in a position to do duty and the thumb will be in readiness to strike the space bar." This method, now self-evident and ubiquitous, was a bold departure. The goal was not so much raw speed as the ability to tran-scribe accurately while looking at the copy rather than the keys.[25]

Mrs. Longley had retired by the mid-1880s, but a Maine shorthand and typing instructor named Bates Torrey published an even more influ-ential manual, *Practical Typewriting,* in 1889. Citing McGurrin, and also blind typists who had mastered the all-finger method, Torrey presented a powerful case for keeping the hands in fixed positions and typing a given word consistently with the same finger motions. It was also Torrey who recognized the importance of touch, and who first applied the word "touch" to all-finger typing. The success of his book, designed for self-instruction and correspondence as well as school use, inspired a wave of other texts, and the touch method spread across the Western states. One author, A. C. Van Sant of Omaha, developed the system of fingering that has been taught ever since and is still the basis of computer typing CD-ROMs. Within about twenty-five years of the modern typewriter's intro-duction, the community of instructors and typists had stabilized a technique.[26]

A TECHNOLOGICAL FOSSIL AND ITS CRITICS

The Remington typewriters used by McGurrin and the early textbook writ-ers arranged their letters in a pattern that by the turn of the century was called the Universal Keyboard. Christopher Sholes had begun with a purely alphabetic arrangement but had revised it because the hammers, which were arranged in a circle in the earliest Remington, tended to jam. Sholes and Densmore worked out a new arrangement of keys for their No. 2 Remington machine, patented in 1878. Just as the musical keyboard was stabilized before 1500, its textual counterpart remains almost identical to the diagram on the patent.

Nobody has been able to reconstruct Sholes's and Densmore's reason-ing completely. It would probably be necessary to find an operating Model 1 or 2 typewriter and experiment with combinations of letters. The QWERTY keyboard, as it came to be known, was clearly a compromise. On the middle row of text there was a nearly alphabetical sequence: DFGHJKLM. The last letter was later moved to the bottom row, where the original C and X were also later reversed. On the top row was a vowel clus-ter (UIO) out of alphabetical order. Sholes and Densmore were both famil-iar with newspaper type cases, arranged not in alphabetical order but roughly according to letter frequency. The QWERTY keyboard did not follow these patterns but was conceived in a similar spirit.[27]

Sholes and Densmore made a fateful assumption about the operator's technique. Compositors used thumb and forefinger and looked at the type case as they worked, and it seemed reasonable to think that typewriter operators would do the same—as, indeed, all but a few initially did. For this style of work the QWERTY keyboard was relatively efficient. Its leading twentieth-century critic, August Dvorak, found that the most frequent letters were typed with the first two fingers of the left hand and the index finger of the right. There seems to be a balance between putting all the most frequent characters near the center of the keyboard and maintaining an order that will make it easier to find keys visually, like keeping O and P as well as the middle-row sequence together.

As early as 1875, proposals circulated for more efficient keyboards. Once touch typing prevailed, it would have been logical to look for even greater speed and comfort by devising a new arrangement. Yet neither McGurrin, nor Mrs. Longley, nor Torrey is known to have proposed any modification. It is unlikely that they feared devaluation of their skill by keyboard reform. McGurrin became a well-to-do banker in the West, Mrs. Longley left teaching, and Torrey and others could have sold new editions of their textbooks. The QWERTY keyboard was not the best imaginable, even in the late nineteenth century, but it was good enough that expert typists did not feel frustrated by it. Remington wisely did not try to monopolize its keyboard arrangement under patent law. And instead of developing their own proprietary systems, Remington's major emerging competitors, Underwood and Royal, kept the Universal Keyboard. The innovators were producers of interchangeable type element machines, notably George C. Blickensderfer. The Blick's home row, DHIATENSOR, was claimed capable of spelling 70 percent of English-language words. But even Blickensderfer offered the Universal Keyboard as an alternative option.[28]

The Universal Keyboard was spreading just as the Jankó piano keyboard was failing in the musical marketplace. Most typists were no longer independent writers, attorneys, or court reporters but part of a large labor pool operating machines they did not own. Just as concert halls needed pianos with standardized keyboards, businesses wanted to be able to engage typists without having either to replace or refit the machines, or retrain their recruits. Typists and typing students likewise wanted to learn the most widely accepted arrangement. Later the economist Paul David called such pressures for standardization "network externalities" and cited the QWERTY keyboard in an influential paper arguing that historical contingency can, paradoxically, lock inferior technology into place.[29]

During the early twentieth century, those who challenged the QWERTY design came from outside the office machine business. Apostles of industrial rationality, they would have found it inconceivable that the dead hand of the past could constrain the invisible hand of the marketplace. Frank and Lillian Gilbreth, who developed new methods for analyzing human skills, took an interest in the typewriter, producing equipment and techniques for people with disabilities. They also took notes on the sequence of motions in typing and made slow-motion study films of championship typists at work. One of Gilbreth's assistants in his typewriter work was William Dealey, who later became a professor of education in Texas. Dealey's brother-in-law, August Dvorak, taught educational psychology at the University of Washington.[30]

Dvorak is best known today as the inventor of the leading alternative keyboard design, but he was interested in everything that could affect typing productivity, including posture and fatigue. In the text that he wrote with Dealey and two typing teachers, Nellie L. Merrick and Gertrude Catherine Ford, and in workbooks for students, Dvorak actually devoted little space to keyboard reform. He was more concerned with every detail of skill regardless of keyboard design, especially developing a rapid, light touch ("ballistic motion") and reducing tension. He was a connoisseur of techniques, showing aspiring typists and typing teachers how they could benefit from the "looseness and lightness" of top performers in fields as different as tournament golf and window washing. All of this he supported with citations from the latest in the growing psychological literature of typing behavior. Dvorak's book, sponsored by the Carnegie Foundation for the Advancement of Teaching, is still impressive for its unpretentious but authoritative manner, and for its belief that typing students could participate in the scientific advancement of their own skills. And Dvorak also fought some dogmas of traditional typing instruction, especially exaggeration of the importance of touch and the "queer bias for petty accuracy."[31]

Keyboard reform was still important to Dvorak and Dealey. They were convinced that the universal keyboard was an obstacle to good technique. It was expert typists who first noted the problems of QWERTY. A speed coach interviewed by the Gilbreths had already remarked that "[t]he young ladies think that it is the mechanical construction of the machine that is

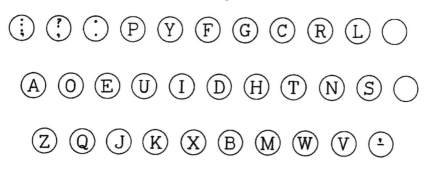

August Dvorak, a disciple of the time and motion expert Frank Gilbreth, devised, with William L. Dealey, one of a number of reform keyboards promising new efficiency in writing, as Paul von Jankó's keyboard did in musical performance. Dvorak's claims for higher speed proved exaggerated. Comfort was the real advantage of the arrangement. From U.S. Patent 2,040,248, May 12, 1936.

retarding their speed." The speed coach went on to note that most of the writing was done with the left hand and that too many words were typed by that hand only. Dvorak went on to estimate that overcoming the defects of QWERTY accounted for half to three quarters of instruction time. Work was divided unevenly among the fingers. Overloading the left hand interrupted the ideal rhythm of alternating use of left and right. Too many combinations of two letters (digraphs) were made by adjoining fingers. Eighty-five percent of digraph combinations had to touch the top row, while only 7 percent could be made on the home row.[32]

Rejecting previous revised keyboards, Dvorak and Dealey developed and patented a new arrangement. Vowels were relocated to the home row of the left hand, and the consonants most often used were on the right, so that hands had to alternate with each syllable. Loads were adjusted to the capacity of each finger. The largest source of errors, digraphs made by adjoining figures, was cut by 90 percent. Much more work could be done on the home row, cutting "hurdling"—the scurrying of fingers over the keyboard. (Only a hundred common words can be made on the home row of the Universal Keyboard, three thousand with the simplified.) Dvorak and Dealey's studies in junior and senior high schools suggested substantial gains: an average of 48 words per minute in the second semester, for example, versus a maximum of 28.4 w.p.m. in comparable studies with a Universal Keyboard.[33]

The Dvorak and Dealey layout, patented in 1936, attracted great interest but appeared at the worst possible time. The typewriter market was

glutted in the Depression. Dealers were even buying and wrecking func-
tioning old machines to suppress the used-typewriter trade, but the major
brands were solidly built and with proper maintenance could last for
decades in normal office use. Schools, the most promising market following the results of the Carnegie-funded experiments, could hardly afford
conversion, especially when there was no sign of acceptance in commerce.
Wages were falling, and there was a surplus of typists. Dvorak still tried to
convince corporations that the new arrangement could reduce their costs
by eliminating typing positions. But typists influenced purchases by their
employers, and the new arrangement seemed to threaten speedups and
layoffs rather than the relaxed high performance that the Dvorak textbook
preached. And Dvorak made yet another strategic mistake, linking the
simplified arrangement with the Remington so-called noiseless typewriter.
Dvorak praised the design for its encouragement of his ideal light staccato
touch; like Frank Lloyd Wright's three-legged chair, it would fail if the
user's body technique was incorrect. But Dvorak overlooked the importance to typing rhythm of the auditory feedback provided by the typebar's
clatter. The Remington also made less crisp impressions, and the mechanism could feel mushy; even Dvorak acknowledged in his text that its feel
"disturb[ed] even expert typists" before they got used to it. Receiving a
U.S. Navy commission in World War II, Dvorak sponsored studies that
appeared to prove that added productivity from the simplified keyboard
would quickly repay the costs of retraining, but his recommendations
were ignored. All 850,000 typewriters bought by the U.S. government
during the war were QWERTY machines.[34]

Even foreign manufacturers failed to challenge the QWERTY arrangement. In France there was strong opposition to the local version, AZERTY.
An influential journal, La Revue Dactylographique et Méchanique, established a commission that proposed a layout designed especially for the
French language. One company, the delightfully named Manufacture des
Armes et des Cycles, adopted it for its own version of a British design in
1909. Unfortunately for its advocates, the French typewriter market
remained small by international standards; in 1913, France produced
4,000 machines and imported 28,000, mainly from U.S. makers who had
standardized on QWERTY. A leading French psychophysiologist and foe
of Taylorism, Jean-Maurice Lahy, likewise tried to challenge American-
style ten-finger typing, but had limited influence. By the end of the 1930s,
the technique of touch typing and its associated technology, the QWERTY
typewriter, had spread with only minor variations throughout the world of

the Roman alphabet and had influenced even non-Roman national keyboards. Reform proposals from as far away as India had been defeated like
alternative musical keyboards, and for similar reasons. Typists formed a
community with hard-won motor reflex skills; they were reluctant, in the
business world, to trade certain proficiency for uncertain gains, no matter
how critical of QWERTY they could be in laboratory settings like the
Gilbreths'. While Dvorak continued to fight for his design through the
1960s, he had been defeated twice: not only in the wartime military market but in the early postwar civilian boom. An incisive analyst and accomplished teacher, he lacked the skills of bureaucratic infighting that might
have implemented his ideas from above, and the knack for public relations
that could have excited enthusiasm from below. There would be one more
window for innovative keyboards in the early microcomputer era. But first
it is time to examine how the keyboard became a computer device in the
first place.[35]

Alphabetic Keyboard Loading by Hand

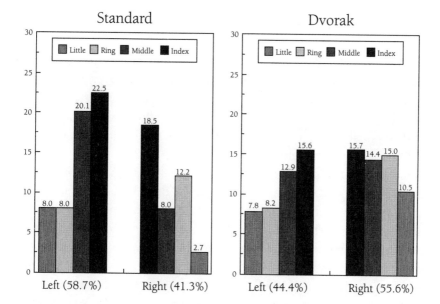

*August Dvorak calculated the data shown here to prove that his key arrangement
evened the distribution of work among the fingers. But his data also show that the
conventional arrangement was well conceived for the original four-finger style of
typing, before typists and typing teachers worked out the five-finger system. Only
the middle finger of the right hand appears underutilized. (Courtesy of Scot Ober)*

THE BODY ELECTRIC

The electrification of the keyboard began at a surprisingly early date. We have already noted the rise of electrified typewriters in telegraphy. In 1902, George Blickensderfer introduced the Blickensderfer Electric, with an interchangeable type element and keyboard control of underscoring, line spacing, carriage return, and tabulation. Hammer force could also be adjusted from the keyboard to produce as many as fifteen carbon copies. The inventor's only mistake was ignoring a different set of network effects: utilities were still supplying electricity only after dark, and the machine's power cord was designed to be screwed into a lightbulb socket. Blickensderfer's death in 1917 left his ideas orphaned by the company's successors, but Remington and Electromatic (absorbed by IBM) began making commercially successful machines in the 1920s.[36]

By the 1930s, IBM was using typewriter-style keyboards for card punching; even early scientific computation used stacks of punched cards, and the keyboard was already an important input device. Yet well into the 1950s, keyboard control was far from universal. Scientists and engineers continued to rely on slide rules for personal computation until inexpensive handheld calculators with keypads appeared in the 1970s. (The ingenious but costly portable mechanical alternatives like the cylindrical Facit used slides and cranks.) The nineteenth and early twentieth century developed wonderful precision instruments for calculating data such as the tides and the areas enclosed by irregular curves. The electrical engineer and scientific administrator Vannevar Bush even built a machine for solving differential equations, applied to problems from atomic structure and ballistics to railroad timetables.[37]

It was digital processing that wedded the QWERTY keyboard to the computer. Early World War II and postwar digital machines used keyboard-encoded punch cards and tape, first paper and then magnetic, and teletype-style printers. More and more computer functions could be managed by typing rather than by rearranging plugged wires and setting dials. Academic scientists began to use time-sharing systems controlled entirely by teletype. The crucial date for the keyboard, though, came in the later 1960s, when the Multics time-sharing system developed by MIT, Bell Laboratories, and General Electric replaced the clacking teletype with a video display terminal. In 1972 the addressable cursor appeared, and keyboard control sequences were standardized. In the late 1970s, when microchip costs had dropped low enough to make hobbyist microcomputers possible, the obvious control hardware was the keyboard plus monitor. The Altair, which employed switches, was soon overtaken by new consumer brands running the BASIC programming language with a video display terminal and keyboard similar to those of the Digital Equipment Corporation VT100, ubiquitous in industry and academia. With the expansion of personal computing in the 1980s, accelerated by improved Web browsers and Internet service from the mid-1990s, the QWERTY keyboard has become a universal interface.[38]

But along with this victory of QWERTY has appeared new hope for reform. Electronic keyboards can be reprogrammed with free software for the Dvorak layout; ever since version 3.1, Microsoft Windows has joined the Macintosh in offering built-in Dvorak support. Keys can easily be relabeled, and moderately priced replacement keyboards offer dual labels and push-button toggling between layouts. From the mid-1980s to the early 1990s the future of the Dvorak layout seemed brighter than ever.

One early hope was healthier keyboarding. As I discussed in *Why Things Bite Back,* the exposure of millions of professional and amateur typists to the new equipment in the 1980s produced an unforeseen epidemic of overuse injuries of the hand, most of them known as carpal tunnel syndrome (CTS). The condition's origins and treatment are still imperfectly understood, including the parts played by physical anomalies and psychological states of users, by working conditions and employer attitudes, by variables such as the response of keys to fingers, and by use of the mouse. The apparent efficiency of the computer keyboard clearly played a part. The manual typewriter had demanded a healthy elevated position for the wrists and had enforced breaks for throwing the carriage and for removing and inserting paper. Computer keyboards permit dangerously rapid and uninterrupted data entry coupled with awkward wrist placement.[39]

The Dvorak layout, which reduces finger travel dramatically, at first seemed an ergonomic blessing. People with CTS and related ailments have praised the arrangement. Yet its medical advantages still have not been proven. Without significant numbers of Dvorak users in organizations it is difficult to gather statistics, and voluntary Dvorak adopters probably are different on average from their QWERTY colleagues: perhaps more affluent or health conscious, but also possibly more concerned about CTS and other conditions after previous episodes. Since total reported cases of CTS dropped from 41,000 in 1993 to 26,300 in 1998, most of them industrial, prevention in the office has seemed less urgent. The schools would be a logical place to test the benefits of the Dvorak layout, as computer-related injuries have been growing among young people and there is no doubt that the Dvorak system is much easier to learn. But few schools even appear to offer Dvorak instruction.[40]

Productivity is another matter. Dvorak's own writings, reflecting the Gilbreths' influence, assumed that adults would be able to sustain the advantage of up to 40 percent in favor of his system that he found among students. Efficiency is easier than health to monitor in controlled experiments. And when a new round of studies began early in the personal computer era, their results were surprising. In 1982 the cognitive psychologists Donald A. Norman and Diane Fisher compared computer simulations of QWERTY, Dvorak, and alphabetical keyboards. Norman and Fisher found that although the Dvorak arrangement did indeed save on motion as its advocates had long claimed, gains in speed were modest: the advantage was only about 5 to 10 percent. They found the long-maligned QWERTY keyboard surprisingly rational in its high number of alternating-hand sequences. The Norman studies and others bolstered an influential 1990 rebuttal to Paul David's analysis by two fellow economists, S. J. Liebowitz and Stephen E. Margolis, who argued that the Dvorak arrangement had lost on its merits.[41]

The critics of the Dvorak layout have a point. Typists' minds are able to manage the additional 37 percent finger travel of the QWERTY keyboard without a corresponding loss of speed. Differentials range from a mere 2.6 percent for Dvorak, to 11 percent. A 1980 Japanese study suggested 15 to 25 percent faster performance in timed writing and 25 to 50 percent faster production of letters, reports, and tables. Scot Ober, a professor of business education rather than a cognitive psychologist, believes that training methods account for some of the difference. Dvorak was a master teacher and developer of instructional materials. To be truly fair, an experimenter would have to provide equally expert textbooks and workbooks, which are not commercially available for the Dvorak program. Even in Seattle's Keytime typing instruction school, known for the innovative methods of its founder, Linda Lewis, the great majority of students choose the QWERTY option. There are also large professional groups, notably computer programmers, for whom neither QWERTY nor Dvorak is an efficient arrangement, especially now that the Internet has revised the importance of letters (like W) and signs (like @).[42]

Innovative keyboard designs have not fared much better than new key patterns. As early as the 1920s, a German industrial health specialist recommended that keyboards be split in the center and tilted like the roof of a tent, to keep the hands in a more relaxed, natural position. In the days of mechanical typewriters this was a prohibitively expensive concept, but the rise of the computer terminal in the 1960s made it feasible. The American ergonomist Karl Kroemer revived it in the 1970s. Since then, a dozen or more manufacturers, including IBM and Microsoft, have offered such designs, but they form a tiny part of the market. The barrier is no longer the technology of manufacture but the investment of users in their technique. Our muscle memories are accurate to within a millimeter, and a new arrangement can be discouraging at first. A few brands, such as Goldtouch, can be adjusted very gradually, but they are rarely offered as original equipment and are sold mainly over the Web and by small independent distributors. One of these, Neal Taslitz, is an ergonomic activist dismayed by the obstacles that small manufacturers face in the marketplace, especially in access to national retail chains. He finds that corporate buyers of computing equipment are so cost-conscious that keyboards are one of the first items on which makers—even prestigious top-tier companies—economize. Thus IBM first spun off and then discontinued using its award-winning buckling-spring technology, famous for a distinctive clicky feel. Most keyboards today are almost disposable. The remaining high-quality, high-price designs seem to be bought by a few affluent individuals and overuse-injury sufferers: the very clientele attracted to the Dvorak key arrangement, though the Dvorak layout and new keyboard contours oddly are seldom combined in the same product. Whatever the reasons, the keyboard is virtually the only part of a computer that has stagnated or declined in performance.[43]

KEYBOARDERS ALL

The response of some computer industry leaders to the keyboard's problems has been the dream of its abolition. When the Macintosh graphic user interface arrived in the 1980s, the mouse at first took over some of the commands of the keyboard, but many users soon clamored for function keys and keyboard shortcuts. In the late 1980s and early 1990s visionaries at Apple and elsewhere called for a new kind of computer that would recognize handwriting on a portable tablet. While the movement ended in a wave of failures, notoriously of the original Apple Newton, pen-based computing never actually died, and both Apple and Microsoft are doing their best to revive it. Yet the smaller the portable device, the more likely it appears that someone will find a way to plug a portable QWERTY keyboard into it. Several are available for the Palm Pilot personal digital assistant (PDA). Even graphically oriented Apple included a detachable keyboard with the last version of its Newton MessagePad—as well as a Unix-based text command option in its latest Macintosh operating system. And PDAs without plug-in keyboards are routinely connected to conventional computers for keyboard-entered data.[44]

Voice control was as exciting in the late 1990s as handwriting recognition had been ten years earlier, and with similar results: a wave of marketing and financial troubles that has failed to shake underlying optimism about the technology's future. While voice recognition software can promote its own overuse injuries, it can now work with natural phrasing and infer spelling from context. But it is unlikely to eliminate the keyboard, because it will still make errors (or users will still fail to enunciate properly), and editing copy orally is even slower and more tedious than correcting it with a keyboard. Optical character recognition data also need checking and editing. Typing will probably be further reduced in familiar applications in the future, but it will also be extended to new tasks.

A new global keyboard order is emerging. Intensive "production" typing is less necessary because more data arrive already digitized and need only formatting and correction. The heavy typing that remains can, like the production of keyboards themselves, be outsourced to low-wage countries. Young employees have no fear that typing skills will trap them in dead-end jobs, as used to happen to women. The generation of senior executives who regarded a keyboard near their desk as a breach of caste purity has long retired. Even professions famous for their attachment to real slate blackboards and hand-lettered transparencies, like pure mathematicians, now are as likely to be typing away like anyone else, thanks to software that sets the most complex equations.

While visionaries still dream of tiny wireless devices, the twenty-first century remains faithful to Sholes's layout. It replaces the card catalogue in libraries and the posted directory in office buildings. It controls industrial processes and home entertainment systems. Many people in large cities actually prefer the numerical keypads and monitors of automatic teller machines to human tellers. Public and private dial telephones are rare, keypads almost universal. Scientific pocket calculators sell for less than slide rules ever did, not just for text production and financial analysis but for the control of everything from industrial processes to entertainment options. Asian countries use combinations of Western keys to form their own scripts, and a new standard called Unicode lets the keyboard generate every character of each important world language.

Keyboards have become more "universal" than ever imagined by the frenetic pitchmen of the nascent typewriter industry, those bold pioneers of high-technology salesmanship. They have fulfilled the dreams of millions of nineteenth- and earlier-twentieth-century schoolchildren, drastically reducing tedious hours of penmanship practice. But thereby they have also deprived millions of adults of a potentially graceful and satisfying skill, reducing much handwriting to illegible cursive and childlike printing. The neurologist Frank R. Wilson has argued persuasively that excessive substitution of computer use for manual activities can impoverish childhood. And just as the rivalry of revisionist keyboards helped maintain QWERTY, the division of handwriting advocates into antagonistic cursive and italic camps has inhibited the reintroduction of handwriting lessons in the United States. There are even calls for laptop computers for all students, as there once were movements to give them typewriters. From the early nineteenth century, the keyboard has been a deliverance for men and women with disabilities, now restoring even speech with voice synthesis. But excessive keyboard use by children has also given them a needless handicap, inability to express themselves rapidly and legibly with a pen or pencil. Researchers of at least one U.S. university are now developing new generations of secure computer-based tests, convinced that word-processing and electronic mail have made handwriting almost unnatural.[45]

We have seen that wearing shoes and sitting in chairs puts us on a technological treadmill: enhancing our comfort and expanding our capabilities, these body technologies also increase our dependence on themselves. Handwriting has become like barefoot walking, satisfying with practice but at first painful to the adult relearner. There has been a cultural ratcheting, too: visual arts, poetry, music, and even intimate personal communications have been subtly influenced by an interface from the counting house. Perhaps the process began when Mark Twain's friend, the humorist Petroleum V. Nasby, watched a typewriting demonstration with Twain in Boston in 1874 and abandoned writing and lecturing to join James Densmore's marketing organization.[46]

Twenty years after Nasby's career change, a typist named Lucy C. Bull published an essay in the *Atlantic Monthly* on the machine's impact. Her remarks are even more profound today then when she wrote them. The device "which is beginning to supplement the labors of clergymen, lecturers, and contributors to the magazines" was, after all, a business tool, "constructed almost entirely in accordance with the demands of commerce." She noted that at least one man had dictated a letter to his fiancée, and that accents and other literary fine points were already in decline. But, most interestingly, she foresaw the universality of the keyboard even before electronics let it overcome its mechanical limits; she understood that the professional keyboard operator would be unnecessary when all had their own machines. The phrase that she used as her title, melding device and operator, might also be a motto for the twentieth and twenty-first centuries: "Being a Typewriter."[47]

CHAPTER NINE

Second Sight

Eyeglasses

S O FAR, we have considered objects widely supposed to enhance human performance: the bottle to improve infant nutrition (although science later proved otherwise), the sandal and shoe for protection from the hostile surfaces other technology has created (although we now know the case for barefootedness), the office chair and reclining chair for optimal posture and health (although slouching is still endemic and the recliner has become an emblem of the sedentary life), the keyboard for more efficient production of both sound and text (although both musical and computer keyboards are still far from ideal).

Eyeglasses are like other body technologies in their potentially far-reaching, even deforming effects. Just as sitting and keyboard affect posture, so do spectacles. (In fact, the satirist Barry Humphries, in the person of his character Dame Edna Everage, calls them "face furniture.") Many people have one ear slightly higher than the other, which tilts their glasses, or a nose that lets the glasses ride downward. Adjustments to compensate for these problems, such as inclining the head sideways or thrusting it forward, and breathing through the mouth, can lead to headaches, neck, and facial joint pain, according to a rehabilitative medicine specialist. Like footwear, poor vision and spectacles can also change walking patterns. The extremely myopic Friedrich Nietzsche took short, high steps by his fifties to avoid stumbling, and in the nineteenth century there were even attempts to train horses in a fashionable, high-stepping gait by fitting them with distorting spectacles. (Conversely, some myopic people, unable to recognize friends and acquaintances at a distance by their features, become expert at identifying their unique carriage.)[1]

Correctly fitted and worn, though, glasses are different from the other technologies we have considered. They are corrections, not deformations or even improvements. Except for certain low-power magnifiers that work like eyeglasses, the optics that extend our senses—binoculars, telescopes, microscopes—are rarely attached to the body. Eyeglass wearers, an ever-increasing part of humanity, become different from other people in appearance so that their performance can be as similar as possible. And despite their corrective purpose, glasses can also affect the body and spirit as well as the face of the wearer, becoming a mask difficult to separate from the person. Sir Arthur Conan Doyle (a well-trained general practitioner and failed ophthalmologist) had Sherlock Holmes observe in "The Adventure of the Golden Pince-nez": "It would be difficult to name any articles which afford a finer field for inference than a pair of glasses." And toward the end of the story the killer's appearance confirms Holmes's deductions from the glasses: thick nose, close-set eyes, peering expression, rounded shoulders. In fact, a pair of eyeglasses can evoke not just a face but a realm of speech and gesture. The elaborate clothing and makeup Humphries wears in character as Dame Edna are not the key to his performance. "As soon as the glasses go on, I cease to be myself," he declared to one writer.[2]

Optics are like the seating we have studied, growing in importance with the spread of information and literacy. We think of reading, like shoes and chairs, as one of the unquestioned goods of civilization. But just as we come to depend on footwear and seating, as we lose the natural protective layers of callused skin on our soles and cease to be able to sit comfortably on the ground, we also lose the unmediated contact with the world that hunter-gatherers enjoy. At least in later life, nearly all of us need artificial aids, and society depends economically on their existence. The type sizes of many of our printed materials presuppose that large numbers of readers will use spectacles to read them. The insides of both mechanical and electronic devices have parts that require some degree of magnification. And while computer monitors can magnify type as conventional printed materials never could, it is also true that new generations of handheld mobile devices set severe limits on the size of type. So eyeglasses, like keyboards, are essential interfaces of the body as well as the mind with the environment.

Spectacles might have been invented two thousand or more years ago, when the principles of lenses were already known. It took a certain kind of society that emerged in medieval Europe to create the conditions for their spread. Respected and mocked for centuries, glasses express both deficiency and superiority. Our techniques of gathering and spreading information rely on the technology of optics, and we are still not completely at ease with the consequences.

The Garden of Eden myth is more than a Judeo-Christian fable. It is also a metaphor of the surprisingly healthy and leisurely lifestyle of preagricultural antiquity, or at least parts of it, that archaeologists and anthropologists have been discovering. Our Paleolithic ancestors were bigger and stronger than we are and had sturdier and healthier teeth. Archaeology can tell us little directly about their vision, but there is reason to think it was acute. Their survival would have depended on their ability to recognize game and identify edible plants at a distance, though the lenses of their eyes no doubt hardened like ours, especially after the age of forty, making it more difficult for the ciliary muscle to focus on near objects. But their perceptual world—*Merkwelt,* as the behavioral biologist Jakob von Uexküll called it—was in the middle range and beyond. And besides, despite their robustness, relatively few survived into their forties.

Whatever their other afflictions, few of our ancestors would have been myopic. Myopia or nearsightedness is the focusing of light in the eye at a point in front of the retina, making far objects indistinct and leading to eye rubbing and squinting. Some people are myopic at birth, but most cases develop in childhood or early adulthood, when the eyeball begins to elongate before the lens and cornea have grown to focus on it.

Myopia appears to have been rare among hunter-gatherers, to judge from the handful of twentieth-century peoples whose vision has been studied. Australian aborigines have an extremely low rate of myopia, for example, and as a group are mildly farsighted; even comparing individuals with normal sight, native Australians can see significantly more sharply than Australians of European descent. Almost equally negligible myopia rates have been found on the South Pacific islands of Vanuatu, Bougainville, and Malaita; among the forest people of Gabon; and among the Eskimos of Greenland. Among Alaskan Eskimo families studied in the 1960s, adults who had always lived traditionally had a 2 percent myopia rate, which increased in their schooled offspring to more than 50 percent. Not that illiteracy guarantees good eyesight. Some old agricultural societies without mass education, like World War I–era Egypt, have been relatively myopic. And the Egyptians' neighbors, migratory Nubians and Bedouins who included many devout readers of the Koran, were rarely nearsighted. One fact remains: no hunter-gatherer people has ever been shown to have a high rate of myopia. While the studies are few and methodologically imperfect, they suggest that some ways of life—

especially urban and literate ones—relax the pressure that usually selects for good eyesight.[3]

Why do the children of visually acute hunter-gatherers begin to show higher rates of myopia soon after they receive a Western education? One recent study suggests that refined starches introduced by Europeans may raise insulin levels, changing the developmental program of children's eyes. Whatever other genetic or physiological influences shape myopia, reading and other detailed visual work probably play some part, though specialists still do not understand the mechanism. Reading—perhaps especially reading of finely detailed ideographic scripts like Chinese and Japanese—involves the central area of the retina to the exclusion of other neurons. The resulting blur in the rest of the visual field, unmarked by the reader, may release neuropeptides that overstimulate the elongation of the eye. The eye may also elongate in keeping close work in focus. This frequent explanation is plausible but incomplete. It does not suggest why the lengthening is not equalized by long-distance focusing that should tend to shorten the eye.[4]

Whatever the details of the mechanism, the technology of writing and the technique of reading appear in most societies to have an unintended and unwanted influence on the developing body. In our efforts to see the fine print of life more distinctly, whether legal niceties or technical drawings, literate peoples are apt to blur the big picture. And indeed, cross-cultural psychologists have found that urban living leads to a superior grasp of detail at the expense of the ability to detect larger patterns. Culture remakes our perceptions as well as our visual apparatus. A spectacle case belongs to the technological baggage of modernization.[5]

BEFORE SPECTACLES

Every literate society has two overlapping populations whose vision needs correction. One is the myopes, who are usually comfortable at reading distance but whose longer sight is indistinct. The other group is the presbyopes, men and women whose eye muscles can no longer focus their lenses properly for near work and require artificial refraction. Almost all men and women over forty need some assistance, yet it took a surprisingly long time after the introduction of writing for humanity to use magnifiers routinely for reading, let alone wear eyeglasses. Artisans and artists in the ancient Mediterranean and early medieval Europe must have used magnifiers on occasion. Jay M. Enoch, a professor of physiological optics and historian of technology, has found a number of objects that were fashioned with converging plano-convex lenses to allow extraordinary detail to be viewed with the naked eye. A bull's-head rhyton (drinking vessel) of about 1550–1500 B.C. has a minute silhouette of a human head visible through the "cornea" of one eye. Greek rings of 350–300 B.C. have gold-foil images that appear to float in the clear glass that covers them. And a supposed splinter of the True Cross in Rheims is sandwiched between two concave magnifying jewels.[6]

The Romans were familiar with the working of rock crystal (clear quartz) and, later, glass. Greeks and Romans made double convex burning glasses, already mentioned in Aristophanes' *Clouds*. But they never developed eyeglasses. As the classical scholar Nicholas Horsfall has observed, they had expert assistance from literate slaves and freedmen: a *lector* read texts, and a *scriba* or *notarius* (secretary or stenographer) took dictation. Roman winters were dark and the lamps crude and smoky by modern standards; Romans also suffered from frequent eye inflammations. Yet the elite could enjoy their own form of mobile voice recognition and word processing, dictating while walking, traveling, or in the baths. Readers and secretaries were part of the entourage of leading political figures, on call around the clock but rarely acknowledged as individuals. The church fathers maintained this pattern, turning to monks and devout young laymen.[7]

Most specialists believe that medieval as well as ancient craftsmen used magnifiers for their fine work. Even today, the detail needs enlargement to be fully appreciated even by the keenest-eyed museumgoer. It is true that none of the many images of medieval scribes in their workshops illustrates such an instrument. Some believe that very myopic young workers might have needed no optical aid, but so many exquisitely detailed masterpieces could hardly have been journeymen's work. Lenses in the form of polished rock crystal and other minerals, or concave mirrors, were probably trade secrets. Artists' use of devices remains controversial even in the generally technophile twenty-first century, as painters and scholars debate whether such giants as Caravaggio, Vermeer, and Ingres employed imaging devices like the camera lucida. An illuminator known to work with a crystal or gemstone might not have been disgraced in the eyes of royal and aristocratic patrons, but his work could have seemed less masterly and less valuable.

Medieval philosophers were excited about the principle that refraction, the bending of light, could make objects appear larger, but until the later thirteenth century they were probably unaware of the lenses their contemporaries were using. It was then that the friar Roger Bacon first wrote of transparent materials like crystal or glass held above text by those with vision problems. But even Bacon did not analyze the design of corrective optics and was probably thinking only of the magnifiers already in use.

WHY GLASSES?

It was not natural philosophers like Bacon who made the first eyeglasses but unnamed artisans in Pisa and Venice, and possibly in France and Germany as well. In 1306, a monk named Giordano da Rivalto, of Pisa, praised eyeglasses in a sermon delivered in Florence, declaring that their manufacture, "one of the best arts and most necessary that the world has," originated twenty years earlier with an unnamed man he knew, probably a Pisan craftsman. True to the practical side that we have noted in medieval religion, another monk in the same Dominican monastery was able to duplicate the secret process by 1313. Even in Venice, where the crystal-workers' guild had been producing crystal magnifiers, the Pisan term *occhiali* (first documented in Giordano da Rivalto's sermon and still the Italian word for eyeglasses) prevailed.[8]

Largely ignored by the learned students of optics, Florentine spectacle making benefited from the presence of a community of artists vividly interested in the laws of perspective, and of a dynasty well known for both lavish patronage and chronic problems. Eyeglasses became a thriving Florentine trade, even an industry by the standards of the day. The princely courts of Italy placed large orders. By 1462, Duke Francesco Sforza of Milan was praising Florentine spectacle workmanship as the best in Europe and ordering through his diplomats three dozen pairs at once for his enthusiastic courtiers. One or more workshops were able to fill this order within two weeks. His son and successor purchased as many as two hundred pairs at once. Nor did Florentine optics demand a princely budget; at a ducat per dozen, they were within the reach even of the skilled artisans of Florence. Further, a dozen of the spectacles the Duke ordered had concave lenses "for distant vision for the young": this is the first clear reference to the production of lenses for nearsightedness. This suggests that at least in Italy, both myopia and the technology for correcting it had been growing. (Both the Duke and his ambassador to Florence, a distinguished humanist, were myopic.) The Florentines were also the first to systematize correction of presbyopia. Reading glasses were classified by five-year age increments, and grades of myopia were recognized. By the seventeenth century, Galileo and other scientists were promoting the development of new, higher grades of optical glass shaped by the regulated heat of oil lamps.[9]

(Few descriptions of European lens making before early modern times survive. According to Dennis Simms, a historian of optical technology, the earliest method of pouring glass into molds for lenses originated in the Islamic world a thousand years ago, and from 1550 to 1950 it was refined by attaching blanks to templates and grinding and polishing them with other templates of opposite curvature. Vincent Ilardi, who has studied the economics of the Renaissance Italian eyeglass trade, adds that this process required little skill once blanks and molds for appropriate correction could be duplicated. Despite automation of the optical industry in the last fifty years, a cottage industry survives in the hills of the Veneto.)[10]

Florence also had an ideal legal climate for a new technological field. Venice guarded spectacle making as an extension of its luxury glass industry, even threatening to imprison relatives of workers who emigrated. In Florence, there was no separate guild of spectacle makers, and artisans openly proclaimed that they would teach their craft to women and even to little boys and girls (*putti*). It was an atmosphere ideal for the transfer of skills and knowledge. But while Florence was outstanding, it was not unique. Other cities and monasteries in Italy also had flourishing opticians, and spectacle making spread to northern Europe.

Was it a coincidence that the spectacle trade was flourishing not just in Florence but in cities and monasteries throughout Italy at the very time when paper mills and printing presses had been spreading through Europe? Printed matter and eyeglasses served the same markets: priests, laymen, clerks, students, merchants, artisans, all avidly reading or doing other skilled detailed work. As early as 1340, up to half the six- to thirteen-year-old children of Florence were in school, and by the end of the fifteenth century, 70 percent of the people of Valenciennes, France, were said to be able to read. While only a small proportion of literate people owned books, those who did often associated them with eyeglasses. At least one early binding has a compartment for spectacles, and the impression of an eyeglass frame has been found in another book. The book crafts also demanded magnification. If manuscript illuminators must have worked with crystals and later with spectacles, so must punch cutters and type-founders. And eyeglasses also benefited printing economically. They helped make possible smaller-format editions that saved high-priced paper. A growing number of scholars and artisans, especially middle-aged and older ones, were able to work longer with finer detail. In fact, the historian Lynn White, Jr., suggests that vision aids had a profound effect on the careers of Europeans, allowing older men to remain in positions of authority long after their eyes could no longer read texts unaided; this expanded readership probably helped create the market for early printing. Eyeglasses and print were thus part of a technological complex.[11]

Useful as they were, spectacles were crude by nineteenth- and twentieth-century standards. The glass was heavier than today's, doubly convex for presbyopia or doubly concave for myopia. It had impurities and, sometimes, coloring based on doubtful medical ideas—or, as today, for protection from the sun. There was no way to fit them securely. The lenses might be close enough to the face to be brushed by the wearer's eyelashes. Many sellers were untrained peddlers who carried an assortment for trial by the customer. No standard for measuring lens strength existed.

With all these shortcomings, it is no wonder that spectacles had an equivocal reputation in their first three or four centuries of use. On one hand, this technology was linked with a prestigious and still uncommon technique: reading. On the other hand, it had no basis in science. Medieval students of optics were uninterested in them and played no part in developing them, although the theologian Nicholas of Cusa, in *De Beryllo* (1458), commended lenses as instruments allowing reason to penetrate further into the world.[12]

AUGUST AND ASININE OPTICS

Lenses divided the Renaissance against itself, appearing in turns wise and silly. When Tomaso da Modena painted the chapter hall of the church of San Nicolò in Treviso, near Venice, in 1352, he depicted forty prominent Dominicans reading, writing, or thinking. Among these were Cardinal Nicholas of Rouen peering through a magnifier, and Cardinal Hugo of Provence writing with hinged eyeglasses on his nose. Early in the next century, Jan van Eyck painted one of the wealthiest men of his community, Canon George van der Paele of Bruges, in the presence of the Virgin, carrying in his hand a small-format breviary and a pair of folding eyeglasses as signs of his learning. For such distinguished clerics, eyeglasses were already two things: a badge of their wearers' station in life, but also an aspect of their personality, as they would be of public figures even in the twentieth century. Thomas More may also have chosen to be depicted wearing them.

And the trend was projected back anachronistically into the past, in representations of learned men of antiquity like Virgil and St. Jerome.[13]

There was also a less flattering alternative convention, as the drawings of Peter Bruegel the Elder (ca. 1525–1569), widely reproduced as engravings, illustrate. The uncomprehending, thin-lipped, philistine connoisseur peering over a painter's shoulder, with hand on purse to complete the deal, wears spectacles; the wild-haired, passionately engaged artist does not. An ass looks into a schoolroom over a shelf with a musical score and a pair of glasses, as though to underscore the uselessness of all aids to learning for those incapable of understanding. Apes loot the wares of a traveling merchant, including glasses among other reputedly idle trinkets such as Jew's harps: a clear reference to the contemporary epithet "spectacle seller" to mean a deceiver of the gullible, the poor quality of his wares obscuring rather than sharpening vision. Far from being enlightened, the simians are only demonstrating their usual larceny and vanity. Yet Bruegel and the buyers of his prints did not dismiss optics entirely. In his allegorical series of the virtues, Temperance is wearing a clock on her head and a bit in her mouth, and holding a bridle in one hand and a pair of eyeglasses in the other—all technologies suggesting measure and self-control.[14]

On balance, Renaissance artists and writers suspected aids to vision. They loved to associate instruments of wisdom with vanity, and to portray the elderly as more foolish than venerable. Eyeglasses advertised not only the infirmities of age but the snares of self-deception and even of deadly sin. In a 1510 woodcut by Hans Baldung Grien, Lust has a bear's muzzle, donkey's ears, horns—and giant spectacles representing what later came to be known as the male gaze. On the opening page of Sebastian Brant's famous *Ship of Fools* (1494) appears a woodcut of an eyeglass-wearing, donkey-eared book lover (bibliofool?) in cap and bells, armed with a fly-swatter to protect the precious library, whose contents he is unable to understand. A figure with a moronic expression is dead center on the title-page woodcut of Rabelais's *Gargantua* (1534). The German physician and alchemist Heinrich Khunrath summarized this suspicion in a tailpiece he had appended to a treatise published in 1599, and that he also used in other works, depicting a bespectacled owl, flanked by two lit candles and holding crossed flambeaux. The caption reads:

> *What use can torch, light, glasses be*
> *To folk determined not to see?*

The symbolism of the bird is still obscure, but the ambiguity of the eyeglasses would have been familiar to all late Renaissance readers.[15]

By the end of the sixteenth century, the multiplication of books and pamphlets, stimulated by neoclassical learning and raging theological and political controversy, made vision aids more important than ever for the learned, especially those over forty. Privileged and demanding presbyopes, like the great book collector and cryptologist Duke August the Younger of Brunswick-Wolfenbüttel (1579–1666), sought out exceptional craftsmen from centers like Augsburg; in contrast to the Florentine workshops, Duke August's optician took a full year to fill his first order for two pair. The Duke's optician later fashioned another pair for eye protection while hunting, complete with rearview mirrors and a set of magnifier glasses called perspective tubes. Despite the Duke's dependence on and enthusiasm for vision aids, though, published surviving portraits omit them. He may have felt sensitive about his eye ailments, but he also might have been reacting to a century of satire. In any case, he was not willing to identify his personality with the technology he used, as Canon van der Paele had evidently been happy to do.[16]

It was only in southern Europe, in Italy and especially Spain, that first the upper classes and then all sections of society openly prized eyeglasses as extensions and amplifications of the self. Physicians were still skeptical about the medical value of spectacles, but wearers were becoming more devoted. A new type of frame appeared later in the century, secured behind the head with straps, or behind the ears with cords, to permit more comfortable continuous wear. Perhaps the growth of reading material was accelerating the progression of myopia, or perhaps fashion was taking its course, but an engraving around 1580 portrays a professor of medicine and philosophy at the University of Padua, Girolamo Capivaccio, wearing corded glasses with great dignity. In Spain, eyeglasses spread to the very highest circles of society. In 1623, a notary of the Inquisition, Benito Daza de Valdés, even published a treatise on eyeglasses and the spiritual and secular rewards they provide. By then, spectacles were firmly established in the aristocracy. During the 1630s, Jusepe de Ribera painted a member of the exclusive Order of Santiago wearing them with sash and armor, and King Philip II himself was said to have a pair with its frame attached to his hat.[17]

Once the elite had given their approval, eyeglasses could cease to be mere prostheses and become articles of fashion and status. The Spaniards admired large frames, such as those that Doña Rodriguez wears in Chapter

48 of *Don Quixote*. And big lenses, fastened behind the ears, became the hallmark of one of El Greco's greatest paintings, the portrait of a cardinal (usually considered to be the Grand Inquisitor Fernando Niño de Guevera) of around 1600. To later generations of non-Spaniards, the glasses have seemed to proclaim the Inquisition's piercing gaze, but contemporary scholarship suggests their real purpose for the artist and sitter were different: to remind the cardinal's enemies at court of his direct appointment by the pope and determination to retain his position against all intrigues. His display of the latest technology was intended to reinforce his skill and resourcefulness. In the end the cardinal lost his inquisitorial position, and spectacles much of their mystique.

In Bartolomé Murillo's *Four Figures on a Step* (ca. 1660), a woman thought to represent a madam is wearing the oversized type, *ocales,* supposedly reserved for the nobility. In the eighteenth century, the Spanish eyewear fashion persisted, but it had become an object of mockery by northern Europeans who saw the glasses as symptoms of pomposity or masks for stupidity.[18]

SCIENCE AND SPECTACLES

Neither the Spaniards nor their critics appreciated the changes that were beginning in the seventeenth century and that would redefine the debate over optics. For the first time, investigators were beginning to understand the physics of vision and the effects of lenses. The most common explanation in antiquity and the Middle Ages had been that the eye sent out rays. Three stunning accomplishments of the early seventeenth century changed this model. Galileo's construction of the telescope and his observation of the earthlike contours of the moon suggested that lenses could multiply the powers of human vision to a previously unimagined degree. Johannes Kepler (1571–1630), in his work *Dioptrice* (1611), showed that images are formed by rays of light on the retina passing through the eye's lens, and René Descartes (1596–1650) presented the law of refraction in his own *Dioptrique* (1637). These discoveries (and the development of the microscope) meant that the grinding of lenses, once exclusively a craft, was on its way to being a science. Glasses lost their former stigma as suspect devices that distorted as they clarified. They became what they are considered today, corrective instruments compensating for the imperfections of the eye. And where the Spanish aristocracy had sought prestige in heavy optics that must have been as unpleasant to wear as Cardinal Niño de

Guevera's friar's chair was to sit on, the scientific revolutionaries gave more thought to comfort and utility. It was Kepler who first used optical theory to propose the meniscus lenses in general use today, which have a curved section that virtually eliminates distortion at the edges of the visual field.

The rise of scientific optics did not transform spectacles into an everyday technology of the body for most Europeans and North Americans in the late seventeenth or eighteenth century. Like armchairs and keyboard instruments, high-quality eyeglasses remained luxury goods, made by skilled artisans for limited markets. Kepler's idea of a curved lens was developed by a Russian theorist in the early eighteenth century, but the design was not widely available until well into the nineteenth. Cheap German spectacles were available in England for as little as fourpence, but optics were crude and frames were cheaply plated. Good glasses, framed in steel or precious metals, cost at least a shilling, and in North America far more.[19]

Despite the price, there was still no standard for dispensing glasses and no scientific equipment for refracting eyes; buyers usually had to try on multiple pairs. The eighteenth century did make one important contribution to the regular wearing of glasses: in the 1720s, English opticians introduced temple spectacles with solid side pieces swinging out on hinges, and sometimes double-hinged for greater compactness. This new design was soon recognized as the best way to keep the lenses properly positioned on the face and remains the standard. Those who could afford effective glasses bought and used them. But outside Spain, vision aids were still slightly embarrassing acknowledgments of infirmity. George Washington, who owned a double-hinged pair, was said to have excused himself when using them in public, explaining that "I have not only grown gray but almost blind in the service of my country." The many eighteenth-century and early-nineteenth-century spectacles on display at the New-York Historical Society with hinged or sliding temples appear as substantial optical instruments, not as fashion. The temples often terminate in circles of a diameter of two or three centimeters, which pressed against the side of the head.[20]

Despite the expense and awkwardness, attitudes were changing. A few politicians were even proud of their vision aids. The Society owns paintings of both Patrick Henry and Aaron Burr wearing glasses, though pushed up on their foreheads. The association of glasses with education must have appealed to both, yet neither felt comfortable with glasses as part of the permanent record of his face. That step was, of course, taken by Benjamin

Franklin, who really did cement halves of concave and convex lenses together in one frame to create the first bifocals and avoid the need for switching between reading and distance glasses. And a few other scientists, inventors, and artists also did not hide their use of glasses. In old age Jean-Baptiste Chardin painted himself wearing a tortoiseshell pair. But it was American scientist-artist siblings who showed the nascent acceptance of vision aids most strikingly. In 1801, the painter Rembrandt Peale depicted his brother, the young botanist Rubens Peale, with the still rare geranium that Rubens had grown from seed for the first time in America. Rembrandt showed Rubens's glasses in his hand. But family and friends believed that because Rubens wore the glasses constantly for his extreme farsightedness, they should be on his head, so Rembrandt added a second pair masterfully to his brother's forehead.[21]

AFFIRMING THE ARTIFICIAL

Rubens Peale, for whom glasses had opened a new world, was a pioneer of new attitudes. Over the 1800s, eyeglasses were transformed. At the beginning of the century they were heavy and costly, or light and almost worthless. By its end, they were stunning examples of the high technology of their day, precision instruments available at least to the lower middle class and the better-off working class. They brought together some of the most advanced chemistry, physics, and medicine of their era, creating new industries and professions. And as in the fifteenth and sixteenth centuries, the technology of printing and the skills of reading reinforced each other. Reading built the demand for vision aids, and high-quality glasses helped expand the reading public. We have seen that many medical authorities welcomed infant formula as a superior alternative to human breast milk, and that countless households began to prefer the performance of a celebrated artist, fixed on a player piano roll, to homemade music. Magazine editors considering submissions started to favor the crude typography of a bar striking through an inked cotton ribbon over the most elegant and legible handwriting. So the eyes were brought into a debate over the future relationship between the human senses and technology. Was the increasing human symbiosis with the mechanical world a source of pride or of alarm? Spectacles, more than any other invention, started and stimulated the debate. If musical and textual keyboards established standard ways to record and transmit information, optics transformed the way Western humanity gathered and absorbed it.

We have already seen how many innovations come from people at the margins of a technology, applying techniques acquired in other contexts. The first true optical-grade glass was developed in 1790 by the Swiss watchmaker Pierre-Louis Guinand, who later worked with the Munich physicist Josef Fraunhofer, himself originally an artisan. In manufacturing, the leader was neither a physician nor an optician but a minister in Rathenow near Berlin: Johann Heinrich August Duncker, an optical hobbyist and inventor with a strong background in physics and mathematics. His treatise on eyeglasses remained a standard work for decades. Duncker received a royal patent for a machine capable of grinding eleven lenses precisely and simultaneously from one drive mechanism. A worldly philanthropist, he was angered by the damage caused by inferior glasses and also concerned about the fate of war victims. Employing disabled veterans and war widows, he provided free glasses for troops and poor children. Rathenow became one of the world's first centers for the large-scale production of precision goods; by the 1820s the city had two hundred factories, making it the world's most advanced optical center. Duncker's invention was said to have achieved for German optics what mechanical looms did for British textiles. So dominant did Germany become that the United States did not even produce optical glass in quantity until supplies were disrupted during World War I.[22]

In the middle and late nineteenth century, university-trained scientists further transformed the optical industry and the dispensing of eyeglasses. In 1862, the Dutch ophthalmologist Herman Snellen developed the chart (with the E at the top) and the scale (with 20/20 as the norm) still in wide use today. In Germany, the chemist Otto Schott and the physicist Ernst Abbe made the Zeiss works in Jena the world technical leaders by the end of the century. Meanwhile, the physicist Hermann von Helmholtz had founded the modern discipline of physiological optics and refined the ophthalmoscope, which permitted direct inspection of the retina and thus helped to establish ophthalmology as a medical specialty. Helmholtz also increased the prestige of optics by demystifying the human eye, arguing powerfully that its imperfections show Darwinian adaptation rather than omnipotent design. Alone, the eye would be an unsatisfactory camera; as used by the mind, it had evolved into a superb instrument of understanding. Since we have been unconsciously manipulating its impressions all along, a reader would conclude, it is perfectly reasonable to improve them further with optics. To determine just which corrections the eyes needed, Edison's electric light made possible new generations of diagnostic equipment, including refracting instruments that could test countless lens combinations for determining prescriptions.[23]

INFORMATION AND MYOPIA

Never had science or technology changed the everyday appearance and behavior of ordinary people so strikingly as in the hundred years following Duncker's high-minded venture. The answer is not just in the new knowledge but in the demand that encouraged them. Mass literacy made vision aids a necessity for tens of millions of people. The governments of northern Europe and North America, especially in the middle and later years of the century, saw increasing economic and military value in extending fluency in reading and writing from the middle class and the cities to the countryside, and later from boys to girls. New industrial processes and military weapons alike made written instructions important throughout the ranks. As we have seen in the history of the typewriter, women were beginning to dominate the clerical workforce.

Prussia, the birthplace of the modern eyeglass industry, led the world in establishing a rigorously controlled school network. Matthew Arnold wrote in his 1866 report to the Schools Inquiry Commission that Germany's educational system "in its completeness and carefulness is such as to excite the foreigner's imagination," and the German victory over France in 1871 was widely credited as much to the Prussian schoolmaster as to the drill sergeant. Prussian male illiteracy was already below 5 percent when the war broke out, but France and England had almost closed the gap by 1900, when virtually everyone in northern and western Europe could read and write—one of the most impressive government programs in history, both the cause and effect of the explosion of inexpensive reading materials thanks to steam presses, wood-pulp paper, and national postal systems.[24]

In the nineteenth century, reading appeared in a different light, no longer as the attribute of a clerical and administrative elite but as a mental skill indispensable to the new industrial world of complex equipment and printed manuals and regulations. Governments also needed literate populations to enforce policies and mobilize opinion, and manufacturers' marketing depended on posters and advertisements. Reading was the mental technique that was essential to the maintenance and operation of nineteenth-century technology.

The Victorians, now considered the staunchest apostles of progress, should have been ecstatic at the spread of reading and writing skills to the masses. Many early-twenty-first-century academics and writers look to the late nineteenth as a golden age of print and educational reform. Of course, there had been warnings against the perils of bookishness since antiquity and, as we have seen, in the Renaissance. Even in the early nineteenth century, in the golden age of German classicism, social critics warned of *Lesewut,* literally "reading mania." The masses, it seems, just could not win; they were either illiterate or foolishly ambitious.[25]

By the 1860s, the old critique of the perils of print had new support from an unexpected quarter, the medical profession itself. Physicians had long been concerned about the effects of reading on the bodies of schoolchildren; we have noted that health reformers were trying to improve classroom seating decades before corresponding innovations in factories and offices. By the 1870s, the new instruments for measuring vision were revealing an unexpected trend: nearsighted children.

Not surprisingly, the first warning appeared in the land that contemporaries identified with educational standards, precision workmanship, and optics: Prussia. And the diagnostician was the leading ophthalmologist of Breslau (now Wrocław, Poland), Dr. Hermann Ludwig Cohn (1836–1906). As a young man he published *An Investigation of the Eyes of 10,060 School Children* (1867), the first major epidemiological study of vision, assailing the dim lighting and cramped seating of traditional schools. In his travels he examined the eyes of nomads like the Bedouins, of huntergatherers, and of the illiterate peasants of some of Germany's remote areas. Their excellent distance vision impressed him, and he blamed the growing prevalence of myopia on the heavy reading loads and detailed work of latenineteenth-century education and industry.[26]

Thanks to Cohn's influence, doctors from St. Petersburg and Tiflis to New York City and Cincinnati were testing schoolchildren for myopia and other conditions; in 1886 Cohn listed forty studies involving hundreds of schools, including Harvard. Evidence was overwhelming that the rate of myopia increased from grade to grade, suggesting that additional reading was changing the shapes of students' eyes permanently. In Cohn's own studies, the proportion of nearsighted students rose from 5 percent in the countryside to fully half the graduating classes of the *Gymnasien.* Book learning was, it seemed, hazardous to visual acuity.[27]

Cohn's warnings were hardly isolated. While the rest of Europe and North America were largely celebrating the Reich's secondary schools, Germany's doctors, politicians, and parents deplored "overburdening," which allegedly was leading to a soaring suicide rate and "nervous debility." To

Cohn and his supporters, myopia was another manifestation of the physical and mental "diseases of civilization" brought on by crowded cities, pollution, business pressures, and lack of fresh air and exercise. The flood of information seemed overwhelming. Literally and figuratively shortsighted, industrial humanity appeared to be walking into an evolutionary trap, as success in life paradoxically produced physical and mental infirmity.[28]

For all this agitation, there were equally noted physicians and critics who virtually welcomed the spread of myopia and of eyeglasses. Cohn was scandalized by his Dutch colleague F. C. Donders, author of the international standard work on eye examinations. The higher degrees of myopia were serious, Donders admitted, but could be prevented with timely spectacle prescriptions; as for the milder forms, they "bring a capacity for delicate hand work and scientific investigation that we would not like to do without." Donders wrote that he would not want to abolish mild myopia even if he could, as it was a sensible accommodation of the eye to practical needs. It would not bother him if it turned out that the scholar and the peasant each developed the eye most suitable for his use. Another German expert flatly declared that education and skills demanded physical and mental sacrifices, and a third compared myopia to the military surrender of life for the Fatherland. Where Cohn saw sickness and danger, and called Donders's "delicate hand work" "generally superfluous," many others found what they considered Darwinian adaptation.[29]

Cohn valiantly advocated better lighting and seating for the schools. He studied typography to find medically optimal fonts and type sizes and even considered requiring tinted schoolbook paper. In fact, many such design reforms were adopted, but in vain. High rates of correctable myopia persisted despite school hygiene programs. Cultural conservatives responded by turning from environmental to hereditary explanations. By 1896, even Cohn had shifted his interest from educational reform to mass testing that would let the German navy identify the part of the German population with unspoiled vision, the better to deploy them in the marine artillery.[30]

In England and the United States, the school crusade against myopia was less strident and anxiety less pronounced. The growing proportion of eyeglass wearers was seen increasingly as a sign of technological progress rather than of physiological degeneracy. In 1893, the president of the ophthalmological section of the British Medical Association declared approvingly at the society's annual meeting that with the progress of popular enlightenment and civilization, in the world of the future "a man who goes about with his eyes naked will be so rare that the sight of him will almost raise a blush." Reporting this speech in the *Atlantic Monthly,* Ernest Hart summed up educated opinion of the time, that research had shown that very few people have ever had perfect vision and that the prevalence of glasses reflected the progress of scientific diagnoses and treatments since the 1870s. There was no shame in improving the eyes technologically.[31]

PROUD PROSTHESES

Hart's report signaled a growing voice against what he called the "croakers of decadence." For the first time since the golden age of Spain, men (though not yet women) were beginning to flaunt their eyewear as manifestations of the latest in science. The growing number of schoolchildren and young adults wearing glasses had provoked popular scorn as well as elite alarm; the earliest taunts of "four-eyes" and "four-eyed" date both in England and the United States from the 1860s and 1870s. Even the overworked youth of Germany had a similar and still less flattering expression, *Brillenschlange,* literally "eyeglass-snake," referring to the pattern on a cobra's hood.[32]

A new generation of assertive myopes did not take such taunts passively. There was a growing sense that eyeglass wearing was not just a scientific correction but an extension and expression of the personality. In 1880, for example, a writer in the London *Saturday Review* remarked that "no artificial adjuncts of the human body are so apparently identical with its nature as spectacles," and that he knew those who could use them to smile, frown, sneer, and even eat. One such person of the day was Theodore Roosevelt, who went to the Dakota Territory as a (gentleman) cowboy to fortify and demonstrate his manliness and succeeded beyond all expectation. And no deed from that time of his life became more celebrated than knocking out an armed ruffian who had used the dread slur "four-eyes." ("As I rose," TR later wrote proudly, "I struck quick and hard with my right just to one side of the point of his jaw, hitting with my left as I straightened out, and then again with my right. . . .") During the Spanish-American War he further turned stigma into asset by leading the charge of his company of Rough Riders up San Juan Hill wearing his signature pince-nez, now immortalized on Mount Rushmore.[33]

In fact, pince-nez, with two lenses mounted on the wearer's nose and held there with a spring or other pressure system, had already become a wildly fashionable accessory. The same *Saturday Review* correspondent

No. 724,151.   PATENTED MAR. 31, 1903.

J. C. ANDERSON.
SPECTACLES.
APPLICATION FILED JAN. 27, 1903.

NO MODEL.

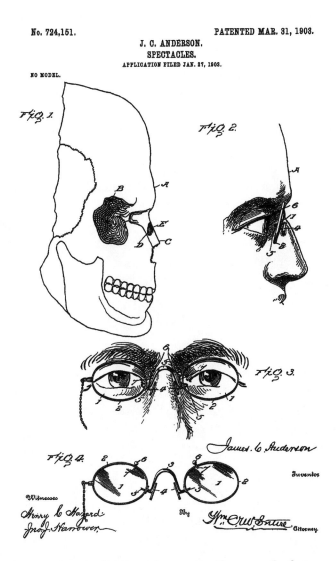

The pince-nez, kept in place by spring pressure alone, was the first eyewear to become a fad among both men and women. Many ophthalmologists and opticians considered it difficult to fit, and the glasses fell off repeatedly. Yet as this patent specification observes, contemporaries thought the design avoided the "elderly appearance" of spectacles with temples. The inventor believed he had found a better way to make the pince-nez stay put.

noted their popularity and marveled at the transformation they produced in the wearer of temple eyeglasses: "You hardly recognize your friend. The face looks but half-clothed, and it wears a rollicking expression which is in strong contrast with the sobriety of its old spectacled days." It hardly mattered that the spring pressure and weight left indentations on the sides of the nose, that pince-nez were nearly always uncomfortable and easily broken, and that many ophthalmologists and opticians denounced them for their failure to maintain a consistent and proper distance from the pupils. By the turn of the twentieth century, whole Harvard and Vassar classes were sporting pince-nez. Snobbery was part of their appeal; they imparted a superior, "quizzing" gaze like the lorgnette before them, and became a double-barreled monocle. With their black ribbon, ostensibly for protection of the optics, but often for dramatic gestures, pince-nez almost encouraged theatricality. Arthur Conan Doyle knew a great clue when he saw one. (Perhaps because of this perceived arrogance, both military officers and civilian employers in the nineteenth century often required subordinates to address them or even to work without their glasses, even if their actual performance suffered.)[34]

Pince-nez soon had bolder competition. In 1913 the Kansas City *Star* detected "an age glorying in infirmity" and noted "owl-like round lenses the size of twin motor lamps" in "bulky tortoise shell and imitation celluloid." New materials were at last beginning to expand both performance and expression, if only in helping revive the look of a traditional frame material that had fallen out of favor. In the 1920s and 1930s new materials attracted a generation of celebrities who discovered that big, round glasses could appear youthful and daring: Harold Lloyd, Buster Keaton, and the (alas, only almost) timeless George Burns. Le Corbusier made the common headlamp spectacles into geometric icons of modernity, and postmodernity. In *Précisions* (1930) he declared the hallmark of the new man of worth to lie "not in the ostrich plumes in his hat, but in his gaze." Seeming not only to gather but to project information, they have remained favorites of Philip Johnson and I. M. Pei. Female fashion authorities like Diana Vreeland and Carrie Donovan found the look equally appealing.[35]

Metallic spectacles and pince-nez remained the choice of many of the Versailles generation: Woodrow Wilson, Georges Clemenceau, and especially David Lloyd George. If oversized hard-rubber, celluloid, or plastic spectacles enlarged the eyes of the wearer and exposed him or her to an admiring countergaze, pince-nez continued to connote an earlier, medieval theory of vision: eyes that emit piercing, searching rays, private eyes that bore into public spaces. The pince-nez thus remained favorites of the twentieth century's own grand inquisitors, Heinrich Himmler and Lavrenti Beria, and of Beria's most celebrated victim, Leon Trotsky. The metal frame also was the choice of the macho myope in the Roosevelt mode. The writer Isaac Babel, hoping to transcend his identity as a Jew "with spectacles on his nose and autumn in his heart," outdid even TR when he rode with the anti-Semitic Cossack cavalry in the Russian civil war. Franklin Roosevelt's lifelong use of pince-nez reflected his youthful admiration of the cousin who initially was thought to overshadow the better-looking Franklin in vitality. And like the other technologies we have studied, they changed FDR's body language early in life, in his case leading him to tilt his head back and literally look down his nose as he spoke to people. It later became Roosevelt's genius to turn this affectation into a token of optimism rather than condescension. (Hitler, on the other hand, resisted the public display of eyeglasses, having speeches and documents prepared with special large-character typewriters.)[36]

GLASSES AND MASSES

World War II was a turning point in the optical modification of the human body. Armies could not afford to exclude soldiers with correctable eyesight, and many Depression-era recruits were properly examined and fitted for the first time in their lives, raising public expectations of visual acuity. In postwar England the spectacle benefit became a symbolic feature of the National Health Insurance system at least until the Thatcher era, and even in fragmented U.S. health care, vision plans remain popular in corporate benefits programs. Once a mark of otherness and later in some quarters of degeneration, eyeglass wearing became the acceptable and even prestigious condition of the majority. Today, six out of ten Americans and Britons wear lenses of some kind.[37]

The postwar relationship of the body and personality to vision aids has taken two directions: expression and invisibility. Thermoset plastics (unlike the rigid Bakelite that prevailed in the interwar years) gave free reign to fantasy and to experimentation with new designs that could complement every facial shape. Women's eyewear, long stigmatized, was first to make this leap. As usual, the pioneer came from outside the industry mainstream. A New York artist and window display designer, Altina Schinasi Miranda, bored by the women's glasses she saw displayed in the city, devised a new pointed look that appealed to Clare Boothe Luce and other stylish and prominent women. Demand exploded in the 1950s, and the fashion press discovered that glasses could be flattering and even sexy. Ever since, celebrities and authorities have turned both men's and women's glasses into a quasi-cosmetic art of matching frames with faces and clothing styles. Finding the correct historic spectacle or a suitable futuristic one is a recognized skill of Hollywood production designers— think of Arnold Schwarzenegger's contoured sunglasses in *Terminator*— and characters' eyewear is often reborn as production models to meet demand. The tradition of flamboyant glasses continues with Elton John and Barry Humphries.[38]

In popular as well as celebrity culture, glasses have been revalued. In the technological boom of the 1990s, plastic-rimmed taped glasses were praised for their "nerd chic." At midcentury, the spectacles of Piggy in William Golding's *Lord of the Flies* (1954) have the vital power to start fires, yet Piggy himself is a born victim. In J. K. Rowling's Harry Potter book series (beginning in 1997), a bullied child instead becomes the protagonist, and Harry Potter–style round glasses have been revived for fans.

Glasses are for protection as well as reading, in large part as a positive unintended consequence of military research in the 1940s. Early plastic lenses made of polymethyl methacrylate (PMMA), a material marketed under the trademarks Perspex, Plexiglas, and Lucite, scratched easily. But during World War II, the sheets of glass in some U.S. bombers were held together by a bonding compound called CR-39, developed by Pittsburgh Plate Glass, to reduce weight and increase the planes' range. An optical researcher, Dr. Robert Graham, was able to experiment with CR-39 after the war and developed technology to turn the difficult, sticky raw materials into lenses. Meanwhile, the U.S. government was mandating higher breakage-resistance standards that added to the weight of optical glass, while chemists in the late 1970s developed long-sought scratch-resistant coatings for plastic. The result was that CR-39 became the preferred material and captured most of the lens market by the 1980s. Glass became a premium material because of the extra step needed to bring it to safety standards. (Graham later used his fortune to establish a eugenically inspired fertility program called the Repository for Germinal Choice, instantly dubbed the Nobel Sperm Bank by the press.) The search for protection did not end with CR-39; polycarbonate, a more expensive material with superior impact resistance, has since become the standard for sports protection. Many factories mandate polycarbonate safety glasses for employees whether or not they need vision correction: hard hats for the eyes.[39]

Military aviation also helped create the modern sunglass industry. Tinted glasses had existed in China, and in the West at least since the Renaissance—a group of friars in *Don Quixote* are wearing them on their travels—but were developed scientifically only in the twentieth century. Bausch & Lomb originally produced Ray-Ban green goggle lenses in the 1920s for U.S. Army aviators troubled by glare reflected from clouds. During the Depression, when plastic sunglasses cost only a quarter, Bausch & Lomb put the new glass into plastic frames and began selling them under the trademark Ray-Ban. In place of the nineteenth century's optimistic valuation of the sun, postwar research substituted far more sobering warnings about ultraviolet light, which have made sun protection almost as important in outdoor eyewear as shatter resistance.[40]

Meanwhile, a second trend was growing. Instead of supplying optics as externally mounted protection for the eyes, it aimed to augment or even reshape the surface of the eye. Like the improvements in worker seating that we have observed, this work originated in wartime necessity. An English ophthalmologist, Harold Ridley, operating on Royal Air Force fighter pilots injured in battle, discovered that their eyes had not rejected the slivers of PMMA from the cockpit hood. He wondered whether this plastic could replace the lenses removed from elderly patients when cataracts, areas of hard and cloudy dead cells, had grown excessively. Since cataracts are one of the world's leading causes of blindness, and conventional cataract surgery required patients to use special thick glasses afterward, Ridley risked his reputation and incurred the scorn of most of his British colleagues to develop a new operation, the intraocular lens transplant. Since the first procedure in 1949, 200 million people worldwide have received artificial lenses. Meanwhile, new plastics developed from the 1930s to the present have made possible a technology proposed three hundred years earlier by René Descartes, the contact lens. For the first time, laypeople were applying medical devices to the surfaces of their own eyes, and the eyeball lost something of its sacred character. Optometrists, long separated from MDs in part by their lack of training in physical contact with the eye, now began to cross a psychological boundary.[41]

It is only a step from the contact lens to surgery that actually reshapes the cornea—most often to treat nearsightedness, but also for farsightedness and astigmatism. The most popular procedure, Lasik ("laser in situ keratomileusis"), was approved by the U.S. Food and Drug Administration in 1995 and now is performed on an estimated million patients a year at a cost of $2.5 billion. A special knife removes a circular flap in the eye, exposing the cornea, which is reshaped by a computer-controlled laser. The paradox of Lasik (and, to a lesser extent, other refractive surgery) is that because it has a generally high rate of satisfaction and offers unusually speedy and visible results to most patients, it has become the most common elective surgery in the United States, so that even a small proportion of complications and disappointing results (such as impaired night vision) affects tens of thousands of people. Yet both this dissatisfaction and the joy of successful results arise from the same unhappiness with the eyeglass as an external technology. The technology of the body seems to follow a trend from heavier to lighter and more flexible external devices, to even lighter ones worn against the body, and finally to the reshaping of the body itself. Thus, as Valerie Steele showed in an exhibition at the Fashion Institute of Technology in 2000, fashion has not so much rejected the corset in the twenty-first century as substituted diet programs and exercise machinery for fabric, whalebone, and steel.[42]

THE REVENGE OF MYOPIA

Even if a safe surgical procedure is developed, there will probably always be a strong market for eyeglasses. Surgical implants to correct presbyopia are still at an early experimental stage and are based on an unconventional theory of the condition's development. Contact lenses for presbyopia are still at an early stage of development, and many people consider even soft contacts uncomfortable, so that eyeglasses—perhaps using new materials but following principles established hundreds of years ago as musical keyboard layouts do—are likely to remain the most common form of vision correction.[43]

Correction will be needed. The twenty-first-century information society, already visible in its outlines a hundred years ago, continues to affect vision. Myopia appears to be spreading in the United States and other industrial nations. It was estimated in the early 1970s that 25 percent of the population between twelve and fifty-four years old is myopic. A 1996 study in Massachusetts, though, revealed a 60 percent rate in the twenty-three- to thirty-four-year-old bracket, declining to 20 percent among state residents older than sixty-five. Contact lenses help conceal the true rate even as they increase the cost of corrected vision.[44]

The causes of myopia are still debated. Genetics influences the likelihood of developing the condition; children with two myopic parents are over six times more likely to be affected than those with at least one nonmyopic parent. But environmental influences are even more striking, and epidemiologists have found strong links to schoolwork. Among children enrolled in U.S. Orthodox Jewish secondary schools, males studied sixteen hours each school day—presumably this includes homework as well as instruction—and had an 81.3 percent myopia rate; females, who studied eight hours, had a 36.2 percent rate. (Among Jewish children attending secular high schools, with a six-hour study load, 27.4 percent of males and 31.7 percent of females were nearsighted.) Asian societies with rigorous school programs are similarly affected. In Hong Kong 75 percent of high school students and 90 percent of college students are now myopic. In Singapore, fully 98 percent of medical students are nearsighted; the national air force can find few visually qualified pilot recruits. Yet there are also puzzles in the environmental history of nearsightedness. Iceland has long been both homogeneous and literate; why should its rate of myopia have increased from 3.6 percent in 1935 to 20.51 percent in only forty years?[45]

Singapore, with its social discipline, high literacy, precision industries, and educational drive, recalls the values of nineteenth-century Prussia. Seemingly repeating, with a vengeance, the causes and effects that so alarmed Dr. Hermann Cohn, it is now believed to have the world's highest rate of myopia. A new complication has been added by the research of a biologist, Joshua Wallman, whose experiments with animals suggest to him that when children wear eyeglasses to correct myopia they may inadvertently be changing the ways in which the eyes grow, actually elongating them further. This idea still has not been proved or disproved.[46]

What is certain is that even a skill as abstract as literacy has an unexpectedly strong physical aspect. In the history of humanity, our attention has shifted from the horizon to the length of our own arms: the printed page or the electronic monitor, or at the farthest the television screen. As with other technologies of the body, in changing our world we have changed ourselves—and not only ourselves. The Western metaphor of the Creator as watchmaker, popular ever since the Enlightenment, suggests even the Deity squinting through a lens, as though remade in our own image.

CHAPTER TEN

Hardheaded Logic

Helmets

I F THERE IS a distinctively twentieth- and early-twenty-first-century body technology, it is paradoxically one of the oldest external modifications of the body, possibly older than the chair (and indeed sometimes used for seating): the helmet. Of all the familiar objects we have observed, it was the helmet that appeared anachronistic for most of the eighteenth and nineteenth centuries. It was never entirely abandoned in military service, but for years it remained the specialty of firefighters and some police forces.

Helmets, more than any other technology, defy conventional chronology. They seem to evolve like metallic and polymeric crustaceans, but not conventionally; a form may disappear for a thousand years and then reappear on a new branch. Another may keep its shape but change materials and habitats. Medievalists have advised the commanders of industrial armies, and armorers from dynasties of European craftsmen have helped tool up for new designs with classic jigs and hammers. Helmets represent risk aversion and aggression, ethnocentrism and cosmopolitanism and sometimes both sides of each dichotomy at once.

In itself the helmet is distinctive because it does not increase comfort or improve performance as some footwear does, and as chairs do for people accustomed to them. It does not simplify learning or communication, as musical and text keyboards do, nor does it augment or extend our senses, as spectacles do. The helmet sacrifices comfort and even performance to protection. It increases the mass of the human head by as much as two or three kilograms, impedes cooling in hot weather, and often limits both peripheral vision and hearing. Human skull shapes and sizes vary so much that fitting troops has always been challenging. And with discomfort comes the risk that soldiers will remove their helmets and expose their heads to the enemy.

The helmet is also radical apparel. It is the only rigid prosthesis that able-bodied men and women now routinely use—an exocranium. Humanity has evolved to discriminate keenly among faces, to see subtle nuances of feeling in changing expressions. Helmets are not just mechanical shields but symbolic frames, like spectacles. A human being wearing one acquires a new persona. The Greek hoplite peering from the depths of his Corinthian helmet, the medieval knight in his sallet, the World War I–era German in his metal helmet, the World War II GI in his almost spherical pot, the astronaut, the patriotic hard hat, the deep-sea diver: heroes and villains almost become their helmets. And even flawed designs can build cohesion and morale by evoking shared values.[1]

TAKING COVER

Peoples all over the world have probably made head protection from organic materials such as shells, vegetable matter, and layers of cloth, but few objects have survived. Some helmets, like those made by the Ibo of West Africa, helped make warfare more a ritual sport than the carnage of most Mediterranean and Near Eastern conflicts. Other peoples must have developed effective armor because their wars were deadly serious. As recently as the early twentieth century, the Gilbert Islanders of the South Pacific were known for their spears studded with sharks' teeth and for correspondingly durable armor, which included the skin of a puffer fish killed while inflated: its natural spikes protected the wearer while intimidating at least the inexperienced foe.[2]

The first metal helmets probably appeared in the third millennium B.C. in the Middle East; they were linked to the development of complex, literate societies that conducted organized warfare. Head protection was part of the earliest version of the arms race, not yet between swords and shields but between maces (weapons with weights at the end of handles) and helmets. The earliest protection consisted of leather and felt caps, which in Homeric times were further protected by boars' tusks. There must have been layers of cushioning under the shells of even the simplest metal helmets, for comfort and heat transfer. An industrial study of the 1960s confirmed the value of head protection—and shows the link between the oldest conflicts and recent industry. A metal helmet spreads the force of an impact over the surface of the indentation the blow makes. On the basis of this research, the military historians Richard A. Gabriel and Karen S. Metz have calculated that a Sumerian helmet consisting of two millimeters of copper over four millimeters of leather could spread the force of an impact so that to knock an opponent unconscious required superhuman strength.[3]

As Gabriel and Metz observe, defensive equipment—helmets and body armor—drove the development of early offensive weaponry, not the other way around. The mace, despite occasional revivals as late as the European Middle Ages, was considered an ineffective weapon in combat against helmeted men, and became the ceremonial object it remains. In the early third millennium, the Sumerians had already developed excellent metal helmets with protection for the ears and the back of the neck as the contemporary NATO helmet has; and it was the same people who developed a new style of ax to pierce such armor, having a sharp, socketed copper blade in which a haft fit securely.[4]

In Mesopotamia, with closely spaced rival city-states and peoples sharing a military culture, these innovations spread quickly. Even Egypt largely abandoned the mace for the ax after invasion by the helmeted Hyksos around 1700 B.C. Throughout the ancient world, the helmet became indispensable for most forms of combat. In the second millennium and the early first millennium B.C., the Assyrians became the first troops to be equipped with helmets on a massive scale. The helmet, usually with a conical shape ideal for deflecting downward blows, was part of the image of the Assyrian army, probably the world's first dreaded fighting machine— all the more remarkable given Assyria's short supplies of metal and fuel. (The Assyrians were also the first ancient army to protect feet as efficiently as heads; they wore tall leather jackboots with iron-studded soles.)[5]

The Greeks were relative latecomers to the manufacture of metal helmets, yet they rather than the peoples of the Middle East are considered the founders of the Western tradition of armor. Aesthetically and technically, their achievements were as impressive as those of the Sumerians and Assyrians. Around the middle of the eighth century B.C., they developed a new style of warfare and distinctive arms and armor to accompany it. The Greek infantry soldier was typically a free farmer-citizen with a heavy civic and financial investment in a panoply of arms. (For the Greeks, armor's essential component was not the helmet but the great shield or *hoplon,* of bronze-reinforced wood.) The best known Greek armor is the Corinthian helmet, first documented about 750 B.C. It reflected exceptional craftsmanship, conforming closely to the head and sweeping forward to form integrated cheekpieces so that only the wearer's eyes and throat were even partially visible.[6]

The protection of helmet and shield was cumbersome as well as costly. The classical historian Victor Davis Hanson has speculated about how the features of the Corinthian helmet, which prevailed from about 750 to 500 B.C., affected ancient Greek tactics, in other words, how a technology helped shape techniques. The helmet had no earholes and allowed only a limited range of vision. These restrictions encouraged the rigidly stylized form that Hanson believes combat took: two formations of heavy infantry several ranks deep, protecting each other with their giant shields, charging in formation until the two opposing masses collided on an open field, pushing each other with their shields and thrusting their spears to strike at throats and other gaps in armor. Only elementary commands could be heard, and all battles were fought in the helmets' claustrophobic darkness, the pressure of comrades on all sides replacing vision. The helmet weighed about 2.2 kilograms—the same as a U.S. infantry helmet of the Vietnam era—but lacked twentieth-century suspension systems. A blow could knock it off or even break a vertebra if the head snapped backward. Battles were fought in the Greek summer heat; hoplites must have perspired profusely, yet martially correct hairstyles were long rather than close-cropped. Horsehair crests were as much for display and morale as for absorbing shocks. They added unwelcome weight and reduced usability by raising the helmets' center of gravity. Only immediately before battle did soldiers lower the Corinthian helmet over their faces.[7]

Many scholars believe there was more individual combat, sometimes without helmets or armor, than Hanson and others acknowledge. Hanson himself suggests that like later soldiers, hoplites modified and personalized helmets and other equipment, and that helmets may have been individually recognizable to members of a unit. But the Corinthian helmet still evokes the terror as well as the glory of combat for good reason.[8]

This uncomfortable, tactically limiting, and costly object remained in use for 250 years, perhaps because of the prestige of its form but also because strategy and tactics became well adapted to it. It was replaced in the fifth century by a smaller, circular head covering of bronze, and sometimes possibly of felt alone, called the *pilos*. Casualties increased; the change evidently marked a riskier, more mobile style of fighting. Other troops began to wear a Thracian helmet, modeled after a cloth cap from the region and also affording far better visibility and hearing, even with the unnecessary weight of simulating a felt hood in bronze.[9]

The Greeks thus were paradoxically masters of the aesthetics and craftsmanship of helmets but unwilling to apply their analytical genius to shielding the brains that produced them. Jacques Ellul believes this attitude was not due to fear or indifference, but suspicion of the potential for brute force and immoderation the Greeks found in all technology. With no such qualms, the eclectic Romans adopted the powerful short sword (*gladius*) of the Spaniards and the bronze helmets and steel-edged shields of the Gauls. The French historian Victor Duruy marveled at the proud Romans learning from their conquered enemies, "ceaselessly improving the science by which they had subdued the world."[10]

The Romans actually used a variety of bronze and iron helmets in the course of the Republic and Empire, showing great attention to battlefield realities. From a kind of reversed metal jockey cap, they developed complex forms with reinforcements, hinged cheekpieces, and effective neck guards, including some of the greatest masterpieces of ancient armor. Even a simple design like the Montefortino helmet of the Republican period had cheek pieces combining side protection with unobstructed vision and hearing. Specialists disagree about just how the Romans used their weapons but admire the coherence of design.[11]

The most common Roman legionary helmet of Hollywood films is known to specialists as the Imperial Gallic (also called Niederbieber from an important find) of bronze-clad iron. With its crossed ridges and brow and neck protection it is as much an icon as its Corinthian predecessor. One expert admirer, the German military historian and reenactor Marcus Junkelmann, calls it "the crowning conclusion, the synthesis and quintessence of the entire evolution of ancient helmets." Junkelmann believes that at 2.5 kilograms it was less comfortable than earlier models but that it provided more protection to the wearer than any other.[12]

The Roman helmet stayed at this high point for only a century or two before repeating the Greek experience of the fifth century B.C., the transition to a simpler, conical form later called the *Spangenhelm* (German scholars often have the last word when Roman texts are silent). It was made of iron riveted in segments rather than hammered from a single sheet of iron or brass; where they existed, cheek and neck pieces were attached with leather rather than with hinges. Private craftsmen had made the Imperial Gallic helmet; the *Spangenhelm* was produced by less skilled artisans in state workshops. It became the favorite of all the warriors of late antiquity and the early Middle Ages who had the means to buy one, including the Vikings. (Nineteenth-century Wagnerian production designers added the horns.)[13]

THE HELMET BETWEEN EMPIRES

European medieval warfare added a weapon relatively uncommon until late antiquity: the bow, which Eurasian steppe nomads had used so fearsomely. And hand-to-hand combat could be brutal. Some foot soldiers may often have had only soft head coverings, and a surprising number of head injuries appear in the remains found in a mass grave from the Battle of Towton (1461) in the Wars of the Roses. Because armor was so valuable, none was found at the site. It is not clear whether helmets had been struck from the casualties' heads or whether they were fighting with no head protection besides the soft hats shown in some depictions of medieval warfare.[14]

Helmets protected the medieval combat elite—nobles, knights, and the professional fighters surrounding them—but probably few foot soldiers. Armor was not only a matter of life or death, but a source of prestige and hazardous recreation in the tournament. A royal warrior-connoisseur like Emperor Charles V of Spain and Austria wore 125 pounds of tilting armor with a helmet weighing forty pounds. Many surviving helmets show evidence of the forces and missiles they deflected; in single combat, a helmet and body armor could withstand blows from the strongest opponent armed with a combat ax that put the old Mesopotamian and Egyptian maces to shame: a three- to five-pound head mounted on a two-pound, five-foot shaft and swung with a two-handed grip.[15]

How did the evolution of helmets interact with the weapons and tactics of medieval warfare? While innovations in helmet construction are well documented from surviving artifacts, the reasons for change are not always apparent. Who wore the single-piece barbute of the fifteenth century; why did it start out as an open-faced helmet and come more and more to resemble the Corinthian helmet? In any case, two medieval helmet styles were to become classics. The kettle hat or chapel-de-fer has a bowl with a wide brim. Taking less time and skill to make than more contoured cavalry models that enclosed the face, it was ideal for equipping foot soldiers in large numbers. The brim must have helped protect them from volleys of arrows, and from missiles hurled from walls during sieges. The chapel-de-fer also was relatively comfortable and did not interfere with vision or hearing. Many knights preferred it; the thirteenth-century Jean

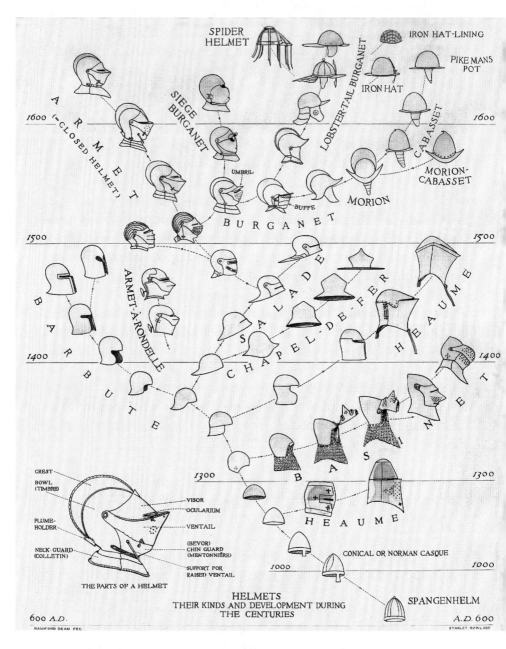

SPIDER
HELMET

IRON HAT-LINING

PIKE MANS
POT

A R M E T
(=CLOSED HELMET)

SIEGE
BURGANET

LOBSTER-TAIL BURGANET

IRON HAT

CABASSET

1600

1600

UMBRIL

MORION-
CABASSET

BUFFE

MORION

B U R G A N E T

1500

1500

S A L A D E

ARMET-À-RONDELLE

H E A U M E

B
A
R
B
U
T
E

C H A P E L - D E - F E R

S A L A D E

1400

1400

B A S I N E T

1300

1300

CREST

BOWL
(TIMBRE)

PLUME-
HOLDER

NECK·GUARD
(COLLETIN)

VISOR

OCULARIUM

VENTAIL

(BEVOR)
CHIN GUARD
(MENTONNIÈRE)

SUPPORT FOR
RAISED VENTAIL

H E A U M E

CONICAL OR NORMAN CASQUE

1000

1000

THE PARTS OF A HELMET

HELMETS
THEIR KINDS AND DEVELOPMENT DURING
THE CENTURIES

SPANGENHELM

600 A.D.

A.D. 600

BASHFORD DEAN FEC.

STANLEY ROWLAND

Bashford Dean, an outstanding ichthyologist of early-twentieth-century New York City, applied his evolutionary outlook to his other passion, arms and armor, preparing this tree of helmet morphology, one of a series of sixteen arms and armor developmental charts he designed in the 1920s. (Courtesy of The Metropolitan Museum of Art, Department of Arms and Armor) de Joinville wrote in his chronicle of lending his kettle hat to King Louis IX for fresh air after the king had spent hours in a massive helm.[16]

The sallet was an equally versatile design. Its brim was contoured back over the neck for added protection, and it often had a pivoted visor with an eye slit, which could be raised and lowered; it also had a removable chin protector called the bevor. The sallet potentially offered the visibility of the kettle hat with the option of protective attachments. Originating in northern Italy, it became popular in England and Germany and appears in Albrecht Dürer's famous engraving *Knight, Death, and the Devil*. Some sallets had built-in eye slits. Others, like the plainer kettle hats, were made for men-at-arms and sweep back directly to the tail without contouring. Some were left unpolished ("rough from the hammer") for economy, not for visual effect, but these "black sallets" now have a sinister look that helped inspire Darth Vader's armor in George Lucas's *Star Wars* series. Most sallets were relatively open, hence airy and popular. Like the Corinthian helmet, a very deep design could be so confining, hot, and disorienting that soldiers put it on at the last moment. Discomfort and thirst could be so unbearable that men removed them during battle, exposing their heads to fatal wounds.[17]

By the early sixteenth century, armorers' skills had reached a peak, but firearms were already starting to threaten their trade. First-class helmets could defy the most massive crossbow that demanding patrons sometimes brought for proving. Recent experiments have shown that a sixteenth-century musket with new-style corned (uniformly grained) powder could achieve a muzzle energy of 4,400 joules, an over twenty-fold increase over the 200 joules of a crossbow bolt and enough to penetrate milled sheet steel two millimeters thick. Bulletproof armor of the later sixteenth century was too heavy to be worn proudly and gracefully. Twentieth-century experiments have proved that late medieval armor had allowed flexibility and comfortable movement; heat, not weight, was its real problem. But the new equipment was often carried by servants as part of a baggage train and could not always be unpacked for unexpected combat. Educated opinion of the time held that the new armor could disable a man by the age of thirty-five.[18]

GRANDEUR AND DECADENCE OF THE HELMET

Not only firearms but new military ideas worked against body armor. Commanders of the early seventeenth century, led by Gustavus Adolphus of Sweden, sought maneuverability and speed and willingly reduced their men's armor. The helmet usually was the last piece to be abandoned. With its reinforcing crest or ridges, permanently mounted visor, nosepiece or face protector, and lobstertail neck protector of riveted plates, the pot helmet was the early modern European (and Ottoman) ancestor of the utilitarian metal helmets of the twentieth century. Homely but effective, it was popular with musketeers.[19]

Arms and armor historians like to point out that protection never disappeared. Bashford Dean, an American ichthyologist and paleontologist who helped establish the technological history of armor as a scholarly field—and founded the Department of Arms and Armor of the Metropolitan Museum of Art—observed that in the seventeenth and eighteenth centuries a number of military authorities still valued armor. A number of important eighteenth-century commanders are known to have worn it; helmets still protected against swords and low-velocity bullets. There is a portrait of Lord Jeffrey Amherst from 1760 with a gleaming helmet beside

The contour of this sallet (ca. 1450–60), still formidable after fire and corrosion, was reborn spectacularly in the twentieth century: in the German steel helmet of World Wars I and II, in the Kevlar helmets of the United States and its allies, and in popular culture. (Courtesy of The Metropolitan Museum of Art, Bashford Dean Memorial Collection, Bequest of Bashford Dean, 1928; 29.150.12)

him. It was more common, though, for officers to wear a small metal skull-cap called a *secrète* under their hats.[20]

While less and less effective within Europe, helmets and body armor helped European armies conquer the world. The historian Geoffrey Parker has called violence the continent's leading export in the early modern age, and the Western exoskeleton and exocranium made terrifying impressions. In the early twentieth century, a Solomon Island informant recalled stories of his ancestors' first encounter with Europeans, probably Spaniards, who appeared to have a removable turtle shell on their bodies and heads and who laughed when spears glanced off their bellies. In the New World there is no known case of Indian arrows piercing European plate armor, and by the seventeenth century armor's price was surprisingly low. After one raid, two thousand helmets were dispatched from English armories to Virginia colonists who had petitioned for help. But in the colonies, too, heat, fatigue, and Indians' acquisition of firearms ultimately all but ended the use of helmets.[21]

Even in Europe, helmets never entirely vanished. Sappers, digging under heavy fire to attack enemy fortifications, had their own head coverings. The French military continued diligently to develop new models, which weighed up to 6.7 kilograms. On the other hand, the cavalry helmets of the nineteenth century, the most glorious survivors of medieval and Renaissance tradition, were far less suitable for the demands of peacetime, let alone battle. In 1865 a correspondent of the *Times* of London called attention to the plight of an orderly of the heavy cavalry who was barely able to control his runaway horse because he had to use one hand to hold down his helmet with its "monstrous tail." Calling for the present model to be retired to a museum, he repeated an "authentic story" of a regiment of heavy cavalry "charging with their swords dangling from their wrists while they held on their helmets."[22]

USER ENGINEERING

Nineteenth-century military helmets may have been better designed than they now appear, but even at their best they did not point the way to the future of head protection. Twentieth-century helmets probably owe more to a popular tradition that had been developing over the centuries. From medieval times, soldiers, firemen, and others without access to prestigious armorers had headgear and other equipment made from leather processed for strength and durability, called *cuir bouilli*—"boiled leather"—and sometimes stitched together in sections to form protective combs or ridges.[23]

A New York firefighter, Jacobus Turck, may have introduced a similar design in North America before 1740; we know only that it had a high crown and narrow brim. Later in the century, the volunteer companies of the newly formed United States of America were proudly civilian and wore what was becoming the mark of Western respectable manhood for most of the nineteenth century, a cylindrical hat. The firefighters' model was made of leather. The most influential craftsman, a firefighter himself, was the New Yorker Henry T. Gratacap, who developed an improved formula for processing leather to hold its shape and withstand heat. He adapted the form with raised stitching and neck protection that had been used for sappers' and fire helmets in Europe, curving the brim to channel water more efficiently and adding a front plate, held by an eagle, for company insignia. Gratacap remained in business from 1836 to 1868, selling up to a hundred helmets a week nationally, and the design remained the U.S. standard through most of the twentieth century.[24]

Eighteenth-century English firefighters, employed by private insurance companies, used similar leather helmets, but ambitious officials in the nineteenth century were not satisfied with these serviceable hats. From the 1700s, French fire services dressed in a military style, and during the Empire and Restoration the *sapeurs-pompiers* of Paris have remained members of the French army. (Marseilles is protected by the battalion of *marins-pompiers,* part of the navy.) French firefighters had a succession of magnificent steel and brass helmets which even with insulation must have conducted heat all too well. The Metropolitan Fire Brigade of London wore a leather helmet in ancient Greek style until 1866, when its new chief officer, the charismatic Eyre Massey Shaw, introduced a French-inspired brass helmet of memorable panache that remained in service until the electrical fires and shock hazards of the twentieth century forced development of a new compacted cork model covered with special insulating cloth. European and American fire companies evoked two different kinds of tradition, one knightly and paramilitary, the other utilitarian and civic. But the glory of the European style may well have made up in morale and in images of popular culture what it lacked in insulation. (Today's European firefighters are likelier than their more conservative American counterparts to have astronaut-style helmets with heat-reflective visors and built-in two-way radios.)[25]

The nineteenth century's villains, like its heroes, saw new possibilities for helmets. The New York City gang called the Plug Uglies earned their name by stuffing their high-crowned felt (plug) hats with cloth as primitive head protection. (The London police are believed to have reinforced their top hats similarly before the adoption of the high cork helmet.) Decades later, the Australian bandit Ned Kelly wore ninety-seven pounds of armor, said to have been made from old plowshares by outback blacksmiths. He capped it with a massive headpiece like a larger version of the medieval great helm. The armor was proof against the Martini rifles used by the police, and Kelly was able to keep them at bay for months. In the end he was shot in the legs, captured, and executed, but he was the only person in the century, civilian or military, to have given a public demonstration of the tactical value of a helmet. (There were many experiments with body armor, including some in the U.S. Civil War, and even a trade of European armorers supplying South America and Africa with chain mail garments, but their only other publicized use was in the Boer War, and few details were known by the time World War I broke out.)[26]

What is surprising about the nineteenth century is the number of hazardous trades without hard headgear: miners, building laborers, workers in heavy industry. Only relatively late in the century did Western people develop the idea of an accident that was not just an individual's misfortune but a symptom of social injustice. Early philanthropies like the Red Cross were intended for the military wounded, and many of their supporters bitterly opposed initiatives to extend help to civilian casualties. Gradually the law began to accept a principle of employer responsibility—without culpability—for worker injuries. Eventually governments, insurers, and other third parties would force or persuade employers to provide protective equipment, but before 1914 there is little evidence that even trade unions made this a high priority.[27]

THE EXOCRANIUM RETURNS

When World War I began, one military item that no nation had on hand in large numbers, or even in planning, was the helmet. To withstand direct rifle fire, the armor of the time would have needed to be as ungainly as Ned Kelly's, no doubt considered a bizarre atavism. But a big surprise in weaponry helped bring back metal headgear. In the 1890s, French officer-inventors showed how a blast's energy could be used to return an artillery piece to its original position, so it could be aimed more accurately on every shot with the help of forward observers, and reloaded four times more often. England alone fired almost a million rounds on a single day in September 1917, and more than five million tons during the entire war. Wounds from artillery fragments accounted for more than half of all casualties, significantly more than rifle or machine-gun bullets. While machine guns and artillery could be equally deadly to troops going over the top in battles like the Somme, barrages were constantly menacing troops still in the trenches, devastating them psychologically as well as physically. Nervous breakdowns were called shell shock for good reason.[28]

At the beginning of the war only the heavy cavalry of the major nations still wore a helmet and breastplate, but they could do little at the front. A French officer who had proposed a helmet was rebuffed by General Joseph Joffre, who did not believe the war would last long enough to put it into production. Like many other innovations we have seen, the first helmet arose by improvisation or lucky accident in the field. A wounded soldier told Intendant (Quartermaster) General August-Louis Adrian that a mess bowl in his hat had saved his life, and the general had a metal skullcap made like the old *secrète* that fit under the service cap. Its favorable reception encouraged him to develop Europe's first new standard-issue helmet since the pot helmet of the early seventeenth century. But Adrian did not return to the pot; he and his staff modified the standard French fireman's helmet of the time, giving it a slightly shallower brim and lower crest but leaving the bowl's shape unchanged. This simplified the tooling for the manufacture of hundreds of thousands.[29]

The bowl was the only thing simplified and rationalized about the Adrian helmet. Its predecessor may well have been based in turn on a helmet from Melos in the Louvre, of a Hellenistic type called the neo-Attic. Despite wartime pressures, it was surprisingly complex to produce, demanding seventy stages even after preparation of the metal components. The slot and crest that added to the time and expense also weakened the helmet's structure and added (with other ornamentation) at least a hundred grams to its weight. Yet Bashford Dean saw France's reasons for keeping it. Its beautiful form reflected the work of the immensely popular military artist Édouard Detaille and helped build the troops' spirit and morale. The wearer, Dean wrote, "becomes fond of his helmet and his feeling toward it is a distinct asset. . . . He is convinced that its shape is excellent, he is accustomed to its lighter weight, and he would gladly wear it under conditions in which he would probably cast aside a heavier and a better helmet." As we saw earlier, the French army also had a style of marching and fighting that emphasized mobility and élan rather than momentum. The Adrian helmet as a technology was well matched to the French technique of war.[30]

The British helmet showed a radically different approach to design. Its shallow bowl permitted the use of relatively thick steel that could be formed in a single pressing while keeping its thickness. Like its predecessor the kettle hat, it protected the shoulders as well as the head from objects falling from above, and offered good protection from direct bullet hits. Simplified production also mattered to the U.S. Army, which placed an initial order for two million, at the time no doubt the largest single helmet project in human history. Like the French soldier, the British Tommy seems to have identified with his headgear. Yet Bashford Dean, observing the effectiveness of the U.S. model based closely on it, was disappointed with its exposure of the back and sides of the head.[31]

While the French and British/American helmets were classics in their own way, the helmet that proclaimed the twentieth-century revival of the exocranium was German. The German military, so renowned for its meticulous planning, had not realized that a leather shortage would force the fabrication of pressed-paper substitutes for the South American hides in its leather spiked helmets. The French innovation arose from a soldier's happy accident and the English from a practical manufacturer's proposal; the German helmet was a scientific project. Friedrich Schwerd, an engineering professor, collaborated with August Bier, who held a chair of medicine and was using powerful magnets to remove metal fragments from wounded soldiers' heads. The surgeon had been heartbroken by the terrible brain damage done by pieces no larger than cherry pits. Many initial survivors were dying later in agony. The army command soon brought Schwerd to Berlin on Bier's recommendation, and Schwerd identified the crucial element for a new metal helmet: "a neck protector which stands away from the neck and extends forward to the temples and over the brow." And he knew his military history, citing the German sallet (*Schallernhelm*) as his model.[32]

Schwerd was firm about avoiding the temptation to strengthen the helmet's front against direct rifle fire; he knew that riveting extra pieces would weaken the helmet and that additional weight would discourage troops from wearing it. Several firms competed to find the best metallurgical formula, and experts even measured ventilation, which turned out to be superior to the leather helmet. For the tradeoff between weight and protection there was an apparently ingenious solution, a detachable plate that could be mounted on the helmet's front with two lugs that doubled as ventilation holes. No other piece of armor in world history had been evaluated by so many experts. For the thickness of the steel, the shape was remarkably deep; a U.S. manufacturer was not able to press steel of a similar formula into the shallower British bowl shape.[33]

The German steel helmet revealed the paradoxes of protection. A sincerely humanitarian device, it was simultaneously a tool to help German troops kill as many of the enemy as possible. It combined an evocative, almost romantically medieval form, with sophisticated medicine and metallurgy. After the war, veterans adopted it or rather recycled it as a paramilitary icon; the largest and most influential of the postwar militias was called the *Stahlhelm,* and even before the appearance of the National Socialist Party, the ultra-rightist Ehrhardt Brigade wore swastika-decorated steel helmets.[34]

The helmet remained a foundation of twentieth-century warfare with remarkably few changes in principle. In World War II the Germans and the British held to improved versions of their designs from the Great War, and Japan chose a British-influenced shape rather than any reflection of their traditional armor or their ally's *Stahlhelm.* America used a semispherical bowl like that of the seventeenth-century pot, with only a slight brim and visor. The design was actually the work of Bashford Dean's years as a major in the U.S. Army, not only evaluating other nations' armor but preparing new designs. Dean especially favored one, No. 5, for its combination of protection—the smooth, round shape maximized the chance that a bullet would bounce off—and simplicity of production. And medieval skills returned in the months before Pearl Harbor, as Dean's successor at the Metropolitan Museum, Stephen V. Grancsay, helped the army prepare an aluminum master of a new model in the museum's fully equipped armorer's shop, which in turn was used in tooling for production. At the suggestion of General George S. Patton, the hexagonal-webbed football helmet suspension recently invented by John T. Riddell was adapted as a separate plastic liner. The resulting M-1 steel helmet was formally adopted on June 9, 1941. The McCord Radiator and Manufacturing Company in Detroit learned how to form the seven-inch-deep bowl in a single pressing—an engineering milestone—and reduced the full production time for all twenty-seven operations from steel blank to finished helmet to only twenty-two minutes. Over four decades 25 million were to be produced, making the M-1 the most widely used helmet model in history.[35]

The M-1 was the elegant answer of American modernism to the ominous angularity of the German model. And like other classic technologies it opened unplanned opportunities for user improvisation. At the end of its production a U.S. master sergeant wrote nostalgically of its uses: as a seat, pillow, washbasin, cooking pot, nutcracker, tent-peg pounder, wheel chock, and even—with the explosive from an unserviceable Claymore mine—popcorn popper. (Since heat degraded steel shell, the army tried to discourage improvisation.) The M-1 prevailed through the Vietnam era and beyond.[36]

During World War II, Germany's specialists were increasingly aware of the *Stahlhelm's* shortcomings, especially its weight, its high exposed front, and the weakened angles between the crown and the visor and neck protector. A new, sloping design was proposed but rejected by Hitler, who shared his generation's attachment to the *Stahlhelm*. It was the army of the new German Democratic Republic, anxious to avoid both Western designs and the helmets of its Soviet masters, that finally adopted Model B/II and used it until reunification. Whether or not any of the parties to the Cold War realized it, this design strongly resembled another U.S. experiment, a "deep salade" (*sic*) that Bashford Dean had developed with the Metropolitan Museum's armorer and the Ford Motor Company in 1917 and that had impressed President Woodrow Wilson himself.[37]

In 1978, Captain Schwerd may have had the last laugh. With the recently invented polymer Kevlar (phenolic polyvinyl butyryl) as reinforcement, patented and trademarked by DuPont, the U.S. armed forces were able to create a laminated helmet no heavier than the M-1 with superior ballistic protection. The army's designers insisted that the new helmet, based on exhaustive scientific measurements of actual heads, was engineered from scratch and that similarities to the *Stahlhelm* were superficial. In any case, the new model proved itself from the Grenada campaign to the Gulf War, and sarcastic nicknames like "Nazi helmet" and "Kevlar Fritz" faded away as the shape became more familiar. Indeed, the sallet was, as we have seen, not originally German at all.[38]

THE INDUSTRIAL FRONT

From the trauma of the Great War a new attitude spread to civilian life, that hard protective head coverings were emblems of rational courage and even extensions of the self.

Some European miners had long reinforced their headgear—Cornish miners wore heavy felt hats treated with resin against rock from tunnel ceilings. Some U.S. miners improvised similar protective head coverings.

But according to a U.S. Bureau of Mines official interviewed in the early 1920s, a surge of new helmet designs had been inspired by miners' wartime appreciation for their helmets. An Oklahoma zinc miner declared, "If the old tin hat would stop shrapnel, I reckon it'll stop these pieces of jack." At least one mine operator handed out military trench hats to its workers at its own expense. The helmets had become common in California and West Virginia as well.[39]

The U.S. Navy also helped civilian innovation. Edward W. Bullard, whose family made miners' carbide lamps, was impressed by the English-style helmet he wore as a doughboy. The navy, alarmed by frequent and severe head injuries in shipyards, asked Bullard to develop head protection for civilian workers, just as munitions plants in England during the war had (as we have seen) installed what may have been the first large-scale industrial posture seating. Bullard designed and produced a hat from alternating sheets of canvas and glue, painted black, with overlapping front and rear brims. Set with steam, it was called the Hard-Boiled Hat and introduced in 1919—in its layered construction, a predecessor of the Kevlar helmet of the 1980s. It was the first headgear made to protect miners from falling objects. Head protection was extended only gradually to other industries, probably through purchases by individual workers. The first construction site to require hard hats was the Golden Gate Bridge; its head engineer, Joseph B. Strauss, was alarmed by injuries from falling rivets, and worked with Bullard, still a San Francisco company, to adapt the Hard-Boiled Hat for heavy construction. In the same year, 1938, Bullard produced the first aluminum hard hat, tough but light.[40]

By America's entry into World War II, there was an active market for industrial safety headgear. On the very day of Pearl Harbor, December 7, 1941, the *New York Times* noted that a Department of Agriculture laboratory had developed "a plastic helmet out of heavy cotton cloth and soybeans" that could shield civilian workers' heads from forces of up to forty pounds. Wartime authorities accelerated industrial protection. A catalogue of fire and police equipment published by a Detroit company during the 1940s features a GI in a pot helmet, furiously firing a machine gun, juxtaposed with a civilian fireman in a Gratacap-style hat, more calmly discharging the company's latest carbon dioxide extinguisher. A lighter version of the doughboy hat is listed with a reminder that "U.S. Service Inspectors require uniformed guards, safety patrolmen and maintenance men to wear steel protective helmets." The company offered "Head Protector" helmets of molded vulcanized fiber, "cradle cushioned against blows," noting that "thousands" were in use around the world "in mines, mills, on construction projects, and where men are subjected to falling objects." The Head Protector had an almost spherical dome and a 360-degree brim set at an angle. There were also "Hedgard Safety Hats" of unspecified material but with a novel suspension cradle, perhaps inspired by the M-1's new system. The catalogue notes that protective headgear would pay for itself in lower insurance rates, a sign that while relatively new it was privately recognized if not yet legally required in most applications.[41]

Industrial helmets flourished in the postwar years and even became emblems of American engineering in the 1956 novel of Helen Marie Newell, *The Hardhats.* The author had grown up in Idaho construction camps and worked as a wartime aircraft mechanic. An *Idaho Voter* review noted that "those metal contraptions that save men's lives" were "familiar objects to Westerners," yet they were also still strange enough to make construction workers look like "creatures from Mars." In November 1958, *Popular Science* ran a two-page spread featuring fifteen styles of work helmet. And civilian head protection became newsworthy. In 1959, under Governor Nelson Rockefeller, the New York State Commerce Department worked with an advertising agency to distribute aluminum safety hats with the state seal to the Empire State's top officials, extolling its citizens as "hard hat doers" and promising companies growth and prosperity "in this hard hat climate."[42]

By the late 1960s, bipartisan opinion in the U.S. Congress favored new legislation to reduce an alarming number of industrial accidents. President Richard M. Nixon called for better personal protection for the nation's workers in August 1969, and by the end of 1970 he had signed the law creating the Occupational Safety and Health Administration (OSHA), a new national authority to make and enforce standards of equipment and practice, of course including work helmets. More than ten years after Rockefeller's failed public relations campaign, there were already many construction projects in New York City with workers wearing mandatory helmets, now widely required by contractors if not by state and local laws. In May 1970, several hundred construction workers came to Wall Street as counterdemonstrators when antiwar students were protesting the shooting of students by National Guardsmen at Kent State University in Ohio. Workers were reported to have pursued the longest-haired students and "swatted" them with their helmets; some "hard hats" countered that they had been provoked by a middle-aged ringleader who spat on an American flag, and others observed that white-collar Wall Streeters were at least as

numerous and forceful as the workers. On May 20, about 100,000 construction and dock workers marched from City Hall to Battery Park with
patriotic songs and slogans. It was almost as though industrial safety,
engendered by the army and navy, was coming to its original patrons' aid.[43]

Whatever the reality, reports of workers attacking student leftists
caught journalists' and social scientists' attention. Political stickers and
American flag decals turned out to reflect a tradition of customizing and
individualizing gear, rather than to be the signature of any new paramilitary force. For protective equipment makers, many of the head coverings
were not properly hard hats at all but hard caps, having only small visors
rather than full brims.[44]

As protestors and counterdemonstrators continued to clash, the police,
too, started wearing helmets more often. In fact, New York patrolmen
assigned to turn off illegally opened hydrants as early as summer 1961
were issued plastic helmet liners to shield them from bricks and bottles
hurled by defiant youths and parents. And as helmeted riot police began to
bear down on protesters in the late 1960s and 1970s, demonstrators in
turn improvised their own head protection. Joschka Fischer, the future
foreign minister of Germany, wore a black motorcycle helmet in his youthful rumbles with white-helmeted officers, while in 1969, a year after the
disastrous Democratic convention in Chicago, the street-fighting Weather-
men faction of Students for a Democratic Society faced Chicago police
again; now the demonstrators wore motorbike helmets and carried batons
of their own. When a new cadre of militant demonstrators emerged in the
antiglobalization protests at the Genoa and Seattle World Trade Organization meetings, head protection had become de rigueur on both sides. In
Los Angeles, the police-centurion image has so prevailed that *LA Weekly*
joked in 2002 that "riot helmets look as natural on cops as mustaches."[45]

Does their new armor help the police? British commentators of different political outlooks, whose government is replacing the traditional high-
crowned helmet with a squatter and more impact-resistant mutant, have
second thoughts. Just as we have seen that the zori sandal encouraged a
certain gait in Japan, the 1864 helmet promoted, according to Andy Beckett of the *Guardian,* a distinctive walk: "very upright, chest out, arms
swinging stiffly, the hat brim down low to stop it wobbling, eyes slightly
narrowed in consequence," setting the wearer apart from lesser security
officers who were "jittery solutions to Britain's current disorders." And the
new helmet might even subtly provoke rather than deter confrontation. To
Tom Utley, a columnist for the *Telegraph,* a new "practical" helmet will say

!WHAT MRS. GRUNDY SAYS ABOUT THE HELMET.

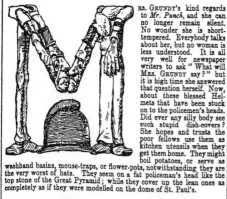

RS. GRUNDY'S kind regards to *Mr. Punch*, and she can no longer remain silent. No wonder she is short-tempered. Everybody talks about her, but no woman is less understood. It is all very well for newspaper writers to ask "What will MRS. GRUNDY say?" but it is high time she answered that question herself. Now, about these blessed Helmets that have been stuck on to the policemen's heads. Did ever any silly body see such stupid dish-covers? She hopes and trusts the poor fellows use them as kitchen utensils when they get them home. They might boil potatoes, or serve as washhand basins, mouse-traps, or flower-pots, notwithstanding they are the very worst of hats. They seem on a fat policeman's head like the top stone of the Great Pyramid; while they cover up the lean ones as completely as if they were modelled on the dome of St. Paul's.

Is there no feeling for Art at Scotland Yard, or is it necessary that the Human (X 32) Form Divine be rendered hideous as well as insolent? Who would like to be "moved on" by such an absurdity? A noble classic Helmet, with vizor, ear flaps, and a highly ornamental cat on top, would have been better; or if COMMISSIONER MAYNE is really and truly wedded to ugliness, he need not have gone further than the "pot-board" in his own back kitchen, the old iron saucepan is always ready to his hand.

There, it serves him right, why did he not seek her aid, instead of taking his own silly, obstinate way. She thinks if he had but asked her opinion, she could have helped him a little. . She has been turning her thoughts to Policemen's hats lately, and with no little success. Seeing that the old Policeman X in his long-faced hat, who was the

image of her glass mustard-pot, has only been replaced by new Policeman X in his helmet, the picture of her pewter pepper-castor, she looked up a few more domestic models.

The Beer-Jug, symbolic of the servant with whom the gallant officer chats at the street corner;
The Flower-Pot, because it suggests nothing in particular;

The Dinner-Bell, to mark the emptiness of his head; and
The China Tea Saucer, as a hint of the shallowness of his brain.

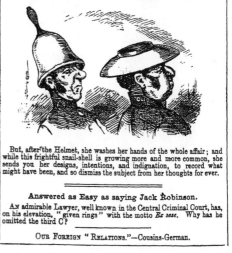

But, after the Helmet, she washes her hands of the whole affair; and while this frightful snail-shell is growing more and more common, she sends you her designs, intentions, and indignation, to record what might have been, and so dismiss the subject from her thoughts for ever.

Answered as Easy as saying Jack Robinson.

AN admirable Lawyer, well known in the Central Criminal Court, has, on his elevation, "given rings" with the motto *Es sese*. Why has he omitted the third C?

OUR FOREIGN "RELATIONS."—Cousins-German.

to many people, " 'Here is a man who is dressed to be thumped.' They will therefore thump him." Protection, whether of police or of demonstrators, can edge into provocation.[46]

Work helmets have long lost the aggressive populism of the 1970s. The grand tradition of improvising, whether in the military or on the job, is discouraged. The armed forces' Kevlar helmets are unsuitable for cooking and their camouflage designs do not invite Vietnam-era embellishment. In civilian as in military life, equipment is better designed than ever, with superior materials, and there are even steps to build in ventilation, relieving one of the helmet's oldest ergonomic problems.

SKID LID KIDS

Once helmets were revived for soldiers and extended to many workers, safety head coverings for athletes almost inevitably followed. Part of the reason was cultural: if a helmet represented the courageous infantryman or miner, it could call attention to the rigor and even danger of sport.

As in mining and police work, reinforced hats were the first safety headgear. The top hat of the nineteenth and twentieth centuries is the descendant of a simple round felt hat—along with the frock coat, originally the plain garb of the common man, as opposed to the wigs and cocked tricorn hats of courtly dress. For hunting and riding, a round hat could be and still is made, reinforced with additional material.

The first documented hard civilian sports helmets appear to have been British flying ("crash") helmets sold in 1923 as Royal Air Force surplus. By the early 1940s specially produced head coverings with vulcanized rubber or pulp shells and web sling suspensions, originally designed for racers, were widely used by soldiers and civilians. Possibly the first scientific study of noncombat helmet effectiveness was published in 1941 by the military neurosurgeon Hugh Cairns. He showed the importance of head injuries in motorcycle casualties and began the scientific evaluation of materials, finding pulp four times more effective than rubber.[47]

Jockeys were the next civilian athletes to use hard head protection after the war. By 1924 lightweight but strong fiber helmets fitting under caps like the old steel skullcaps were introduced in Australia, and the next year they were made mandatory in U.S. steeplechase events. They spread to Thoroughbred racing. In 1927 a similar helmet saved the life of a jockey rolled over and kicked in the head by his runaway colt. Head protection is now required in public riding schools and events in many U.S. states.[48]

The most influential and controversial of the rigid sports helmets was developed by the late 1930s for American football. Originally bareheaded, players of the 1890s had began to use homemade mohair-cushioned leather helmets as play got rougher until manufacturers started to offer what was probably the first ready-made athletic head protection, around 1900. For better impact absorption, the head was isolated from the shell by a web of fabric straps in 1917, a system marginally improved by innovations of the 1920s and 1930s. The breakthrough was a plastic shell with a new web suspension, adjustable for head size. Unlike leather, the lighter plastic allowed the riveting of the suspension, and did not mildew. But even before U.S. entry into the war and the suspension's adoption by the army as the foundation of the M-1 helmet system, it was kept off the market by materials shortages.[49]

After some initial failures following the war, the National Football League admitted the helmet in 1949. By the early 1950s it had virtually replaced the leather helmet. In *Why Things Bite Back* I told the story of its paradoxically catastrophic effect on injuries. It reduced some head damage but was held responsible for tripled neck injuries and a doubling of deaths from cervical spine injuries. Such casualties have resulted in lawsuits that have continued to plague the U.S. sporting goods industry. In 1954, *Sports Illustrated* called them "Martian-looking headpieces" and "the most lethal weapon" for wearer and opposition alike. But the real problem was not in the technology but in the technique. There is no evidence that the Riddell company, any more than so many other innovators, foresaw the change in user behavior that its products helped bring about. Coaches had once instructed players to tackle ball carriers by wrapping their arms around them. Their new technique instead used the helmet as a battering ram not only to stop the carrier but to dislodge the ball—ultimately by aiming below the victim's chin, hoping to knock him out. While helmet technology has continued to evolve since the early 1990s, and safer designs have been promised, the underlying problem remains from football's early years as a spectator sport: fans enjoy and encourage violent plays. For the determined developer of new aggressive techniques, the challenge is finding the loopholes in the new rules.[50]

After the war, pulp and rubber were succeeded by generations of new plastics. Aerodynamic contours replaced the old pudding-bowl design. Top automotive racing models incorporated ventilators and antilift design, dramatizing rather than concealing risks. In fact, racing helmets can be so heavy and forces so great that drivers may need a supplementary restraint such as Peter Hubbard's HANS (Head and Neck System) to keep the helmet from whipping the neck fatally in a crash. While the death of a prominent amateur auto racer, William "Pete" Snell, in 1958 converted most of his fellow competitors to the idea of protection and led to the establishment of a leading safety equipment research foundation in his memory, most organized U.S. motorcycle enthusiasts question mandatory helmet legislation, observing that it has reduced fatalities mainly by discouraging motorcycle riding. In accidents at road speeds above fifteen miles per hour, some U.S. opponents of helmets add, the additional mass of the helmet only helps substitute neck for head injuries. Yet emergency-room surgeons treating motorcycle accident victims remain among the strongest helmet advocates, and many cyclists never ride without head protection.[51]

In hockey and most other sports, the helmet is a tool, not a totem. In baseball and cricket, it is worn only by batters, when at bat. In the early twenty-first century all hockey players wear helmets; the handful of holdouts, grandfathered when protection became mandatory in 1994, are now retired. Even the mocking adoption of flimsy helmets designed for less hazardous ice sports appears to have vanished. The hockey player's helmet is but another element of specialty clothing, a routine necessity like automotive seat belts. Controversy in most sports is not about whether helmets should be worn but about the best balance between comfort and protection. The American Society for Testing and Materials (ASTM), the Snell Memorial Foundation, and other national and private laboratories constantly develop new tests for impact attenuation, just as medieval helmets and armor were proved.[52]

Thanks to a constantly expanding armory of energy-absorbent materials, construction features, and simulation equipment, the helmet now represents both the triumphs and the paradoxes of the technological reduction of risk. There is no universal sports helmet, and the U.S. Consumer Product Safety Commission (CPSC) advises purchasing a separate, purpose-designed model for each sport. By now the list includes not only football and hockey but lacrosse, skateboarding, snowboarding, in-line skating, BMX cycling, equestrian sports, extreme sports, and boxing. After a number of celebrity fatalities, skiing helmets have gained support. Many medical researchers now favor helmets even for youth soccer, where concussions and brain damage have turned out to be more common than most parents and coaches had realized. And the middle-aged are even more likely to suffer soccer and bicycling injuries than their children, partly because they are less flexible about adopting protection. (Paradoxically, the rate of injury from in-line skating is lower than that from bicy-cling and less than half the injury rate from softball, because in-line skat-ing looks so dangerous that protective equipment has been the norm.)[53]

Among these varieties, the most ubiquitous symbol of the rebirth of head protection remains the bicycle helmet, the commonest, lightest, most colorful, and most controversial of all. Yet it is also relatively recent. It began with the surge of mountain bicycles in the mid-1980s, and was led by Bell Sports, the largest maker of auto racing helmets. Through the 1980s and 1990s, the size of the potential market encouraged Bell and others to increase the appeal of protection by using new and lighter mate-rials and ventilated aerodynamic shapes to reduce weight without degrad-ing performance, although tooling for some of the most striking models requires (like so many other high technologies) time-consuming and high-priced hand craftsmanship.[54]

Thanks to stylish design and growing markets, a once stodgy idea became sexy and youthful. In 2001, the CPSC reported that 69 percent of child cyclists and 43 percent of adults in the United States wore helmets. Yet this apparent success has turned up a paradox. In the decade from 1991 to 2001 the surge in helmets was accompanied by a decline in rider-ship and an increase in cyclist accidents, resulting in 51 percent more head injuries per bicyclist. To some opponents of mandatory helmet laws, these statistics are the result of what is called risk compensation, bolder behav-ior arising from the feeling of enhanced security. And there are indeed peo-ple who will push the envelope of any safety device. But most bicycle safety advocates reject an increase of risk taking as an explanation. They believe the CPSC statistics of helmet use may be exaggerated, and they point to adverse traffic conditions, more aggressive behavior by motorists (attributed in turn by risk compensation theorists to the reassurance of seat belts and air bags) and faster bicycles (which the same theorists would probably explain by the perceived safety of helmets). One major point on which both sides appear to agree is that helmets offer limited protection in automobile-bicycle collisions.[55]

The most powerful case against mandatory bicycle helmet laws may be that they ignore the primary role of motorists in cycling injuries and trans-fer too much of the burden of safety to the cyclist. There is evidence that they also discourage cycling, especially because effective helmets are not cheap relative to the price of bicycles and because they must be replaced after absorbing shocks and as children's heads grow. Thus the British Med-ical Association has endorsed voluntary bicycle helmet use but opposed a legal requirement, on the grounds that the health benefits of bicycle exercise far outweigh risks of injury.[56]

Today's sports helmet does the same job as the ones introduced in the ancient Near East. It spares the wearer a skull fracture or concussion by spreading the energy of an impact, no longer with a mace tip but with an asphalt surface. And, like the ancient helmet, it can build group spirit. Ancient, medieval, and twentieth-century military helmets were designed in part to foster morale; in the view of critics, the sporting helmet may create an excess of confidence. And many wearers of today, like Greek hoplites and medieval soldiers before them, complain of stuffiness, heat, and sensory barriers. The similarities between warlike and recreational helmets go even further, in the view of at least one commentator. In 1984, a twenty-nine-year-old writer and amateur hockey player, Gregory Bayan, defended his bareheaded ways in *Newsweek*. Lamenting the loss of daring individuality from Olympic and professional hockey, regretting the appearance of televised players as "identical automatons," he compared the helmet to "a dorsal fin that breaks water and heralds the presence of a leviathan . . . something deeply wrong in America." To Bayan, the helmet's enthusiasts were "Hardheads," determined to eradicate individuality and risk in the name of safety.[57]

Yet the most curious result of the avoidance of risk was to appear only in the late 1980s, as the result of a successful medical campaign to reduce sudden infant death syndrome (SIDS). Parents were urged to put babies to sleep on their backs so as to lower the risk of their suffocating. The unintended consequence was that many infants spent so much time on their backs that their heads became flattened, a condition called plagiocephaly. While mild cases can self-correct if parents vary their babies' positions while the babies are supervised or awake, thousands of infants have been wearing custom-made orthopedic helmets to correct the condition. There are already seventeen companies making them in the United States alone. This therapy is a small price to pay for occasional and reversible mishaps of a lifesaving technique, but it also shows that one is never too young to become a Hardhead. The helmet was born in ancient warfare, and in wearing it we have become men, women, and children from Mars.[58]

Epilogue

Thumbs Up

E VERYDAY TECHNOLOGY sometimes reshapes the body; the feet of
shod people are different from the feet of those who have always
walked barefoot. More important, it helps shape how we use our
bodies. Technology, and the techniques of using it, have coevolved over
millennia.

The most challenging open question is whether mind, body, and
machine will fuse in some radical new way over the next generation. For
over fifty years, waves of enthusiasts have proclaimed the dawning of a
new age of augmented humanity. But they are far from agreed on the
mechanism. For some, the intimacy is limited to even more portable and
powerful versions of the devices we take with us now: computers that
might be as easily carried as cell phones and personal digital assistants
(PDAs) now are, and viewed through special eyeglass displays. Spectacles
could also transmit the emotional states of users so that a speaker, for
example, could detect a group's interest or boredom. There are already
sneakers that can transmit or record information on a runner's performance, and even civilian motorcycle helmets with intercoms and navigation
aids built in. Other enthusiasts scorn mere wearability; they are having
sensors and transmitters surgically implanted in their bodies, as some deaf
children and adults have been fitted with cochlear implants that restore
hearing. The cyborg or human-machine is probably the most powerful
and persistent idea, perhaps because it seems a logical next step from technological symbiosis. Politically the cyborg movement extends from Paul
Verhoeven's original *RoboCop* film to the work of Donna Haraway and

Chris Hables Gray, who see the connection between human and machine as an emancipatory strategy against rigid economic and gender roles.[1]

But is the body really becoming more mechanized? George Washington never wore wooden teeth, and his last set of dentures—made of gold plates inset with hippopotamus teeth, human teeth, and elephant and hippo ivory, hinged with a gold spring—was as good as the craftsmen of his time could produce. He still suffered great discomfort, eating and speaking with difficulty. Yet perhaps this reserve only enhanced his dignity. In any case, it should not be surprising that one in ten Americans had an implant of some kind, not including dentures, by 2002; the nation was founded by a cyborg. Nor was Washington an isolated case. Thomas Jefferson's polygraph (not a lie detector but a linkage and second pen automatically tracing a duplicate of his writing) and semireclining work chair and Benjamin Franklin's bifocals were also giant steps in human-mechanical hybridization. (John F. Kennedy continued the cyborg tradition as one of the first political adopters of the robotic signature machine, a giant and distinctively American step in the cloning of gesture.)[2]

Without antiseptics or antibiotics, wounds in the U.S. Civil War demanded extensive amputation and created an innovative artificial-limb industry. Today, responsive advanced prosthetics, wheelchairs, vision implants, and other assistive devices exceed the nineteenth century's wildest dreams; there is even litigation in the United States on whether a teenage swimmer with an artificial leg was unfairly barred from wearing a flipper on it. But the first choice of medicine and dentistry is still the conservation of natural materials and abilities. The trend has gone from spectacles to contact lenses to laser surgery; dentistry has moved steadily from dentures to prophylaxis and conservation of endangered natural teeth. And some dental researchers believe that adults may be able to grow replacement teeth naturally, just as children teethe. Other regeneration, including the recovery of function for paraplegics and quadriplegics, may follow. And no robot can match the performance of a trained assistance dog.[3]

The body remains surprisingly and reassuringly conservative. The zori design is still used for some of the most stylish sandals for women and men. Even the majority of athletic shoes with the most technically advanced uppers and soles still use a system of lacing at least two hundred years old. The most advanced new office chairs still rely on the hundred-year-old principle of a spring-mounted lumbar support, for all their additional adjustments. Recliners still place the body in the same contours as the library chairs of the nineteenth century; interest in built-in data ports and other technological enhancements is fading, according to industry sources. Not only has the QWERTY arrangement resisted all reform movements, but even alternatives to the flat conventional keyboard are expensive niche products, partly because of innovators' marketing difficulties, but also because few users are willing to relearn techniques in the absence of discomfort. A century after the piano started to lose its prestige and markets, it remains the master instrument with its familiar keyboard. Computers help produce advanced progressive eyeglasses without the apparent breaks of bifocals, but wearers still hold glasses on their heads with the folding temples introduced in the eighteenth century. The latest NATO helmet echoes the outlines of the medieval sallet. But then, our foot bones and vertebrae and fingers and eyes and ears and skulls have not changed much, either. Even automatic transmissions rely on a familiar tactile principle, a knob or handle and lever; push-button shifting is largely a memory of the Edsel. And the twenty-first century's automobiles are still directed and controlled by wheels and pedals—familiar from early modern sailing ships and wagons—rather than the alternative interfaces that appeared in patents and experimental cars. Meanwhile, many technological professionals study body techniques that may need few or no external devices: yoga, martial arts, and the Alexander Technique.

Even the Christopher Columbus of wearable computing has misgivings about integrating himself with today's "smart" technology. Steve Mann holds an MIT doctorate in computer science and teaches in Canada; he was photographed wearing a helmet equipped with a video camera and a rabbit-ears antenna as early as 1980. In his book *Cyborg*, he acknowledges being "increasingly uncomfortable with the idea of a cyborg future," where privacy is sacrificed for pleasure and convenience to a degree he compares to drug addiction.[4]

Today's real advanced cyborg technology is actually neither utopian nor apocalyptic. Virtual-reality helmets are still not playthings but professional tools demanding rigorous training in physical and mental techniques to prevent disorientation and lapses in judgment. At the other extreme of complexity, miniature keyboards on cellular telephones and other devices are surprisingly influential. They have been shifting the balance of power of the human hand from the index finger to the thumb. We have seen that C. P. E. Bach elevated the thumb's role in playing the musical keyboard 250 years ago, and that the pioneers of touch typing rediscovered the fourth and fifth fingers but banished the thumb to tripping the space bar. Now the thumb is enjoying a second renaissance.

The thumb is coming back to computing with pen- and pencil-like devices, such as the styli used with PDAs. Even a radical new mouse, developed by the Swedish physician and ergonomist Johan Ullman, is gripped and moved around the desk with a pen-shaped stick, using the precision muscles of the thumb and fingers rather than twisting the hand and tiring the forearm, in Dr. Ullman's analysis. As for pencils themselves, they have had some of their strongest unit sales yet, increasing by over 50 percent in the United States in the 1990s.[5]

The biggest surprise for the thumb is electronic. In Japan today, there are so many new data entry devices that young people are called *oyayubi sedai*, the Thumb Generation. In Asia and Europe, these users have turned voice recognition on its head, sending short text messages to friends, thumbs jumping around their cellular keyboards. By 2002, there were over 1.4 billion of these transmissions each month in the United Kingdom alone. One British researcher, Sadie Plant, has found that thumbs all around the world are becoming stronger and more skillful. Some young Japanese are now even pointing and ringing doorbells with them. As Plant told the *Wall Street Journal*, "The relationship between technology and the users of technology is mutual. We are changing each other." The major laboratories did not predestine the thumb as the successor to the index finger, though they did help make the thumb's usefulness possible. The full capacities of the digit were discovered through the joint experimentation of users, designers, and manufacturers. And its new role is an expression of the intimate relationship described by Frank Wilson, writing of the "twenty-four-karat thumb": "The brain keeps giving the hand new things to do and new ways of doing what it already knows how to do. In turn, the hand affords the brain new ways of approaching old tasks and the possibility of understanding and mastering new tasks."[6]

In learning new body skills, we have sometimes neglected others. Scores of resting positions known to anthropologists are being replaced by a single style of sitting; humanity has been losing not only languages but body techniques. Infant feeders compete with the emotional and physiological rewards of nursing. Sneakers assist fitness programs, but some may interfere with older people's sense of balance. The reclining chair, originally sold partly as a health device, has become an emblem of the dangers of sedentary living. The piano's high plateau of development in the nineteenth century prepared the way for the player piano and ultimately for recorded music. Typewriter and computer keyboards have eliminated much of the grind of learning penmanship, but also the pleasure of a personal hand. The growth of public education has been accompanied by an increase of myopia, and of eyeglass wearing. And the helmet can promote as well as ward off danger. While augmenting our powers, these devices increase their power over us.

One challenge of advanced industrial societies is a degree of standardization that threatens to choke off both new technologies and new techniques. The remedy is a return to the collaboration between user and maker that marked so many of the great innovations, whether the shaping of the classic American fire helmet or the development of the touch method by expert typists and typing teachers. And research even in the most advanced technical processes confirms the importance of users. In the 1980s, the economist Eric von Hippel studied change in high-technology industries such as the manufacture of scientific instruments and semiconductors and the assembly of printed circuit boards. He found that up to 77 percent of innovations were initiated by users and recommended that manufacturers identify and work with a vanguard of "lead users." We have seen how nineteenth-century musicians worked with piano manufacturers, and how the typewriter entrepreneur James Densmore tested his ideas with the court reporter James O. Clephane. Even rank-and-file operators are able to help modify equipment and systems in the interest of safety and productivity. Cognitive psychologists who study work are rejecting an older model of a single, best set of procedures and learning from the way in which workers modify equipment to help do their jobs better. As Kim J. Vicente puts it, "Workers finish the design."[7]

Design should be not only user-friendly but user-challenging. The piano keyboard is rightly celebrated as an interface manageable for the novice and inexhaustible for the expert. Information interfaces should also invite the beginner while offering the experienced user the chance to develop new techniques rather than attempting to anticipate every desire. Participatory design, introduced by the mathematician and computer scientist Kristen Nygaard, began with Norwegian workers who wanted a say in the development of technology in their industries, but was ultimately embraced by corporations worldwide.[8]

The thumb keyboard has a sad aspect, the threat to handwriting traditions, whether Asian or Western. But it also has a positive side as a mark of human resourucefulness. Thus the thumb may be the best symbol for a new technological optimism based on user self-reliance, a proletarian digit resurgent in the digital age. The index finger signifies authority, marking regulations and warnings in texts, wagging and lecturing in person: the

Rules. The thumb connotes the practical knowledge men and women have worked out for themselves: the Rules of Thumb. And it represents tacit knowledge, the skills we can't always explain: as in the Green Thumb. Most of all, when extended in the almost lost art of hitchhiking, the thumb shows the right attitude toward the future, open and cooperative but with a firm sense of direction.

Notes

PREFACE

1. Michael M. Weinstein, "A Test You're Apt to Flunk," *New York Times,* March 28, 1993.

2. Donald A. Norman, *The Psychology of Everyday Things* (New York: Basic Books, 1988); Henry Petroski, *The Evolution of Useful Things* (New York: Alfred A. Knopf, 1993).

3. Dave Duehls, "Fit to Be Tried," *Runner's World,* vol. 28, no. 10 (October 1993), 26–27; Nicholson Baker, *The Mezzanine* (New York: Vintage Books, 1990), 16–18; Jerry Fodor, "The Appeal to Tacit Knowledge in Psychological Explanation," *Journal of Philosophy,* vol. 65, no. 20 (October 24, 1968), 627–28.

4. Robert D. Richardson, Jr., *Henry Thoreau: A Life of the Mind* (Berkeley: University of California Press, 1986), 290–91; Geoffrey Wolff, *The Duke of Deception: Memories of My Father* (New York: Vintage Books, 1990), 52; Ed Dentry, "Hikers Need Lessons in Tying the Knot," *Rocky Mountain News,* August 1, 1997; Melvyn P. Cheskin, *The Complete Handbook of Athletic Footwear* (New York: Fairchild Publications, 1987), 203–4; Joe Ellis, "Lacing Lessons," *Runner's World,* vol. 21, no. 4 (April 1986), 59; Jane E. Brody, "When the Elderly Fall, Shoes May Be to Blame," *New York Times,* February 26, 1998; Molly Martin, "Fine-Tuning the Fit: It's OK to Play with Your Shoelaces," *Seattle Times,* March 19, 1995.

5. Evelyn cited in Susan Swann, *Shoes* (London: B. T. Batsford, 1982), 20; Suman Bandrapalli, "Where Do Sneakers Come From?" *Christian Science Monitor,* December 1, 1998; Bill Taylor, "Not Fit to Be Tied," *Toronto Star,* February 14, 1998; Lord Baden-Powell of Gilwell, *Scouting for Boys* (n.p.: Boy Scouts of America, 1946), 223; British army boot lace requirement described by Phillip Nutt, telephone interview, November 15, 1999; Ian Stewart, *The Magical Maze* (New York: John Wiley, 1997), 198–203.

6. Helena de Bertodano, "Billionaire Scarred by Poverty," *Sunday Telegraph* (London), December 13, 1998; Jeff Meyers, "He Has Become Phantom of Marathons," *Los Angeles Times,* August 19, 1990; Patrick Reusse, "Cuba Dumps U.S. in L-o-n-g, B-o-r-i-n-g, Sloppy Game," *Star Tribune* (Minneapolis), July 30, 1992.

7. " 'Miracle': Rescuer Describes Man Forced to Cut Off Leg," *St. Louis Post-Dispatch,* July 23, 1993; Carolyn Hughes Crowley, "College Cribbers," *Washington Post,* January 6, 1992.

8. J. H. Thornton, *Textbook of Footwear Materials* (London: National Trade Press,

1955), 55–56, 210–14; Jeff Bailey, "Unfit to Be Tied: It Really Isn't You, It's Your Shoelaces," *Wall Street Journal,* January 28, 1998.

9. Stewart Brand, *How Buildings Learn* (New York: Penguin Books, 1995), 12–23.

10. Hans Fantel, "Portable CD Players Advance," *New York Times,* May 17, 1987; Jon Van, "Teletubby Infatuation Gives Fermilab Inspiration," *Chicago Tribune,* September 13, 1999; "Totally Random," *Scientific American,* vol. 278, no. 5 (November 1997), 28.

CHAPTER ONE

1. Sally Holloway, *London's Noble Fire Brigades* (London: Cassell, 1973), 51.

2. William Booth, "3 Bears Too Clever to Live; At Yosemite, Sharing Human Tastes Can Be Deadly," *Washington Post,* December 11, 1997; John J. Fialka, "Yosemite Bears Prefer Toyotas and Hondas for Late-Night Snacks," *Wall Street Journal,* January 13, 1999.

3. Sandy Bauers, "Philadelphia's Simian Version of the Great Escape," *Philadelphia Inquirer,* June 13, 1999.

4. Ruben Castaneda, "FBI Probing Canine Unit," *Washington Post,* April 4, 1999.

5. Jacques Ellul, *The Technological Society,* trans. John Wilkinson (New York: Vintage Books, 1964), 42–44, 134.

6. Ibid., 28.

7. See William H. McNeill, *The Pursuit of Power: Technology, Armed Force, and Society Since A.D. 1000* (Chicago: University of Chicago Press, 1982), 126–39; William H. McNeill, *Keeping Together in Time: Dance and Drill in Human History* (Cambridge, Mass.: Harvard University Press, 1995), 127–29; Geoffrey Parker, *The Military Revolution of the Seventeenth Century,* 2nd ed. (Cambridge, Eng.: Cambridge University Press, 1996), 6–24. John Keegan, in faulting McNeill for presenting the de Gheyn book as a drill book, argued that it is instead "an industrial safety manual" designed to keep musketeers out of each other's range of fire. But of course it had to be both, even if drill did not include modern cadenced marching. See John Keegan, "Keeping in Time," *Times Literary Supplement,* July 12, 1996, 3–4, on which I have also relied for some details of the preparation of *Weapon Handling.*

8. David Heathcote, "In Every Home an Architect," *Eye,* vol. 8, issue 31 (Spring 1999), 38–39.

9. Ellul, *Technological Society,* 13–15.

10. Marcel Mauss, "Body Techniques," in *Sociology and Psychology: Essays* (London: Routledge & Kegan Paul, 1979), 95–105.

11. Ibid., 99.

12. Ibid., 99–100; H. L. Brock, "The Gentle Art of Walking," in George D. Trent, ed., *The Gentle Art of Walking: A Compilation from the New York Times* (New York: Arno Press/Random House, 1971), 1–2.

13. McNeill, *Keeping Together,* 1–11; Mauss, "Body Techniques," 99, 114–15.

14. On the history of Prussian and German drill, see Mary Mosher Flesher, "Repetitive Order and the Human Walking Apparatus: Prussian Military Science versus the Webers' Locomotion Research," *Annals of Science,* vol. 54, no. 5 (September 1997), 463–87.

15. Jane E. Brody, "Baby Walkers May Slow Infants' Development," *New York Times,* October 14, 1997; Connell quoted by William Hamilton, *Morning Edition,* National Public Radio, September 14, 1998; Joyce Carol Oates, "To Invigorate Literary Mind, Start Moving Literary Feet," *New York Times,* July 19, 1999.

16. Flesher, "Repetitive Order," 468; Jan Bremmer, "Walking, Standing, and Sitting in Ancient Greek Culture," in Jan Bremmer and Herman Roodenburg, eds., *A Cultural History of Gesture* (Ithaca: Cornell University Press, 1992), 16–23; Fritz Graf, "The Gestures of Roman Actors and Orators," in ibid., 47, 49.

17. Nirad C. Chaudhuri, *The Continent of Circe* (London: Chatto & Windus, 1965), 226–27; Nicholas D. Kristof, "Walk This Way, or How the Japanese Kept in Step," *New York Times*, April 18, 1999; Tim Ingold, "Situating Action V: The History and Evolution of Body Skills," *Ecological Psychology*, vol. 8, no. 2 (1996), 171–82; Steele F. Stewart, "Footgear—Its History, Uses and Abuses," *Clinical Orthopaedics and Related Research*, vol. 88 (October 1972), 127; Koichi Kishimoto, "Japan Skipped Along to Power—and Owes Debt to Foreigners for It," *Daily Yomiuri* (Tokyo), November 5, 1997.

18. N. C. Heglund et al., "Energy-Saving Gait Mechanics with Head-Supported Loads," *Nature*, vol. 375, no. 6526 (May 4, 1995), 52–54; Eugenie Samuel, "Walk Like a Pendulum," *New Scientist*, vol. 169, no. 2273 (January 13, 2001), 38.

19. Mauss, *Sociology and Psychology*, 116; Thomas V. DiBacco, "Way Cool! Today's Common Sense Was a Long Time Coming," *Washington Post*, November 4, 1997.

20. Robert McKay, "The Amazing Dr. Henry Heimlich," *Saturday Evening Post*, November 1986, 42–45; Candy Purdy, "When People Choke," *Current Health 2*, vol. 16, no. 6 (February 1990), 24–25; Fenly [sic], "Simply Brilliant: 'Maneuver' Is Just One of His Important Medical Discoveries," *San Diego Union-Tribune*, February 12, 1990; Henry Heimlich, "Don't Slap a Choker on the Back," *New York Times*, July 12, 1988; Pamela Warrick, "Heimlich's Audacious Maneuver," *Los Angeles Times*, October 30, 1994.

21. Fenly, "Simply Brilliant"; Warrick, "Heimlich's Audacious Maneuver."

22. Cathie Viksjo, "Art of Flintknapping Carved in Stone," *Trenton* (N.J.) *Times*, June 12, 1999; Arnold Pacey, *Technology in World Civilization* (Cambridge, Mass.: MIT Press, 1991), 166–67.

23. Donald A. Norman, *The Invisible Computer: Why Good Products Can Fail, the Personal Computer Is So Complex, and Information Appliances Are the Solution* (Cambridge, Mass.: MIT Press, 1998), 162, 170–73.

24. Edwin Gabler, *The American Telegrapher: A Social History, 1860–1900* (New Brunswick: Rutgers University Press, 1988), 79–80, 82–83; Claude S. Fischer, *America Calling: A Social History of the Telephone to 1940* (Berkeley: University of California Press, 1992), 48, 155.

25. Stephanie Faul, "New Directions for Steering," *Car and Road* (American Automobile Assn.), September/October 1996, 6–7; Eric C. Evarts, "Buckling Up Isn't as Easy as It Sounds," *Christian Science Monitor*, March 10, 1999.

26. "Often Outgunned, Police Are Bolstering Firepower," *New York Times*, September 27, 1987; "Armed and Unready: City Pays for Failure to Train Officers with Sophisticated Weapon," *Washington Post*, November 18, 1998.

27. Thomas A. P. van Leeuwen, *The Springboard in the Pond: An Intimate History of the Swimming Pool*, ed. Helen Searing (Cambridge, Mass.: MIT Press, 1998), 27–36.

28. Charles Sprawson, *Haunts of the Black Masseur: The Swimmer as Hero* (New York: Pantheon Books, 1992), 19–23; van Leeuwen, *Springboard in the Pond*, 36–39; Cecil M. Colwin, *Swimming into the 21st Century* (Champaign, Ill.: Leisure Press, 1992), 1–49, 69–73; Frank Litsky, "Allen Stack, 71, a Swimmer Who Broke 6 World Records," *New York Times*, September 19, 1999.

29. James Fallows, "Throwing Like a Girl," *Atlantic Monthly*, August 1996, 84; John Thorn and John B. Holway, *The Pitcher* (New York: Prentice-Hall Press, 1987), 147.

30. Thorn and Holway, *The Pitcher*, 4; Mark Heisler, "It Can Be One Pitch from Over," *Los Angeles Times*, April 9, 1990.

31. Mike Marqusee, "Getting Cricket Straight," *New York Review of Books*, vol. 44, no. 7 (April 24, 1997), 65.

32. Thorn and Holway, *The Pitcher*, 149–54; "Sports," *Encyclopaedia Britannica*, 15th ed. (1998), vol. 28, 158–59; Jeff Lyon, "Outer Limits," *Chicago Tribune*, October 6, 1991, *Good Health Magazine*, 14.

33. Wiebe E. Bijker, *Of Bicycles, Bakelite, and Bulbs: Toward a Theory of Sociotechnical Change* (Cambridge, Mass.: MIT Press, 1995), 30–45, 54–100.

34. John R. Hale, "The Lost Technology of Ancient Greek Rowing," *Scientific American*, vol. 274, no. 5 (May 1996), 82–85.

35. Neil Wigglesworth, *A Social History of English Rowing* (London: Frank Cass, 1992), 87–89; Thomas C. Mendenhall, *The Harvard-Yale Boat Race, 1852–1924 and the Coming of Sport to the American College* (Mystic, Conn.: Mystic Seaport Museum, 1993), 47–49, 86; Eric Halladay, *Rowing in England: A Social History* (Manchester, Eng.: Manchester University Press, 1990), 204–9; Thomas C. Mendenhall, *A Short History of American Rowing* (Boston: Charles River Books, n.d.), 29–37.

36. Ian Fairbairn, ed., *Steve Fairbairn on Rowing* (London: Nicholas Kaye, 1951), 17–45, 81–85, 93–102; Christopher Dodd, *The Story of World Rowing* (London: Stanley Paul, 1992), 155–68; Walter Wülfing, " 'Der Ruderprofessor,' " in Hans Lenk, ed., *Handlungsmuster Leistungssport: Karl Adam zum Gedenken* (Schorndorf, West Germany: Verlag Karl Hofmann, 1977), 20–37.

37. Nick Evangelista, *The Encyclopedia of the Sword* (Westport, Conn.: Greenwood Press, 1995), 26, 491–92, 546–47, 233; Marvin Nelson, *Winning Fencing* (Chicago: Henry Regnery, 1975), 4–5.

38. Evangelista, *Encyclopedia of the Sword*, 254–55, 208–11.

39. Ibid., 197–201; Nelson, *Winning Fencing*, 5, 109–12.

40. Nelson, *Winning Fencing*, 111; Evangelista, *Encyclopedia of the Sword*, 447–48, citing E. D. Morton, *Martini A–Z of Fencing* (London: Queen Anne Press, 1992).

41. Jim Gorant, "Hinge Benefits," *Popular Mechanics*, vol. 175, no. 2 (February 1998), 52–53; "Gerrit Jan van Ingen Schenau," *San Diego Union-Tribune*, April 7, 1999; "Geerit Jan van Ingen Schenau, Dutch Designer of Revolutionary Clap Skate," *Pittsburgh Post-Gazette*, April 13, 1998. These sources were also used in the following paragraphs.

42. David Chung, "Slap Skates Just Link in Chain to Faster Times," *Japan Economic Newswire*, January 15, 1998.

43. Ibid.

44. Amy Shipley, " 'Slapskates' Melt Records, Anger Purists," *Washington Post*, November 30, 1997; Vahe Gregorian, "Clap Skates Don't Get Applause of All the Skaters," *St. Louis Post-Dispatch*, February 9, 1998.

45. Beverley Smith, "New Technology Creates Icy Feud in Speedskating," *Globe and Mail* (Toronto), September 30, 1997.

46. Thomas C. Kouros, *Par Bowling: The Challenge* (Palatine, Ill.: Pin-Count Enterprises, 1993), 177.

47. Weiskopf, *Perfect Game*, 162–75; Chip Zielke, *Revolutions* (Matteson, Ill.: Revolutions International, 1995), 28–32; James Brooke, "900 Reasons for Making the Bowlers' Record Book," *New York Times*, February 9, 1997; John Maher, "Spare Change: With High-Tech Bowling Balls, the Premium on Talent Diminishes in the Professional Bowlers," *Austin American-Statesman*, July 29, 1995; Dick Evans, "Once

Spoiled by Success, Holman Is Starting Over," *Portland Oregonian,* January 30, 1995; Kevin Sherrington, "Bowlers Worry About PBA Tour's Decline," *Seattle Times,* March 12, 1995.

48. Dan Herbst, "Is Practice a Lost Art?" *Bowlers Journal International,* vol. 82, no. 6 (June 1995), 80, 84.

49. Rone Tempest, "China Puts Bowling in New League," *Los Angeles Times,* October 5, 1997; Steve James, "The Helicopter Technique: It Isn't Funny Anymore," *Bowling,* vol. 58, no. 3 (December 1991–January 1992), 28–32; Zakri Baharudin, "Taiwan's Spinning Takes Them Places," *New Straits Times* (Malaysia), March 29, 1999; Akirako Yamaguchi, "Taiwanese Bowler Rolls Gold with 'Helicopter,' " *Daily Yomiuri,* October 6, 1994; Hildegarde Chambers, "Bowler Aims for World Title," *Calgary Herald,* December 10, 1998.

50. Chris Cooper, "How Much Does a Strike Weigh?" *Bowling,* vol. 61, no. 3 (December 1994–January 1995), 34–36.

51. James, "Helicopter Technique," 32.

52. David Warsh, "Toward a Better Understanding of Economic Growth," *Boston Globe,* August 7, 1994; Paul M. Romer, "Beyond Classical and Keynesian Macro-economic Policy," *Policy Options,* July–August 1994, posted at http://www.stanford.edu/~promer; A. Lund, broadcast e-mail message, October 28, 1995, "Feedback on Ratings of Usability Rules of Thumb," cited in Aaron Marcus, "Graphical User Interfaces," in Martin G. Helander et al., eds., *Handbook of Human-Computer Interaction,* 2nd ed. (Amsterdam: Elsevier, 1997), 438.

53. On the colonial deployment of automatic weapons, see John Ellis, *The Social History of the Machine Gun* (London: Pimlico, 1993), 79–110.

CHAPTER TWO

1. Henry Petroski, *To Engineer Is Human: The Role of Failure in Successful Design* (New York: St. Martin's Press, 1985), and *Engineers of Dreams: Great Bridge Builders and the Spanning of America* (New York: Alfred A. Knopf, 1995).

2. J. Bostock, "Evolutionary Approaches to Infant Care," *Lancet,* vol. 1, no. 1038 (1962), 1033–35; cited in Katherine A. Dettwyler and Claudia Fishman, "Infant Feeding Practices and Growth," *Annual Review of Anthropology,* vol. 21 (1992), 180.

3. Huntly Collins, "Low-Tech Breast-Feeding Aid," *Philadelphia Inquirer,* May 24, 1999; Sue Armstrong, "Choice Would Be a Fine Thing: An Increasingly Urban Lifestyle Is Encouraging Women to Abandon Breast-Feeding," *New Scientist,* vol. 148, no. 1998 (October 7, 1995), 44 ff.

4. Valerie A. Fildes, *Breasts, Bottles and Babies* (Edinburgh: Edinburgh University Press, 1986), 17–36, 45–52, 399–401; Valérie Lastinger, "Re-Defining Motherhood: Breast-Feeding and the French Enlightenment," *Women's Studies,* vol. 25, no. 6 (1996), 603–17; Joan Sherwood, "The Milk Factor: The Ideology of Breast-feeding and Post-partum Illnesses, 1750–1850," *Canadian Bulletin of the History of Medicine,* vol. 10 (1993), 25–47; Sally McMillen, "Mothers' Sacred Duty: Breast-Feeding Patterns Among Middle- and Upper-Class Women in the Antebellum South," *Journal of Southern History,* vol. 51, no. 3 (August 1985), 333–56.

5. Fildes, *Breasts,* 262–65.

6. Ibid., 345; Naomi Baumslag and Dia L. Michels, *Milk, Money, and Madness: The Culture and Politics of Breastfeeding* (Westport, Conn.: Bergin & Garvey, 1995), 135–38; M. Michelle Jarrett, " 'An Act of Flagrant Rebellion Against Nature,' " *Winterthur Portfolio,* vol. 30, no. 4 (Winter 1995), 279–88; Christina Hardyment, *Dream*

Babies: Child Care from Locke to Spock (London: Jonathan Cape, 1983), 50–51; Anita Golo, "Infant Feeders Worth Collecting," *San Diego Union-Tribune,* June 10, 1984.

7. Vern L. Bullough, "Bottle Feeding: An Amplification," *Bulletin of the History of Medicine,* vol. 55, no. 2 (Summer 1981), 257–59; Baumslag and Michels, *Milk,* 135–38; Hardyment, *Dream Babies,* 50–51; Golo, "Infant Feeders"; Gabrielle Palmer, *The Politics of Breastfeeding* (London: Pandora, 1988), 174.

8. Dr. Darroll Erickson, telephone interview, June 9, 1999; James B. Meadow, "Time in a Bottle," *Rocky Mountain News,* March 6, 1997; Debra Lee Baldwin, "We've Come a Long Way, Baby," *San Diego Union-Tribune,* October 18, 1989; William S. Walbridge, *American Bottles Old and New* (Toledo: n.p., 1920), 55–72; Diane Ostrander, *A Guide to American Nursing Bottles,* 2nd ed. (York, Pa.: ACIF Publications, 1992), C-VI to C-XI, 171–72.

9. Patricia Harris and David Lyon, "The Power of Plastic," *Boston Globe Magazine,* November 27, 1988, 22 ff.; Dawn Clayton and Anne Maier, "For Babies Who Want to Get a Grip . . . ," *People,* March 3, 1986, 41–42; Baumslag and Michels, *Milk,* 136.

10. Clayton and Maier, "For Babies," 41–42; Baumslag and Michels, *Milk,* 80–83; Derrick B. Jelliffe and E. F. Patrice Jelliffe, *Human Milk in the Modern World* (Oxford: Oxford University Press, 1978), 205.

11. Joy Melnikov and Joan M. Bedinghaus, "Management of Common Breast-feeding Problems," *Journal of Family Practice,* vol. 39, no. 1 (July 1994), 56 ff.; *The Womanly Art of Breastfeeding,* 5th ed. (New York: Plume Books, 1991), 432; Huguette Rugeon-O'Brien et al., "Nutritive and Nonnutrivive Sucking Habits: A Review," *Journal of Dentistry for Children,* vol. 63, no. 5 (September–October 1996), 321–27.

12. Patricia Stuart-Macadam, "Biocultural Perspectives on Breastfeeding," in Patricia Stuart-Macadam and Katherine A. Dettwyler, eds., *Breastfeeding: Biocultural Perspectives* (New York: Aldine de Gruyter, 1995), 16–17; Lisa Cloat, "Caring for Tiny Teeth: The Very Bottle That Nourishes a Child Can Destroy Teeth," *Peoria Journal-Star,* February 10, 1999.

13. Cited in Jelliffe and Jelliffe, *Human Milk,* 209; Dettwyler and Fishman, "Infant Feeding Practices," 181; Max Kaufman, "What's in Infant Formula?" *Washington Post,* June 1, 1999.

14. Randolph M. Nesse and George C. Williams, *Why We Get Sick: The New Science of Darwinian Medicine* (New York: Times Books, 1994), 202; Fildes, *Breasts,* 278–79; Janet Golden, *A Social History of Wet Nursing in America: From Breast to Bottle* (Cambridge, Eng.: Cambridge University Press, 1996), 132; P. J. Atkins, "Sophistication Detected: Or, the Adulteration of the Milk Supply, 1850–1914," *Social History,* vol. 16, no. 3 (October 1991), 317–37.

15. T. P. Mepham, " 'Humanizing' Milk: The Formulation of Artificial Feeds for Infants (1850–1910)," *Medical History,* vol. 37, no. 3 (July 1993), 225–49; William H. Brock, *Justus von Liebig: The Chemical Gatekeeper* (Cambridge, Eng.: Cambridge University Press, 1997), 243–45.

16. Brock, *Justus von Liebig,* 245–49; Palmer, *Politics of Breastfeeding,* 163–65; Carolyn Crowley Hughes, "The Man Who Invented Elsie, the Borden Cow," *Smithsonian,* vol. 30, no. 6 (September 1999), 32–33.

17. Palmer, *Politics of Breastfeeding,* 163–65; Rima D. Apple, *Mothers and Medicine: A Social History of Infant Feeding, 1890–1950* (Madison: University of Wisconsin Press, 1987), 8–13.

18. Apple, *Mothers,* 23–34; Palmer, *Politics of Breastfeeding,* 165–66; P. J. Atkins, "White Poison? The Social Consequences of Milk Consumption, 1850–1930," *Social*

History of Medicine, vol. 5, no. 2 (August 1992), 207–27; Golden, *Wet Nursing,* 132–33.

19. Apple, *Mothers,* 159–60, 173–76; Jelliffe and Jelliffe, *Human Milk,* 191.

20. Jonas Frykman and Orvar Löfgren, *Building Cultures: A Historical Anthropology of Middle-Class Life,* trans. Alan Crozier (New Brunswick: Rutgers University Press, 1987), 232; Jelliffe and Jelliffe, *Human Milk,* 188–93; Apple, *Mothers,* 152–57.

21. Apple, *Mothers,* 169–72; Jelliffe and Jelliffe, *Human Milk,* 195–99.

22. "Food Processing," *Encyclopaedia Britannica,* 15th ed. (1998), vol. 19, 386; Jelliffe and Jelliffe, *Human Milk,* 211–40; Baumslag and Michels, *Milk,* 147–50.

23. Baumslag and Michels, *Milk,* 152–45; Palmer, *Politics of Breastfeeding,* 186–88; Jelliffe and Jelliffe, *Human Milk,* 234, 271–91.

24. Jelliffe and Jelliffe, *Human Milk,* 231; Baumslag and Michels, *Milk,* 154–88.

25. "Breastfeeding and the Use of Human Milk," *Pediatrics,* vol. 100, no. 6 (December 1997), 1035–39; Natalie Angier, *Woman: An Intimate Geography* (Boston: Houghton Mifflin, 1999), 153; Jack Newman, "How Breast Milk Protects Newborns," *Scientific American,* vol. 273, no. 6 (December 1995), 76–79; Erickson, personal communication, June 9, 1999; Allan S. Cunningham, "Breastfeeding: Adaptive Behavior for Child Health and Longevity," in Stuart-Macadam and Dettwyler, eds., *Breastfeeding,* 244–45.

26. Cunningham, "Adaptive Behavior," 253–55; L. J. Horwood and D. M. Fergusson, "Breastfeeding and Later Cognitive and Academic Outcomes," *Pediatrics,* vol. 101, no. 1 (January 1998), E9; A. Lucas et al., "Breast Milk and Subsequent Intelligence Quotient in Children Born Preterm," *Lancet,* vol. 339, no. 8788 (February 1, 1992), 261–64; for recent criticism of the intelligence studies, see Shari Roan, "Analysis Questions Link Between Breast Milk, IQ," *Los Angeles Times,* June 10, 2002.

27. Alan Lucas, "Programming by Early Nutrition: An Experimental Approach," *Journal of Nutrition,* vol. 128, no. 2 (February 1998), 401 ff.

28. Cunningham, "Adaptive Behavior," 249–55; W. H. Oddy et al., "Association Between Breast Feeding and Asthma in 6 Year Old Children," *British Medical Journal,* vol. 319, no. 7213 (September 25, 1999), 815–19.

29. *Womanly Art of Breastfeeding,* 72–73, 161–75.

30. Katherine A. Dettwyler, "A Time to Wean: The Hominid Blueprint for the Natural Age of Weaning in Modern Human Populations," in Stuart-Macadam and Dettwyler, eds., *Breastfeeding,* 39–74; Al Podgorski and Richard A. Chapman, "Breastfeeding: Babies, Yes. But Toddlers and Even Older Kids Too?" *Chicago Sun-Times,* September 26, 1999.

31. Jo L. Freudenheim, "Lactation History and Breast Cancer Risk," *American Journal of Epidemiology,* vol. 148, no. 11 (December 1, 1997), 932–38; Marc S. Micozzi, "Breast Cancer, Reproductive Biology, and Breastfeeding," in Stuart-Macadam and Dettwyler, eds., *Breastfeeding,* 347–84; Patricia Stuart-Macadam, "Biocultural Perspectives on Breastfeeding," in ibid., 11; Palmer, *Politics of Breastfeeding,* 72–73; Jelliffe and Jelliffe, *Human Milk,* 115–17.

32. *Womanly Art,* 123; Sara A. Quandt, "Sociocultural Aspects of the Lactation Process," in Stuart-Macadam and Dettwyler, eds., *Breastfeeding,* 127–43; Peter T. Ellison, "Breastfeeding, Fertility, and Maternal Condition," in ibid., 305–45; Jelliffe and Jelliffe, *Human Milk,* 117–19.

33. *Womanly Art of Breastfeeding,* 161–73; Baumslag and Michels, *Milk,* 31–32, 23–26; Lastinger, "Re-Defining Motherhood," 613; Marylynn Salmon, "The Cultural Significance of Breastfeeding and Infant Care in Early Modern England and America," *Journal of Social History,* vol. 28, no. 2 (Winter 1994), 247–69.

34. Alan S. Ryan, "The Resurgence of Breastfeeding in the United States," *Pediatrics,* vol. 99, no. 4 (April 1997), E12–E16; Marc Kaufman, "What's in Infant Formula?" *Washington Post,* June 1, 1999.

35. Kaufman, "Infant Formula," 13; Marc Kaufman, "Baby Formula Fight Puts Fat Under Fire," *Washington Post,* June 1, 1999.

36. Karen Goldberg Goff, "To Baby's Health," *Washington Times,* August 18, 2002; Isadora B. Steylin, "Infant Formula: Second Best but Good Enough," *FDA Consumer,* vol. 30, no. 5 (June 1996), 17–20.

37. Alan Lucas, letter, *British Medical Journal,* vol. 317, no. 7174 (August 1, 1998), 337–38; Glen E. Mott et al., "Programming of Cholesterol Metabolism by Breast or Formula Feeding," in *The Childhood Environment and Adult Disease* (Chichester, Eng.: John Wiley & Sons, 1991), 56–76; Golden, *Wet Nursing,* 206; Janet Raloff, "Breast Milk: A Leading Source of PCBs," *Science News,* vol. 152, no. 22 (November 29, 1997), 344.

38. Jill Nelson, "Mr. Mom Finally Gets It All," *Washington Post,* September 27, 1987.

39. On the technology and politics of contemporary breast-feeding, see Linda M. Blum, *At the Breast: Ideologies of Breastfeeding and Motherhood in the Contemporary United States* (Boston: Beacon Books, 1999); for a new feminist interpretation, see Alison Bartlett, "Breastfeeding as Headwork: Corporeal Feminism and Meanings for Breastfeeding," *Women's Studies International Forum,* vol. 25, no. 3 (May–June 2002), 373–82.

40. Terence Chea, "Martek to Feed Market for Fortified Formula; Do Additives Make Babies Smarter? Studies Differ," *Washington Post,* January 24, 2002; Eric Nagourney, "Vital Signs: Nutrition, Extra Fortification for Baby Formulas," *New York Times,* January 29, 2002; Nicholas B. Kristof, "Interview with a Humanoid," *New York Times,* July 23, 2002.

CHAPTER THREE

1. William Rossi, "Back to Basics—Again and Again," *Footwear News,* vol. 54, no. 36 (September 7, 1998), 24; Eunice Wilson, *A History of Shoe Fashions* (London: Pitman, 1974), 36–37; Harold E. Driver and William C. Massey, *Comparative Studies of North American Indians* (Philadelphia: American Philosophical Society, 1957) (= *Transactions of the American Philosophical Society,* n.s. vol. 47, part 2); Alika Podolinsky Webber, *North American Indian and Eskimo Footwear: A Typology and Glossary* (Toronto: Bata Shoe Museum, 1989).

2. Phillip Nutt, electronic mail to author, August 20, 2002; David Foster Wallace, "Shipping Out," *Harper's,* vol. 292, no. 1748 (January 1996), 33; Brian Dibble, "Synonyms for *Zori,*" *American Speech,* vol. 54, no. 1 (Spring 1979), 79.

3. "Put Your Best Foot Forward," *Current Health* 2, vol. 27, no. 4 (December 1997), 28–29; Todd R. Olson and Michael R. Seidel, "The Evolutionary Basis of Some Clinical Disorders of the Human Foot: A Comparative Survey of the Living Primates," *Foot and Ankle,* vol. 3, no. 6 (May–June 1983), 322–41; Frank R. Wilson, *The Hand* (New York: Pantheon Books, 1998), 318n2; Félix Regnault, "Le Pied Préhensile chez l'Homme," *Bulletins et Mémoires* (Société Anthropologique de Paris), 5th ser., vol. 10 (1909), 41–42.

4. William A. Rossi, "The Foot's Arches: Myth vs. Fact: Do Arch Inserts Play the Best Supporting Role?" *Footwear News,* vol. 52, no. 20 (May 13, 1996), 14; Kelley Ann Hays-Gilpin, Ann Cordy Deegan, and Elizabeth Ann Morris, *Prehistoric Sandals*

from Northeastern Arizona: The Earl H. Morris and Ann Axtell Morris Research, Anthropological Papers of the University of Arizona, 62 (Tucson: University of Arizona Press, 1998), 38; Earle T. Engle and Dudley J. Morton, "Notes on Foot Disorder Among Natives of the Belgian Congo," *Journal of Bone and Joint Surgery*, vol. 13 (April 1931), 311–18; Shanghai data from Rossi, above.

5. "Discalced Orders," *New Catholic Encyclopedia* (New York: McGraw-Hill, 1967), vol. 4, 893; Friedrich Engels, *The Condition of the Working Class in England*, trans. and ed. W. O. Henderson and W. H. Chaloner (Stanford: Stanford University Press, 1968), 80; Richard Keith Frazine, *The Barefoot Hiker* (Berkeley: Ten Speed Press, 1993), 15–16; David Holmstrom, "Hiking Shoes Are Getting the Boot," *Christian Science Monitor*, June 9, 1997, 15.

6. Frazine, *Barefoot Hiker*, 31–33; Liz Halloran, "Shoeless in the Forest, Hikers Discover a World of Unexpected Sensations," reprinted in ibid., 90–91; David Holmstrom, "Hiking Shoes," 15.

7. M. Douglas Baker and Randi E. Bell, "The Role of Footwear in Childhood Injuries," *Pediatric Emergency Care*, vol. 7, no. 6 (December 1991), 353–55. Sneakers and other rough-soled shoes, worn with socks, are the best preparation against loss of footing, according to the study.

8. Sander Gilman, "The Jewish Foot: A Foot-Note to the Jewish Body," in Sander Gilman, *The Jew's Body* (New York: Routledge, 1991), 38–59; Patricia Vertinsky, "Body Matters," in Nobert Finzsch and Dietmar Schirmer, eds., *Identity and Intolerance: Nationalism, Racism, and Xenophobia in Germany and the United States* (Washington, D.C.: German Historical Institute, 1998), 331–57; Patricia Vertinsky: "The 'Racial' Body and the Anatomy of Difference: Anti-Semitism, Physical Culture, and the Jew's Foot," *Sport Science Review*, vol. 4, no. 1 (January 1995), 38–59; Lynn T. Staheli, "Shoes for Children: A Review," *Pediatrics*, vol. 88, no. 2 (August 1991), 371–75; William A. Rossi, "Dr. Scholl: From Humble Beginnings," *Footwear News*, vol. 54, no. 27 (July 6, 1998), 14; Rossi, "Foot's Arches," 14.

9. David King, "The Way We Were: World War II Presented Baseball with Its Ultimate Challenge," *Houston Chronicle*, July 23, 1995; Howard Seiden, "Flat Feet Don't Automatically Mean Bad Feet," *Montreal Gazette*, October 17, 1992; Elisabeth Rosenthal, "The Maligned Flat Foot: Some See an Advantage," *New York Times*, November 22, 1990; David N. Cowan et al., "Foot Morphologic Characteristics and Risk of Exercise-Related Injury," *Archives of Family Medicine*, vol. 2, no. 7 (July 1993), 773–77 (paper originally presented in 1989).

10. E. E. Bleck, "The Shoeing of Children: Sham or Science?" *Developmental Medicine and Child Neurology*, vol. 13, no. 2 (April 1971), 188–95; Udaya Bahaskara Rao and Benjamin Joseph, "The Influence of Footwear on the Prevalence of Flat Feet," *Journal of Bone and Joint Surgery*, vol. 74-B, no. 4 (July 1992), 525–27; Staheli, "Shoes for Children," 371–75.

11. "White Man's Burden," *Economist*, May 7, 1983, 104; Thomas V. DiBacco, "Hookworm's Strange History: How a Yankee's Research Saved the South," *Washington Post*, June 30, 1992; Adrian Gwin, "Looking Back: No Shoes Serves This Barefoot Fan," *Charleston Daily Mail*, June 7, 1997; Asa C. Chandler, *Hookworm Disease: Its Distribution, Biology, Epidemiology, Pathology, Treatment and Control* (New York: Macmillan, 1929), 174–75, 208–11, 380–83; John Ettling, *The Germ of Laziness* (Cambridge, Mass.: Harvard University Press, 1981), 130; Mary A. Dempsey, "Henry Ford's Amazonian Suburbia," *Américas* (English edition), vol. 48, no. 2 (March–April 1996), 49. Shoes as well as bare feet can transmit lethal infections. A sixty-one-year-old English tourist, who remained carefully shod during a brief holiday in Thailand, stepped on a

thorn while gardening on the day after his return, and thereby injected into his heel the bacillus *Burkholderia pseudomallei,* which had probably colonized his shoes and feet at his Thai resort. Released after two weeks of intravenous antibiotics, he died three months later of a ruptured abdominal aorta. See James K. Torrens et al., "A Deadly Thorn: A Case of Imported Melioidosis," *Lancet,* vol. 353, no. 9157 (March 20, 1999), 1016.

12. Lam Sim-Fook and A. B. Hodgson, "Foot Forms Among the Non-Shoe[sic] and Shoe-Wearing Chinese Population," *Journal of Bone and Joint Surgery,* vol. 40-A, no. 5 (October 1958), 1058–62.

13. William C. Hayes, *The Scepter of Egypt: A Background for the Study of the Egyptian Antiquities in the Metropolitan Museum of Art,* 2 vols. (New York: Metropolitan Museum of Art, 1990), vol. 2, 188; Laurie Lawlor, *Where Will This Shoe Take You?* (New York: Walker, 1996), 8–12.

14. B. J. de Lateur et al., "Footwear and Posture: Compensatory Strategies for Heel Height," *American Journal of Physical Medicine and Rehabilitation,* vol. 70, no. 5 (October 1991), 246–54, presents evidence that high-heeled shoes do not increase the lordosis, or curvature, of the spine.

15. Victoria Nelson, *My Time in Hawaii: A Polynesian Memoir* (New York: St. Martin's Press, 1989), 9–10; Steele F. Stewart, "Footgear—Its History, Uses and Abuses," *Clinical Orthopaedics and Related Research,* vol. 88 (October 1972), 119–22; K. Ashizawa et al., "Relative Foot Size and Shape to General Body Size in Javanese, Filipinas and Japanese with Special Reference to Habitual Footwear Types," *Annals of Human Biology,* vol. 24, no. 2 (1997), 117–29; electronic posting to H-ASIA list, June 14, 1998.

16. Nicholas Wade, "Shoes That Walked the Earth 8,000 Years Ago," *New York Times,* July 7, 1998; Heather Pringle, "Eight Millennia of Fashion Footwear," *Science,* vol. 281, no. 5373 (July 3, 1998), 23–25; Jenna T. Kuttruff et al., "7500 Years of Prehistoric Footwear from Arnold Research Cave, Missouri," *Science,* vol. 281, no. 5373 (July 3, 1998), 72–75; Kathy Kankainen, "Sandal Styles, Materials, and Techniques," in Kathy Kankainen, ed., *Treading in the Past: Sandals of the Anasazi* (Salt Lake City: Utah Museum of Natural History in association with the University of Utah Press, 1995), 21–30.

17. Louis Jacobson, "Ancient Sandal-Makers Were a Step Ahead," *Washington Post,* March 23, 1998; Hays-Gilpin, Deegan, and Morris, *Prehistoric Sandals,* 38–39, 121–22.

18. Stewart, "Footgear," 121.

19. Jonathan Norton Leonard, *Early Japan* (New York: Time-Life Books, 1972), 110; Bernard Rudofsky, *The Kimono Mind* (London: Victor Gollancz, 1965), 49–50; Richard J. Bowring, "Geta," in *A Hundred Things Japanese* (Tokyo: Japan Cultural Institute, 1975), 42–43; *Traditional Japanese Footwear* (Toronto: Bata Shoe Museum, 1999), n.p.; John Fee Embree, *Suye Mura: A Japanese Village* (Chicago: University of Chicago Press, 1939), 100; *We Japanese* (Fujiya Hotel, n.d.), 144; Junichi Saga, *Memories of Silk and Straw: A Self-Portrait of Small-Town Japan,* trans. Garry O. Evans (Tokyo: Kodansha, 1987), 230–32; Liza Crihfield Dalby, *Kimono: Fashioning Culture* (New Haven: Yale University Press, 1993), 169.

20. "Straw Ware," *Kodansha Encyclopedia of Japan* (Tokyo: Kodansha, 1983), vol. 7, 249; "Zori," ibid., vol. 8, 379; "Waraji," ibid., vol. 8, 223; Basil Hall Chamberlain, *Things Japanese,* 5th ed. (London: Kegan Paul, 1927), 38; Susan B. Hanley, *Everyday Things in Premodern Japan: The Hidden Legacy of Material Culture* (Berkeley: University

of California Press, 1997), 51–76; Koizumi Takeo, "Japan's Rich Rice Culture," *Japan Quarterly,* January 1999, 58 ff.; Leonard, *Early Japan,* 112; *Traditional Japanese Footwear,* n.p.; "Big Foot: A Sandal Made of Straws Symbolising Japan's Protection Is Dedicated to the Sensoji Temple, Tokyo [photograph]," *The Independent* (London), November 1, 1998, 2; Dalby, *Kimono,* 169.

21. Wilson, *A History of Shoe Fashions,* 32–33.

22. Nelson, *Hawaii,* 9; K. Nishio, "Die Häufigkeit der Fußmykosen in Japan," *Archiv für Klinische und Experimentelle Dermatologie,* vol. 227, no. 1 (1966), 581–83.

23. Tadashi Kato and Showri Watanabe, "The Etiology of Hallux Valgus in Japan," *Clinical Orthopedics and Related Research,* vol. 157 (June 1981), 78–81.

24. Edward S. Morse, *Japanese Homes and Their Surroundings* (New York: Dover Publications, 1961), 238–39; Dalby, *Kimono,* 86–87.

25. Alice Mabel Bacon, *Japanese Girls and Women* (Boston: Houghton Mifflin, 1891), 15.

26. Nancy Stedman, "Learning to Put the Best Shoe Forward," *New York Times,* October 27, 1998; Ko Tada, "Comparative Analysis of Walking Patterns with Flip-flops Between American and Japanese Males," M.A. thesis, Western Michigan University, 1997, 63–64; Mikiyoshi Ae and Toshiharu Yamamoto, "Soryoku wo kyoka suru" ("Strength of the Sprinter"), *Training Journal,* no. 159 (January 1993), 20–25; no. 160 (January 1993), 42–45 (I am indebted to Kiyoko Heineken of the Gest Oriental Library at Princeton University for her summary of this article, cited in the thesis of Ko Tada); Michael Cooper, ed., *They Came to Japan: An Anthology of European Reports on Japan, 1543–1640* (Berkeley: University of California Press, 1965), 210–11.

27. Stedman, "Best Shoe"; Tada, "Walking Patterns," 1–2, 13–15; Ae and Yamamoto, "Soryoku wo kyoka suru," 159–60, 20–25, 42–45.

28. Barbara F. Kawakami, *Japanese Immigrant Clothing in Hawaii, 1885–1941* (Honolulu: University of Hawaii Press, 1993), 153–61.

29. http://ScottHawaii.com; Steve Scott, telephone interview, September 14, 1999; Charles Rippin, personal communication, February 9, 2003; Kathryn Bold, "Summer Flops: Say Goodby to 'Blowouts' with Beach Sandals That Borrow the High-Tech Features of Athletic Shoes," *Los Angeles Times,* July 13, 1995.

30. Robert T. Elson, *Prelude to War* (Alexandria, Va.: Time-Life Books, 1977), 144–45; Ashizawa et al., "Relative Foot Size," 118; Bernard Rudofsky, *Sparta/Sybaris* (Vienna: Residenz Verlag, 1987), 130–31.

31. John Henry Thornton, *Textbook of Footwear Materials* (London: National Trade Press, 1955), 219–33; Phillip Nutt, personal communication, June 15, 1998.

32. Yoshiaki Shimizu, conversation, February 17, 1999.

33. Bold, "Summer Flops"; Miriam Jordan and Terry Agins, "Fashion Flip-flop: Lowly Sandal Leaves the Shower Behind," *Wall Street Journal,* August 8, 2002.

34. "From an Oriental Teahouse," advertisement, *House Beautiful,* May 1958, 97; "Dress Zori Sandals," advertisement, *House Beautiful,* April 1964, 90.

35. "Swan Thong," *Sydney Morning Herald,* December 27, 1991; Geoffrey Blainey, *Jumping Over the Wheel* (St. Leonards, N.S.W.: Allen & Unwin, 1993), 299–300; Adam Edwards, "The Flip-Side of a Fashion Frontier," *Evening Standard* (London), February 17, 1993; Jan Freeman, "Hear My Thong," *Sunday Age* (Melbourne), December 20, 1992; Sue Hewitt, "Hits and Memories at the Top of the Thong Parade," *Sunday Age,* December 17, 1995.

36. Robert W. Mann, "Three Examples of Vietnamese Footwear from the Vietnam Conflict," *Journal of the American Podiatric Medical Association,* vol. 85, no. 1

(January 1995), 61; Nguyen Bay and Thai Bang, "Vietnam Life: Story of a Bodyguard," *Saigon Times Magazine,* June 1, 1997, n.p.; Ed Timms, "South Vietnamese Veterans Faced Different Struggles After War," *Dallas Morning News,* April 25, 1995.

37. Steve Scott, telephone interview, September 14, 1999; Nelson, *My Time in Hawaii,* 8–9.

38. Phillip Nutt, personal communication, June 15, 1998; Lu-Lin Cheng, "Embedded Competitiveness: Taiwan's Shifting Role in International Footwear Sourcing Networks" (Ph.D. diss., Duke University, 1996), 2–20, 77–80.

39. *Small-Scale Manufacture of Footwear* (Geneva, Switzerland: International Labour Office, 1982), 13–71, 117–19; Cheng, "Embedded Competitiveness," 81–82.

40. These data are drawn from telephone interviews with Phillip Nutt and unpublished reports he kindly supplied; also, "Japanese in Kenya Work for a Healthier Africa," *Daily Yomiuri,* May 1, 1997, n.p.; Robin Eveleigh, "Flip-flops Get to the Soul of the Nation," *Financial Times,* November 20–21, 1999; Jordan and Agins, "Fashion Flip-flop."

41. Philip Williams, "Eritrean Rebel Campaign Backed by Hidden Factories, Ethiopian POWs," *Los Angeles Times,* January 1, 1989.

42. Jen Nessel, "Why I Love Shoes: A Social Code for Sandal Wearers," *Self,* May 1997, 146.

43. Paul Trachtman, "Hands-on Toys," *Smithsonian,* vol. 28, no. 3 (December 1993), 128–33; Suzanne Seriff, "Folk Art from the Global Scrap Heap: The Place of Irony in the Politics of Poverty," in Charlene Cerny and Suzanne Seriff, eds., *Recycled: Folk Art from the Global Scrap Heap* (New York: Abrams, 1996), 12; Paul Webster, "Danger on the Beach: World Polluters Beware," *Observer,* November 24, 1996; N. G. Willoughby et al., "Beach Litter: An Increasing and Changing Problem for Indonesia," *Marine Pollution Bulletin,* vol. 34, no. 6 (June 1997), 469–78; Ian Anderson, "This Beach Ain't Big Enough for the Both of Us," *New Scientist,* vol. 151, no. 2043 (August 17, 1996), 12.

44. Will Englund and Gary Cohn, "A Third World Dump for America's Ships?" *Baltimore Sun,* December 9, 1997; William Gordon, "Christmas Comes to Port Newark," *Newark Star-Ledger,* December 23, 1998; Paul Watson, "Calcutta's Homeless: A City unto Themselves," *Toronto Star,* May 4, 1996; Jennifer Lin, "Poor, Middle Class Left Hungry, Angry as Indonesia's Economic Crisis Worsens," *Dallas Morning News,* October 9, 1998; "Iraq: Surviving Sanctions," *Economist,* December 12, 1998, 47.

45. Bold, "Summer Flops."

CHAPTER FOUR

1. Valerie Steele, *Shoes: A Lexicon of Style* (New York: Rizzoli, 1999), 169.

2. David Orr, "Slow Knowledge," *Designer/Builder,* vol. 4, no. 8 (December 1997), 5–9.

3. Gordon James Knowles, "Dealing Crack Cocaine: A View from the Streets of Honolulu," *FBI Law Enforcement Bulletin,* vol. 65, no. 7 (July 1996), 7.

4. Andrew Maykuth, "Apartheid Prison Now Memorial to Human Spirit," *Times-Picayune* (New Orleans), March 23, 1997; Jeff Testerman, "Inmate Caught After Fleeing Courtroom," *St. Petersburg* (Florida) *Times,* July 22, 1989; Chris Sosnowski and Rochelle Killingbeck, "Manhunt Nabs Escapees," *The Post and Courier* (Charleston, S.C.), December 28, 1985; David Cannella, "New Shoes Trip Up Fugitive," *Arizona Republic,* May 29, 1990; James D. Henderson, W. Hardy Rauch, and Richard L.

Phillips, *Guidelines for the Development of the Security Program,* 2nd ed. (Lanham, Md.: American Correctional Association, 1997), 109–10.

5. Charles L. Perrin, "Sneaker FAQ and Glossary," http://www.pair.com/ sneakers/; Sylvia Rubin, "Cops, Courts, Crooks and Creeps Featured in New Syndicated Shows," *San Francisco Chronicle,* September 8, 1996; Mary Jane Fine, "Unlikely Outlaws," *Sunday Record* (Bergen County, N.J.), November 22, 1998; Stephanie A. Stanley, Ralph Vigod, and Joseph Cambardello, "Police Come to Admire the Area's 'Burglar to the Stars,' " *Philadelphia Inquirer,* November 22, 1998; William J. Bodziak, *Footwear Impression Evidence* (New York: Elsevier, 1990); Andrea Codrington, "Technology and Design Run Wild in the Soles of the Newest Sneakers," *New York Times,* November 6, 1997.

6. John W. Fountain, "Noted with . . . Pride; Way Black When," *Washington Post,* July 25, 1999; Tom Wolfe, *Bonfire of the Vanities* (New York: Farrar, Straus and Giroux, 1987), 110; William Leith, "Pump It Up," *The Independent* (London), July 8, 1990.

7. Michelle Higgins, "The Ballet Shoe Gets a Makeover, but Few Yet See the Pointe," *Wall Street Journal,* August 18, 1998.

8. Wolfgang Decker, "Die Lauf-Stele des Königs Taharka," *Kölner Beiträge zur Sportwissenschaft,* vol. 13 (1984), 7–37; E. Norman Gardiner, *Athletics of the Ancient World* (Oxford: Clarendon Press, 1930), 128–33, fig. 87; E. Norman Gardiner, *Greek Athletic Sports and Festivals* (London: Macmillan, 1910), 271–73; H. A. Harris, *Greek Athletes and Athletics* (London: Hutchinson, 1964), 66–77; Marc Bloom, "Judging a Path by Its Cover," *Runner's World,* vol. 32, no. 3 (March 1997), 54–62; Melvin P. Cheskin, *The Complete Handbook of Athletic Footwear* (New York: Fairchild Publications, 1987), 2–3.

9. Joseph B. Oxendine, *American Indian Sports Heritage* (Lincoln: University of Nebraska Press, 1995), 67–89; John Weyler, "They're Sole Survivors in Race for Life," *Los Angeles Times,* October 16, 1997; Nancy Nusser, "Indian Runners Trounce World's Ultra-Marathoners," *Orange County Register,* April 23, 1995; Graciela Sevilla, "Running with 'the Light-Footed Ones,' " *Arizona Republic,* June 4, 1995; Peter Severance, "The Legend of the Tarahumara," *Runner's World,* vol. 28, no. 2 (December 1993), 74–80; Peter Nabokov, *Indian Running* (Santa Barbara: Capra Press, 1981), 179–82; "Mohave," *The Gale Encyclopedia of North American Tribes* (Detroit: Gale Research, 1998), vol. 2, 206–11.

10. Michael Olmert, "Points of Origin," *Smithsonian,* vol. 13, no. 1 (April 1982), 38–42; Earl R. Anderson, "Footnotes More Pedestrian Than Sublime: A Historical Background for the Foot-Races in *Evelina* and *Humphrey Clinker,*" *Eighteenth-Century Studies,* vol. 14, no. 1 (Autumn 1980), 56–68; Phyllis Cunnington, *Costume of Household Servants: From the Middle Ages to 1900* (London: Adam and Charles Black, 1974), 100–103.

11. Anne D. Wallace, *Walking, Literature, and English Culture: The Origins and Uses of Peripatetic in the Nineteenth Century* (Oxford: Clarendon Press, 1993), 17–66; George Moss, "The Long Distance Runners of Ante-Bellum America," *Journal of Popular Culture,* vol. 8, no. 2 (Fall 1974), 370–82; Walter Bernstein and Milton Meltzer, "A Walking Fever Has Set In," *Virginia Quarterly Review,* vol. 56, no. 4 (Autumn 1980), 698–731; Don Watson, "Popular Athletics on Victorian Tyneside," *International Journal of the History of Sport,* vol. 11, no. 3 (December 1994), 485–94.

12. Bernstein and Meltzer, "Walking Fever," 700–701; Edward Lamb, " 'Weston the Walker' Made Pedestrianism a Way of Life," *Smithsonian,* vol. 16, no. 4 (July 1979), 89; Moss, "Long Distance Runners," 380; Cavanagh, *Running Shoe Book,* 20–23.

13. Bernstein and Meltzer, "Walking Fever," 701; Sue Moore, "Fresh Air on a High-Tech Shoestring," *The Times* (London), March 6, 1991; Anne Caborn, "Reebok Head's Finest Feat," *Sunday Telegraph* (London), December 17, 1989; *1897 Sears Roebuck Catalog* (New York: Chelsea House, 1968), 35, 203, 206, 190–201; Cavanagh, *Running Shoe Book,* 16–19.

14. Hal Higdon, *A Century of Running: Celebrating the 100th Anniversary of the Boston Athletic Association Marathon* (Emmaus, Pa.: Rodale Press, 1995), 71, 73; Cavanagh, *Running Shoe Book,* 27–33.

15. Ralph F. Wolf, *India Rubber Man: The Story of Charles Goodyear* (Caldwell, Idaho: The Caxton Printers, 1939), 40–54; Samuel Americus Walker, *Sneakers* (New York: Workman, 1978), 15, 18–20.

16. George R. Vila, *The Story of Uniroyal: 75 Years of Progress* (New York: The Newcomen Society, 1968), 8–11; John R. Stilgoe, *Alongshore* (New Haven: Yale University Press, 1994), 73–76; Melanie Rickey, "Flash of Genius," *The Independent* (London), June 29, 1996.

17. Rochelle Chadakoff, "Sneaking in Style," *Rocky Mountain News,* May 19, 1994; Laurie Lawlor, *Where Will This Shoe Take You?* (New York: Walker, 1996), 105; Cavanagh, *Running Shoe Book,* 17; Cheskin, *Athletic Footwear,* 6–7.

18. "Tools of the Track," *Toronto Star,* September 3, 1996; Fila, *Story of Uniroyal,* 10–12; Cheskin, *Athletic Footwear,* 7–12; Walker, *Sneakers,* 21, 26, 50–51.

19. Cheskin, *Athletic Footwear,* 12–14, 60–61; Cavanagh, *Running Shoe Book,* 33; Phillip Nutt, telephone interview, November 15, 1999.

20. William Ecenbarger, "A Trend Afoot," *Chicago Tribune Sunday Magazine,* April 18, 1993; Walker, *Sneakers,* 78–79, 84; Tom Vanderbilt, *The Sneaker Book: Anatomy of an Industry and an Icon* (New York: New Press, 1998),13–14; Richard Cohen, "Sneakers as Metaphor," *Washington Post Magazine,* September 8, 1991.

21. Strasser and Becklund, *Swoosh,* 72–77; Kathleen Low, "In the Days When Sports Shoes Weren't Fashionable," *Footwear News,* vol. 41 (October 6, 1985), 42 ff.

22. Cavanagh, *Running Shoe Book,* 33–46; Ronald P. Dore, *Shinohata: A Portrait of a Japanese Village* (New York: Pantheon Books, 1978), 77–78; Strasser and Becklund, *Swoosh,* 77.

23. Walker, *Sneakers,* 29; Peter Levine, *A. G. Spalding and the Rise of Baseball: The Promise of American Sport* (New York: Oxford University Press, 1985), 88; Cheskin, *Athletic Footwear,* 55–57; Phillip Nutt, telephone interview, November 15, 1999.

24. Strasser and Becklund, *Swoosh,* 97; Vanderbilt, *Sneaker Book,* 29–30; Kenny Moore, "An Outrageous Stand," *Sports Illustrated,* vol. 75, no. 6 (August 5, 1991), 60 ff.; Kenny Moore, "The Eye of the Storm," *Sports Illustrated,* vol. 75, no. 7 (August 12, 1991), 60 ff.; Steele, *Shoes,* 172, 183; William Ecenbarger, "A Trend Afoot," 27 ff.

25. Cheskin, *Athletic Footwear,* 129.

26. Strasser and Becklund, *Swoosh,* 350–51, 357–59.

27. Vanderbilt, *Sneaker Book,* 84, 111.

28. Phillip Nutt, telephone interview, November 15, 1999; Ed Nardoza, "A Once Proud Industry Is Now Down to Its Last Three," *Footwear News,* vol. 39 (April 25, 1983), S1.

29. Vanderbilt, *Sneaker Book,* 57–60; Melissa Dallal, "Tinker Hatfield/Nike," *I.D.,* vol. 44, no. 1 (January–February 1997), 63; "Turbo-Charged Shoes," *SGB UK,* May 20, 1999, 50; Patricia Leigh Brown, "Once-Lowly Sneaker Is Pedestrian No More," *New York Times,* May 28, 1992; Tinker Hatfield, "Inspired Design: How Nike Puts Emotion in Its Shoes," *Harvard Business Review,* July–August 1992, 93; Michele Golden, "The Design Guy: It's Definitely the Shoes," *SportsTech,* vol. 1, no. 1 (October

1997), 40–44; Strasser and Becklund, *Swoosh,* 355–56; Peta Bee, "Planet Trainer," *The Times* (London), February 28, 1999.

30. Golden, "Design Guy," 40, 43.

31. Bruce Tulloh, "Sole Searching," *Peak Performance,* March 1991, 4; Cavanagh, *Running Shoe Book,* 86–88, 94–95, 358–60; Ian Hawkey, "Running Shoes Need Not Be Your Achilles' Heel," *The Times* (London), November 23, 1997; Owen Anderson, "The Shoe Scene: Are the Best Shoes Your Own Feet?" *Running Research News,* May–June 1991, 9–10; Adam Turnball, "The Race for a Better Running Shoe," *New Scientist,* vol. 123, no. 1673 (July 15, 1989), 42–44; R. McNeill Alexander and Michael Bennett, "How Elastic Is a Running Shoe?" *New Scientist,* vol. 123, no. 1673 (July 15, 1989), 45–46; R. McNeill Alexander, "The Spring in Your Step," *New Scientist,* vol. 114, no. 1558 (April 30, 1987), 42–44.

32. Cavanagh, *Running Shoe Book,* 166–71; Tom Yulsman, "Anatomy of the High-tech Running Shoe," *Science Digest,* April 1985, 46 ff.

33. Cavanagh, *Running Shoe Book,* 46–49; Tony Baer, "How Long Can This Go On?" *Runner's World,* vol. 21, no. 4 (April 1986), 44.

34. Paul Carrozza, "Inside Cushioning Technologies," *Runner's World,* vol. 33, no. 9 (September 1998), 56–57; James Braham, "High Tech Afoot," *Machine Design,* vol. 63, no. 11 (June 6, 1991), 80–84; Strasser and Becklund, *Swoosh,* 343–49, 355–57, 563–66, 619–22; Mark Hyman, "Bracing for Life Without 'Air,' " *Baltimore Sun,* August 19, 1996. The original Air shoe patents are available at http://www.uspto.gov: U.S. Patents 4,183,156 of January 15, 1980, and 4,219,945 of September 2, 1980.

35. D. R. Martin, "How to Steer Patients Toward the Right Sport Shoe," *Physician and Sportsmedicine,* vol. 29 no. 9 (September 1, 1997), 138 ff.; Adam Bryant, "What Packers and Builders Can Learn from the Bees," *New York Times,* October 6, 1991; Charles Leerhsen, "Now, Running on Empty," *Newsweek,* December 3, 1990; "Jordan Shaking Up the Shoe Market," *St. Louis Post-Dispatch,* November 5, 1989; "Letting You Down Easy," *Consumer Reports,* vol. 57, no. 5 (May 1992), 308; "For Athletes, More Bounce from a 'Space Age' Shoe," *BusinessWeek,* September 16, 1985, 133; "Designers Put the Boot In," *The Engineer,* June 5, 1998, 20 ff.; Carrozza, "Inside Cushioning Technologies," 57.

36. Strasser and Becklund, *Swoosh,* 366.

37. Dallal, "Tinker Hatfield," 63; Vanderbilt, *Sneaker Book,* 111; Tom Hawthorn, "High-tech Shoes Aim to Reduce Sports Injuries," *Globe and Mail* (Toronto), April 10, 1987. Frank Rudy has observed that the envelope holding the compressed gas in Air shoes is actually under maximum load when it is sitting, not when it is being pounded by activity. See Braham, "High Tech Afoot," 80 ff.

38. Athletic Footwear Association, *The U.S. Athletic Footwear Market Today* (1998 and 1999); Ann Wozencraft, "Sports Retailing Chains Can't Seem to Get It Right," *New York Times,* November 28, 1999.

39. Suzanne Kapner, "Market Savvy," *Los Angeles Times,* July 4, 1998; William A. Rossi, "The Shoe Industry's 20-Year Snooze," *Footwear News,* vol. 54, no. 25 (July 22, 1998), 12; William A. Rossi, "Athletic Footwear: Facing Hard Realities," *Footwear News,* vol. 54, no. 21 (May 25, 1998), 32; Aaron Donovan, "A Long Walk, and a Gift of New Shoes," *New York Times,* December 7, 1999 (on an unemployed engineer enabled by the newspaper's Neediest Cases Fund to replace his worn-out sneakers with Timberland boots); Tasha Zemke, "Foot Race," *Pittsburgh Post-Gazette,* October 6, 1998; Antony Ramirez, "The Pedestrian Sneaker Makes a Comeback," *New York Times,* October 14, 1990; "Know It All: Footwear Questions Asked and Answered,"

Footwear News, June 13, 1994, 14; "Nike Set to Sell Cheap Shoes," Associated Press report, September 15, 1999 (source: http://www.abcnews.com).

40. Luella Bartley, "Death of the Trainer," *Vogue* (London), August 1998, n.p.; Steele, *Shoes,* 175–79; Deyan Sudjic, "Nike Air Zoom Seismic," *Weekend Financial Times, How to Spend It* magazine, no. 36 (May 1999), 69.

41. Timothy Egan, "The Swoon of the Swoosh," *New York Times Magazine,* September 13, 1998, 66; Mitchell Raphael, "Corporate Perversion," *Toronto Star,* February 7, 1998; Kathryn Kranhold and Suzanne Vranica, "Amid Heavy Dot-Com Spending on Ads, 1999 Saw the Good, the Bad and the Bizarre," *Wall Street Journal,* December 31, 1999; "Scuffle of the Week," *The Lawyer,* June 21, 1999, 56; Lisa de Moraes, "Putting On the Dogs," *Washington Post,* February 8, 1999; Bob Garfield, "Chauvinist Pigskin," *Advertising Age,* February 1, 1999; Nita Lelyveld, "Disparate Grievances Land at Same Doorstep," *Philadelphia Inquirer,* December 2, 1999.

42. Athletic Footwear Association, *Footwear Market* (1999), 3; Ellen Simon, "Fading Footprints," *Newark Star-Ledger,* September 26, 1999; Cragg Hines, "Republican National Convention," *Houston Chronicle,* August 12, 1996; Dan Morain, "Stanley Mosk: Will Dean of High Court Hang It Up?" *Los Angeles Times,* January 26, 1986; "Foot Notes," *New York Times,* April 6, 1980.

43. Steven E. Robbins and Gerard J. Gouw, "Athletic Footwear: Unsafe Due to Perceptual Illusions," *Medicine and Science in Sports and Exercise,* vol. 23, no. 2 (February 1991), 217–24; E. C. Frederick and Peter R. Cavanagh, letter, *Medicine and Science and Sports and Exercise,* vol. 24, no. 1 (January 1992), 144–45, and reply by Steven E. Robbins and Gerard J. Gouw, 145–47, and literature cited; Steven E. Robbins and Adel M. Hanna, "Running-Related Injury Prevention Through Barefoot Adaptations," *Medicine and Science in Sports and Exercise,* vol. 19, no. 2 (February 1987), 148–56; Steven Robbins, Gerard J. Gouw, and Jacqueline McClaran, "Shoe Sole Thickness and Hardness Influence Balance in Older Men," *Journal of the American Geriatrics Society,* vol. 40, no. 11 (November 1992), 1089–94; Steven Robbins and Edward Waked, "Balance and Vertical Impact in Sports: Role of Shoe Sole Materials," *Archives of Physical Medicine and Rehabilitation,* vol. 78, no. 5 (May 1987), 463–67; "Like to Walk? Put Away Your Walking Shoes," *Tufts University Health & Nutrition Letter,* April 1997, 1, 6; Steven Robbins and Edward Waked, "Hazard of Deceptive Advertising of Athletic Footwear," *Journal of the American Medical Association,* vol. 279, no. 13 (April 1998), 976F; Tom Carter, "Do Barefoot Athletes Have a Leg Up?" *Washington Times,* September 5, 1991; "Canadian Doctor Advises Distance Runners to Go Barefootin'," *Star-Tribune* (Minneapolis), August 23, 1987.

44. Frederick and Cavanagh, letter, 144–45; Beverley Smith, "Pricy Shoes Overrated, Report Says," *Globe and Mail* (Toronto), December 11, 1997; David Israelson, "Shoemakers Stomp Sole Study," *Toronto Star,* December 12, 1997; Joe Henderson, "Shoes and Feet," *Runner's World,* vol. 21, no. 10 (November 1986), 8.

45. Steven Robbins, Edward Waked, and Nicholas Krouglicof, "Improving Balance," *Journal of the American Geriatrics Society,* vol. 46, no. 11 (November 1998), 1363–70; Robbins, Gouw, and McClaran, "Sole Thickness," 1093.

46. John McCarry, "The Promise of Pakistan," *National Geographic,* vol. 192, no. 4 (October 1997), 60–61, 65–67.

CHAPTER FIVE

1. Andy Rooney, "A Chair That Fits, That's What We Need," *San Diego Union-Tribune,* May 23, 1989.

2. Ada S. Ballin, *The Science of Dress in Theory and Practice* (London: Sampson Low, Marston, Searle & Rivington, 1883), 229; Wilhelm Thomsen, *Die Geschichte der Schuhreform Hermann von Meyer's und ihre Beziehungen zur Gegenwart* (Stuttgart: Ferdinand Enke Verlag, 1940), 1–14; *Modern Chairs, 1918–1970* (London: Victoria and Albert Museum, 1970), 90; Alice Rawsthorn, "Tate Modern Take-away," *Financial Times,* May 6, 2000.

3. Gordon W. Hewes, "The Anthropology of Posture," *Scientific American,* vol. 196, no. 2 (February 1957), 122–32; Walter B. Pitkin, *Take It Easy: The Art of Relaxation* (New York: Simon & Schuster, 1935), 34–35; William C. Hayes, *The Scepter of Egypt,* 2 parts (New York: Metropolitan Museum of Art, 1990), pt. 1, 258–59, 262–65; pt. 2, 200–202.

4. Bernard Lewis, *The Middle East: A Brief History of the Last 2,000 Years* (New York: Scribners, 1995), 7, unnumbered illustrations.

5. Li Xing, "Furniture Took Ages to Grow Legs," *China Daily* (Beijing), October 31, 1988 (thanks to Karl Kroemer for calling this article to my attention); Wang Shixiang, *Classic Chinese Furniture: Ming and Early Qing Dynasties,* trans. Sarah Handler and the author (London: Han-Shan Tang, 1986), 14–16, 19–21; Sarah Handler, *Austere Luminosity of Chinese Classical Furniture* (Berkeley: University of California Press, 2001), 9–24.

6. K. H. E. Kroemer, H. B. Kroemer, and K. E. Kroemer-Elbert, *Ergonomics: How to Design for Ease and Efficiency* (Englewood Cliffs, N.J.: Prentice-Hall, 1994), 365–69; Thierry Bardini, *Bootstrapping: Douglas Engelbart, Coevolution, and the Origins of Personal Computing* (Stanford: Stanford University Press, 2000), illustration section, n.p.

7. Ralph S. Hattox, *Coffee and Coffeehouses: The Origins of a Social Beverage in the Medieval Near East* (Seattle: University of Washington Press, 1985), plate 12.

8. See Lorenz Homberger and Piet Meyer, "Concerning African Objects," and Roy Sieber, "African Furniture Between Tradition and Colonization," in Sandro Bocola, ed., *African Seats* (Munich: Prestel, 1996), 22–29 and 30–37, respectively.

9. Li, "Furniture," 8; Diana Fane, ed., *Converging Cultures: Art and Identity in Spanish America* (Brooklyn, N.Y.: Brooklyn Museum of Art, 1996), 284–85, 119; *The New York Times Capsule,* exhibition label, American Museum of Natural History, New York, visited April 6, 2000.

10. Andrew Pollack, "In a Painful Situation, Japanese Choose Chairs," *New York Times,* August 25, 1995.

11. "Researcher: Japanese-Style Toilets Produce Strong Thighs," *Report from Japan,* May 21, 1993, Nippon Notes section, n.p.; Takuo Fujita and Masaaki Fukase, "Comparison of Osteoporosis and Calcium Intake Between Japan and the United States," *Proceedings of the Society for Experimental Biology and Medicine,* vol. 200 (1992), 149–52; Emily Yoffe, "Maybe All That Milk You've Been Drinking Is to Blame," http://www.slate.com, August 2, 1999.

12. Galen Cranz, *The Chair: Rethinking Culture, Body, and Design* (New York: Norton, 1998), 94–101.

13. Paul Saenger, *Space Between Words: The Origins of Silent Reading* (Stanford: Stanford University Press, 1997), 48–49; Bruce M. Metzger, "When Did Scribes Begin to Use Writing Desks?" in his *Historical and Literary Studies: Pagan, Jewish, and Christian* (Grand Rapids, Mich.: Eerdmans, 1968), 123–37.

14. Saenger, *Space Between Words,* 48, 315 n. 104; Cranz, *Chair,* 154–55, 189; Siegfried Giedion, *Mechanization Takes Command: A Contribution to Anonymous History* (New York: Norton, 1969 [1948]), 264–65.

15. Giedion, *Mechanization Takes Command,* 282–87; Dora Thornton, *The*

Scholar in His Study: Ownership and Experience in Renaissance Italy (New Haven: Yale University Press, 1997), 53–59.

16. Fane, *Converging Cultures,* 278–79; Herman Roodenburg, "The 'Hand of Friendship': Shaking Hands and Other Gestures in the Dutch Republic," in Jan Bremmer and Herman Roodenburg, eds., *A Cultural History of Gesture* (Ithaca: Cornell University Press, 1991), 158–59. On the chair as frame, see Edgar Kaufmann, Jr., "Have a Seat," *Art News,* vol. 48, no. 5 (September 1949), 29–36.

17. David Nickerson, *English Furniture of the Eighteenth Century* (New York: Putnam, 1965), 8–9; Geneviève Souchal, *French Eighteenth-Century Furniture,* trans. Simon Watson Taylor (New York: Putnam, n.d.), 108, 112; John Kassay, *The Book of American Windsor Furniture* (Amherst: University of Massachusetts Press, 1998), 124–36.

18. Giedion, *Mechanization Takes Command,* 286–87, 405, 407.

19. Thomas E. Hill, *Hill's Manual of Social and Business Forms: A Guide to Correct Writing* (Chicago: Hill Standard Book Co., 1882), 28; H. Lachmayer, "Le Bureau du Chef," in *L'Empire du Bureau, 1900–2000* (Paris: Berger-Levrault and C.N.A.P., 1984), 59.

20. David Yosifon and Peter N. Stearns, "The Rise and Fall of American Posture," *American Historical Review,* vol. 103, no. 4 (October 1998), 1069; Gabriel A. Bobrick, *School Furniture: A Treatise on Its Construction in Compliance with Hygienic Requirements* (Boston: Rockwell and Churchill, 1887), 3–17.

21. Giedion, *Mechanization Takes Command,* 404–7.

22. Kenneth L. Ames, *Death in the Dining Room and Other Tales of Victorian Culture* (Philadelphia: Temple University Press, 1992), 185–232; Giedion, *Mechanization Takes Command,* 401.

23. Katherine C. Grier, *Culture and Comfort: Parlor Making and Middle-Class Identity, 1850–1930* (Washington: Smithsonian Institution Press, 1997), 124–29; Kenneth L. Ames, "The Rocking Chair in Nineteenth-Century America," *Antiques,* vol. 103, no. 2 (February 1973), 322–27; Ames, *Death in the Dining Room,* 216–32.

24. Cranz, *Chair,* 101–5; photograph reproduced in John Szarkowski, ed., *The Photographer's Eye* (New York: Museum of Modern Art, 1966), 139, and in George Talbot, ed., *At Home: Domestic Life in the Post-Centennial Era, 1876–1920* (Madison: State Historical Society of Wisconsin, 1976), 6–7.

25. Cranz, *Chair,* 155, 206; Adrian Forty, *Objects of Desire* (New York: Pantheon Books, 1986), 123.

26. Anson Rabinbach, *The Human Motor: Energy, Fatigue, and the Origins of Modernity* (New York: Basic Books, 1990); Richard Gillespie, "Industrial Fatigue and the Discipline of Physiology," in Gerald L. Geison, ed., *Physiology in the American Context, 1850–1940* (Bethesda, Md.: American Physiological Society, 1987), 237–62.

27. See Roger Cooter and Bill Luckin, eds., *Accidents in History: Injuries, Fatalities, and Social Relations* (Amsterdam: Rodopi, 1997); "The Worker's Chair," *Times* (London) *Trade Supplement,* September 18, 1920, 10; Irving Salomon, "How Correct Seating Affects Manufacturing Profits," *Industrial Management,* vol. 63, no. 6 (June 1922), 332.

28. Diana Condell, *Working for Victory: Images of Women in the First World War* (London: Routledge & Kegan Paul, 1987), 104–5, 112; E. Matthias, "Getting More from Workers by Giving Them More," *System* (London), vol. 33. no. 5 (May 1918), 308–14; "Working Hours That Prevent Fatigue," *System,* vol. 34, no. 2 (August 1918), 157–64.

29. "The Worker's Chair," advertisement, *System* (London), vol. 39, no. 6 (June 1921), 671; G. A. Bettcher, telephone conversation with author, April 5, 2000.

30. "The Worker's Chair," 671; "Scientific Seating," advertisement, *Success* (London), vol. 40, no. 5 (November 1921), 427.

31. "The Worker's Chair," 671; "The Worker's Chair," advertisement, *Success* (London), vol. 39, no. 4 (April 1921), 455; "Scientific Seating," advertisement, *Success* (London), vol. 40, no. 5 (November 1921), 427; "The Tan-Sad Chair for Workers," advertisement, *Success* (London), vol. 40, no. 6 (December 1921), 455; "Prices Which Defy Competition," advertisement, *Success* (London), vol. 40, no. 6 (December 1921), 463; "Efficient Seating," advertisement, *Success* (London), vol. 43, no. 1 (January 1923), 45.

32. "Worker's Chair," *Times* (London) *Trade Supplement,* November 27, 1920, 262; Salomon, "Correct Seating," 328–30; R. M. Bowen, "Fatigue and Production—Equipment That Eliminates Waste Effort," *Success* (London), vol. 48, no. 3 (September 1925), 150.

33. Forty, *Objects of Desire,* 140–42; Yosifon and Stearns, "American Posture," 1073–78; Yolande Amic, "Sièges de Bureau," in *L'Empire du Bureau,* 72–74; "Fortschritt-Stuhl," advertising sheet, 1928 or 1929 (Switzerland), in the Tschichold Collection, Museum of Modern Art (the author thanks Christopher Mount of the Department of Design for making a photocopy available).

34. Lachmayer, "Bureau du Chef," 60–61; G. A. Bettcher, telephone conversation, April 5, 2000; "Evolution and Revolution," advertisement, *Success* (London), vol. 48, no. 3 (September 1925), 150.

35. "Domore Chair Company—1920's," promotional material from company archives by G. A. Bettcher; G. A. Bettcher, telephone conversation, April 19, 2000.

36. "Domore Chair Company—1930s," promotional material supplied by G. A. Bettcher; Jonathan Lipman, *Frank Lloyd Wright and the Johnson Wax Buildings* (New York: Rizzoli, 1986), 85–91; Jack Quinan, *Frank Lloyd Wright's Larkin Building: Myth and Fact* (New York: Architectural History Foundation; Cambridge, Mass.: MIT Press, 1987), 62.

37. G. E. Macherle to Do/More Chair Company, May 10, 1939.

38. Jennifer Thiele, "Sit Down Before You Read This," *Contract Design,* vol. 32, no. 11 (November 1990), 59.

39. Thiele, "Sit Down," 61; Gillespie, "Industrial Fatigue," 253–57.

40. Eric Larrabee, *Knoll Design* (New York: Abrams, 1989), 152–53; "Portfolio: Domore 800, Seen by Raymond Loewy Associates."

41. "Sitting Down on the Job: Not as Easy as It Seems," *Occupational Health and Safety,* vol. 50, no. 10 (October 1981), 24–28; G. A. Bettcher, telephone conversation, April 19, 2000; Thiele, "Sit Down," 61; "The Big Stakes in Designing a Place to Sit," *BusinessWeek,* April 26, 1976, 46J.

42. "Machines à s'asseoir," *Progressive Architecture,* vol. 61, no. 5 (May 1980), 126–31.

43. David Morton, "Poetic Pragmatics," *Progressive Architecture,* vol. 59, no. 9 (September 1978), 98–101.

44. "You're Looking at the Control Core of the Most Advanced Task Chair Ever Developed," brochure (Washington, D.C.: Rudd International, 1982).

45. United States National Aeronautics and Space Administration, Scientific and Technical Information Office, *Anthropometric Source Book,* 3 vols. (Washington, D.C.: National Technical Information Service, 1978); J. Jay Keegan, "Alterations of the Lumbar Curve Related to Posture and Seating," *Journal of Bone and Joint Surgery,* vol. 35-A, no. 3 (July 1953), 589–603; "The Science of Easy Chairs," *Nature,* vol. 18, no. 468 (October 17, 1878), 637–38; A. C. Mandal, "The Seated Man (Homo Sedens),"

Applied Ergonomics, vol. 12, no. 3 (March 1981), 19–26; A. C. Mandal, "Balanced Seating," *Interior Design,* vol. 58, no. 15 (December 1987), 178–79.

46. Cranz, *Chair,* 170–76.

47. "Can Your Chair Do This?" *Managing Office Technology,* vol. 41, no. 1 (January 1996), 38 ff.

48. Lawrence Shames, "Seats of Power," *New York Times Magazine,* June 11, 1989, 65–66, 78–79.

49. William Houseman, "The Preface to a Serious Chair," *Design Quarterly,* no. 126 (1984), 5–23; "An Automatic Winner?" *Design,* no. 453 (September 1986), 40–41; Alix M. Freedman, "Today's Office Chair Promises Happiness, Has Lots of Knobs," *Wall Street Journal,* June 18, 1986, n.p.; Jura Koncius, "Furniture: High Style Comes to the Workplace," *Washington Post,* June 19, 1986, Washington Home sec., 9.

50. Vernon Mays, "The Ultimate Office Chair," *Progressive Architecture,* vol. 69, no. 5 (May 1988), 98 ff.

51. Robert Bishop, *Centuries and Styles of the American Chair* (New York: Dutton, 1972), 373.

52. Janet Travell, *Office Hours: Day and Night* (New York: World, 1968), 274.

53. Terril Yue Jones, "Face Off. Sit on It," *Forbes,* July 5, 1999, 53; Scott Leith, "Chair vs. Chair," *Grand Rapids Press,* June 13, 1999.

54. Sue Emily Martin, *Flexible Bodies: Tracking Immunity in American Culture from the Days of Polio to the Age of AIDS* (Boston: Beacon Press, 1994).

55. Dan Logan, "Home Office Thrones," *Los Angeles Times,* January 9, 1999.

56. "Science of Easy Chairs," 638.

CHAPTER SIX

1. A. Roger Ekirch, "Sleep We Have Lost: Preindustrial Slumber in the British Isles," *American Historical Review,* vol. 105, no. 2 (April 2000), 343–87; Peter N. Stearns et al., "Children's Sleep: Sketching Historical Change," *Journal of Social History,* vol. 30, no. 2 (Winter 1996), 345–66; Jerome A. Hirschfeld, "The 'Back-to-Sleep' Campaign Against SIDS," *American Family Physician,* vol. 51, no. 3 (February 15, 1995), 622 ff.; Bruce Bower, "Slumber's Unexplored Landscape," *Science News,* vol. 156, no. 13 (September 25, 1999), 205.

2. Lawrence Wright, *Warm and Snug: The History of the Bed* (London: Routledge & Kegan Paul, 1962), 177–78.

3. Robert Gifford and Robert Sommer, "The Desk or the Bed?" *Personnel and Guidance Journal,* vol. 46, no. 9 (May 1968), 876–78.

4. Oswyn Murray, "Sympotic History," in Oswyn Murray, ed., *Sympotica: A Symposium on the* Symposium (Oxford: Clarendon Press, 1990), 3–13; Margaret Visser, *The Rituals of Dinner: The Origins, Evolution, Eccentricities, and Meaning of Table Manners* (Toronto: HarperCollins, 1991), 152–55.

5. Clive Edwards, "Reclining Chairs Surveyed: Health, Comfort and Fashion in Evolving Markets," *Studies in the Decorative Arts,* vol. 6, no. 1 (Fall–Winter 1998–1999), 32–67; Pamela Tudor-Craig, "Times and Tides," *History Today,* vol. 48, no. 1 (January 1998), 2–5. On sixteenth-century information overload, see Geoffrey Parker, *The Grand Strategy of Philip II* (New Haven: Yale University Press, 1998), 42–45.

6. Edwards, "Reclining Chairs Surveyed," 36–37; *The Collection of Arne Schlesch* (New York: Sotheby's, 2000), 179; Wendy Moonan, "Scandinavian Dealer Sells His

Treasures," *New York Times,* March 31, 2000. Despite special mention in the *Times,* the piece failed to sell, perhaps because the estimate of $10,000 to $15,000 seemed high for what Moonan described as "an 18th-century precursor to the La-Z-Boy."

7. Margaret Campbell, "From Cure Chair to *Chaise Longue:* Medical Treatment and the Form of the Modern Recliner," *Journal of Design History,* vol. 12, no. 4 (1999), 327–28; Edwards, "Reclining Chairs Surveyed," 36–39.

8. Tudor-Craig, "Times and Tides," 3.

9. *Dictionnaire des Lettres Françaises: Le Dix-septième Siècle* (Paris: Fayard, 1954), 861–64; William L. Hamilton, "The Long . . . and Short of It," *New York Times,* March 9, 1995; Madeleine Jarry, *Le Siège Français* (Fribourg: Office du Livre, 1973), 53–56.

10. Jarry, *Siège Français,* 67, 78–81; John Gloag, *The Englishman's Chair* (London: Allen & Unwin, 1964), 139–42.

11. Orlando Sabertash, *The Art of Conversation, with Remarks on Fashion and Address* (London, 1842), 148, quoted in Campbell, "Cure Chair to *Chaise Longue,*" 328.

12. Katherine C. Grier, *Culture and Comfort: Parlor Making and Middle-Class Identity, 1850–1930* (Washington, D.C.: Smithsonian Institution Press, 1997), 117–30.

13. Ibid., 117–42, 193–210.

14. Edwards, "Reclining Chairs Surveyed," 42–45.

15. U.S. patent 9,449 (December 1852), issued to J. T. Hammitt; Sharon Darling, *Chicago Furniture: Art, Craft, & Industry, 1833–1983* (New York: Norton, 1984), 86–88; *Asher and Adams's Pictorial Album of American Industry, 1876* (New York: Rutledge Books, 1976), 69; Edwards, "Reclining Chairs Surveyed," 51–54.

16. Giedion, *Mechanization Takes Command,* 418–22.

17. Edwards, "Reclining Chairs Surveyed," 54–56; Kevin Oderman, "How Things Fit Together," *Southwest Review,* vol. 81, no. 2 (Spring 1996), 235–46; Bruce E. Johnson, "The Morris Chair," *Country Living,* January 1988, 23–24; *Sears, Roebuck & Co. 1908 Catalogue No. 117,* ed. Joseph J. Schroeder, Jr. (Chicago: Follett, 1969), 448–49.

18. ". . . and Morris Chair Makes a Comeback," *Petersburg Times,* February 14, 1988.

19. *Sears 1908 Catalogue,* 449.

20. Ibid., 444.

21. Albert B. Jannsen, *La-Z-Boy: I Remember When* (Monroe, Mich.: La-Z-Boy Chair Co., n.d.), 2–7; U.S. Patent 1,789,337, January 20, 1931, issued to Edward M. Knabusch and Edwin J. Shoemaker.

22. Jannsen, *La-Z-Boy,* 7–8; "La-Z-Boy, Specialist in Reclining Comfort," *Furniture South,* vol. 35, no. 3 (March 1956), unpaginated copy; Dave Hoekstra, "A Pilgrimage for the Armchair Traveler," *Chicago Sun-Times,* September 6, 1998; Sears, Roebuck & Co. catalogue, Fall–Winter 1939–40, 620–21.

23. Alexander von Vegesack, *Thonet: Classic Furniture in Bent Wood and Tubular Steel* (London: Hazar, 1996), 90–91, 118–19; Campbell, "From Cure Chair to *Chaise Longue,*" 335.

24. *The Dictionary of Art,* s.v. "Klint, Kaare (Jensen); Alexander von Vegesack et al., *100 Masterpieces from the Vitra Design Museum* (Weil am Rhein: Vitra Design Museum, 1996), 72–73; Renato de Fusco, *Le Corbusier, Designer Furniture, 1929* (Woodbury, N.Y.: Barron's, 1977), 10–22, 30–37, 68; Mary McLeod, "Furniture and Femininity," *Architectural Review,* vol. 181, no. 1079 (January 1987), 43–46.

25. Von Vegesack et al., *100 Masterpieces,* 80.

26. Unless otherwise noted, information on Anton Lorenz derives from inter-

views with his associate, the engineer Peter Fletcher, who collaborated on chair designs with Lorenz until Lorenz's death in 1964, and purchased intellectual property and papers (hereafter cited as "Lorenz papers") from the Lorenz estate; Henry Conston, interview, July 7, 2000.

27. Christopher Wilk, *Marcel Breuer: Furniture and Interiors* (New York: Museum of Modern Art, 1981), 73–78; Werner Möller and Otakar Máčel, *Ein Stuhl Macht Geschichte* (Munich: Prestel, 1992), 85–92.

28. Von Vegesack et al., *100 Masterpieces,* 80; Gunther Lehmann, "Zur Physiologie des Liegens," *Arbeitsphysiologie,* vol. 11 (1941), 253–58.

29. Wilk, *Thonet,* 112–13; Möller and Máčel, *Ein Stuhl Macht Geschichte,* 98.

30. Möller and Máčel, *Ein Stuhl Macht Geschichte,* 98. Many details of Lorenz's relations with Barcalo and other U.S. manufacturers also are given in: 1) unsigned and undated history of Barcalo, Burchfield-Penney Art Center, Buffalo State College; 2) memorandum of company history, Barcalounger Company files; and 3) Ernest F. Becher, "Development of Automotive Seating and Mechanical Chairs" (1983), Barcalounger Company files.

31. "NOW Available for Selective Retail Distribution," advertisement, *Retailing,* September 10, 1945, 4.

32. "Recline-Relax-Recuperate Chairs" (Floral City, Mich.: Floral City Furniture, n.d.); La-Z-Boy archives; Walter B. Pitkin, *Take It Easy: The Art of Relaxation* (New York: Simon & Schuster, 1935), 234–44; "Chair," *The New Yorker,* December 10, 1949, 32–33.

33. George W. Kinker, "Learn to Relax," *Today's Health,* vol. 29, no. 11 (November 1951), 22 ff.; Joseph L. Fetterman, M.D., "Roads to Relaxation," *Today's Health,* vol. 31, no. 11 (November 1953), 18 ff.

34. Peter Blake and Jane Fiske McCullough, "Very Significant Chair," *Harper's Magazine,* vol. 217, no. 1299 (August 1958), 66–71; Joseph Giovannini, "For Long, Leisurely Sitting, Traditional Chairs Are Best," *New York Times,* April 18, 1986.

35. "Leisure, or How Does the American Relax?" *Life,* July 12, 1955, 72–73; Sylvia Wright, "Do We Sit? No, We Collapse," *New York Times Magazine,* April 19, 1964, 26 ff.

36. "Only Stratolounger Gives You a Complete! Profitable! Reclining Chair Department," advertisement, *Home Furnishings Daily,* July 29, 1958, sec. 1, 7; "It's Here!" advertisement, *Home Furnishings Daily,* June 29, 1959, 29; "There Is Only One Leader," advertisement, *Home Furnishings Daily,* June 17, 1963, sec. 1, 3.

37. "La-Z-Boy, Specialist in Reclining Comfort," n.p.

38. Hortense Herman, "New Mechanisms, Forms Make Chairs Big News," *Home Furnishings Daily,* June 16, 1959, 1.

39. Gary Garrity, "Style of Reclining Loungers Creating Demand in Topeka," *Home Furnishings Daily,* June 15, 1964, 44; Clark Rogers, telephone interview, August 5, 2000.

40. John A. Byrne, "Sittin' and Rockin'," *Forbes,* November 7, 1983, 124; "The Evolution of the La-Z-Boy Recliner," 2.

41. Joan Kron, "The *New York* Magazine Comfortable Chair Competition," *New York,* July 22, 1974, 43–49.

42. "Sunny Side Up," *People Weekly,* April 10, 2000, 200–202.

43. J. Jay Keegan, "Alterations of the Lumbar Curve Related to Posture and Seating," *Journal of Bone and Joint Surgery,* vol. 35-A, no. 3 (July 1953), 589–603, with no reference to Lehmann's paper on the Lorenz experiments; National Aeronautics and Space Administration, *Anthropometric Source Book,* vol. 1: *Anthropometry for Designers*

(NASA Reference Publication 1024), 1978, IV-21; Susan M. Andrews, "Recliners: Good for Your Health," *Furniture/Today,* June 12, 2000, 8–9; "Selling the Spanish Siesta," *Economist,* February 6, 1999, 54; K.B., "Siesta Time," *Forbes,* January 31, 1994, 119.

44. American Furniture Manufacturers Association, *AFMA 1999 Sales Planning Guide* (n.p.: AFMA, 1999), 9, 11; Nancy Butler, "Stagnant Recliner Market Offset by Higher Tickets," *Furniture/Today,* November 17, 1997, 8–10.

45. Charles Allen, *Raj: A Scrapbook of British India, 1877–1947* (New York: St. Martin's Press, 1977), 60–61, 84–85.

CHAPTER SEVEN

1. "Keyboard," *New Grove Encyclopedia of Music,* 2nd ed. (London: Macmillan, 2001), vol. 13, 510–13; Max Weber, *The Rational and Social Foundations of Music,* trans. and ed. Don Martindale et al. (Carbondale: Southern Illinois University Press, 1958), 117.

2. "Keyboard," *New Grove Encyclopedia.*

3. Weber, *Foundations of Music,* 114–17; Thomas Levenson, *Measure for Measure: A Musical History of Science* (New York: Simon & Schuster, 1994), 27–38.

4. Paul Saenger, *Space Between Words: The Origin of Silent Reading* (Stanford: Stanford University Press, 1997), 82–99; Henri-Jean Martin, *The History and Power of Writing,* trans. Lydia G. Cochrane (Chicago: University of Chicago Press), 153–54; David Gelernter, "Bound to Succeed," *New York Times Magazine,* April 18, 1999, 132.

5. Joel Mokyr, *The Lever of Riches: Technological Creativity and Economic Progress* (New York: Oxford University Press, 1990), 48–51.

6. Edmund A. Bowles, "On the Origins of the Keyboard Mechanism in the Late Middle Ages," *Technology and Culture,* vol. 7, no. 2 (Spring 1966), 152–62.

7. See Edward Rothstein, "Every Piano Is a Society," *New York Times,* August 24, 1986; "Tuning," *Encyclopedia of the Piano* (New York and London: Garland, 1996), 410–13. For a fascinating history of this technical but vital subject, see Stuart Isacoff, *Temperament: The Idea That Solved Music's Greatest Riddle* (New York: Alfred A. Knopf, 2001).

8. "Cristofori, Bartolomeo," *Encyclopedia of the Piano,* 97–105.

9. Ibid.

10. Brent Gillespie, "Haptic Display of Systems with Changing Kinematic Constraints: The Virtual Piano Action" (Ph.D. diss., Stanford University, 1996), 7–16.

11. Sheryl Maureen P. Mueller, "Concepts of Nineteenth-Century Piano Pedagogy in the United States" (Ph.D. diss., University of Colorado, 1995), 26.

12. "Beethoven, Ludwig van," *Encyclopedia of the Piano,* 44–47.

13. Arthur Loesser, *Men, Women, and Pianos: A Social History* (New York: Simon & Schuster, 1954), 366–67; Mick Hamer, "Smashing Performance Franz, Play It Again," *New Scientist,* vol. 108, no. 1487–88 (December 19/26, 1985), 46.

14. "Frames," *Encyclopedia of the Piano,* 137–39.

15. Cyril Ehrlich, *The Piano: A History,* rev. ed. (Oxford: Clarendon Press, 1990), 107, 138–40; Loesser, *Men, Women, and Pianos,* 520–31, 611.

16. Craig H. Roell, *The Piano in America, 1800–1940* (Chapel Hill: University of North Carolina Press, 1989), 8–10.

17. Mueller, *Concepts of Piano Pedagogy,* 104–33.

18. Loesser, *Men, Women, and Pianos,* 291–92.

19. "Barrel Piano," *Encyclopedia of the Piano,* 41–42; Arthur W. J. G. Ord-Hume, *Pianola: The History of the Self-Playing Piano* (London: Allen & Unwin, 1984), 9–22.

20. Mick Hamer, "Rave from the Grave," *New Scientist,* vol. 160, no. 2165-66-67 (December 19/26, 1998/January 2, 1999), 18–19; Ord-Hume, *Pianola,* 31–35.

21. Roell, *Piano in America,* 14–15; Hamer, "Rave from the Grave," 18.

22. Roell, *Piano in America,* 185–221; Loesser, *Men, Women, and Pianos,* 602–3; Ehrlich, *Piano,* 186–87, 222.

23. Rudolph Chelminski, "In Praise of Pianos . . . and the Artists Who Play Them," *Smithsonian,* vol. 30, no. 12 (March 2000), 71–72.

24. "Clutsam, Ferdinand" and "Keyboards," *Encyclopedia of the Piano,* 82, 202–4; Ernest Hutchenson, *The Literature of the Piano: A Guide for Amateur and Student,* 2nd ed., rev. Rudolph Ganz (London: Hutchinson, 1969), 14–15.

25. Ibid.

26. "Paul von Jankó," *Die Musik in Geschichte und Gegenwart* (Kassel: Bärenreiter, 1957) vol. 6, 1710–13; "Jankó, Paul von," *Encyclopedia of the Piano,* 185–86; Edwin L. Good, *Giraffes, Black Dragons, and Other Pianos: A Technological History from Cristofori to the Modern Concert Grand,* 2nd ed. (Stanford: Stanford University Press, 2001), 258–62.

27. Alfred Dolge, *Pianos and Their Makers* (New York: Dover, 1972), 78–83; John S. Allen, telephone interview, April 25, 2001.

28. "Jankó," *Musik in Geschichte und Gegenwart,* vol. 6, 1713.

29. Dolge, *Pianos and Their Makers,* 80.

30. Constantin Sternberg, "Ad Vocem, Jankó Keyboard," *The Musical Courier,* vol. 22, no. 2 (January 14, 1891), 39; Good, *Giraffes,* 262–64.

31. Dolge, *Pianos and Their Makers,* 80.

32. "Keyboard Revamp Could Take Terror Out of Piano Practice," *New Scientist,* vol. 132, no. 1793 (November 9, 1991), 28; Tara Patel, "Mathematical Piano Sounds a Logical Note," *New Scientist,* vol. 134, no. 1821 (May 16, 1992), 20.

33. Albert Glinsky, *Theremin: Ether Music and Espionage* (Urbana: University of Illinois Press, 2000), 22–49.

34. Ibid, 119–20, 138, 142, 145–48.

35. Jean Laurendeau, *Maurice Martenot, Luthier de l'Électronique* (Montreal: Louise Courteau, 1990), 45–57, figs. 18–55.

36. Hutcheson, *Literature of the Piano,* 18; on current and older piano construction and maintenance, see Larry Fine, *The Piano Book: Buying and Owning a New or Used Piano,* 4th ed. (Boston: Brookside Press, 2001).

37. Larry Fine, *The Piano Book: Buying and Owning a New or Used Piano,* 3rd ed. (Boston: Brookside Press, 1994), 91–93; http://www.overspianos.com.au and articles reprinted there.

38. Trevor Pinch and Frank Trocco, "The Social Construction of the Early Electronic Music Synthesizer," *Icon: Journal of the International Committee for the History of Technology,* vol. 4 (1998), 9–19.

39. Ibid., 22; Frank Houston, "Gather Round the Electronic Piano," *New York Times,* December 16, 1999.

40. Mark Tucker, "The Piano in Jazz," in James Parakilas, ed., *Piano Roles: Three Hundred Years of Life with the Piano* (New Haven: Yale University Press, 1999), 374; James Parakilas, "New Specialties in the Musical Marketplace," in ibid., 391.

41. See Allen's Web site: http://www.bikexprt.com/music/index.htm.

42. Robert A. Moog and Thomas L. Rhea, "Evolution of the Keyboard Interface:

The Bösendorfer 290 SE Recording Piano and the Moog Multiply-Touch-Sensitive Keyboards," *Computer Music Journal,* vol. 14, no. 2 (Summer 1990), 52–60.

43. Mick Hamer, "Physics Under the Keyboard," *New Scientist,* vol. 108, no. 1487–88 (December 19/26, 1985), 44–47; Edwin M. Good, telephone interview, May 8, 2001.

44. Moog and Rhea, "Evolution of the Keyboard Interface," 52, 58–89.

45. Gillespie, "Haptic Display," 29–30.

46. See ibid. for an exposition of the paradox of piano actions and for ideas on future haptic keyboards.

CHAPTER EIGHT

1. Christopher de Hamel, *Scribes and Illuminators* (Toronto: University of Toronto Press, 1992), 7, 29, 35–39.

2. Henri-Jean Martin, *The History and Power of Writing,* trans. Lydia G. Cochrane (Chicago: University of Chicago Press, 1994), 230–31; de Hamel, *Scribes and Illuminators,* 37–39.

3. Donald M. Anderson, *Calligraphy: The Art of Written Forms* (New York: Dover, 1992), 134–37.

4. Ibid., 140–52.

5. Asa Briggs, *Victorian Things* (Chicago: University of Chicago Press, 1988), 182–86.

6. Anderson, *Calligraphy,* 272; George Pratt, "A Pen of Steel," in Thomas R. Lounsbury, ed., *Yale Book of American Verse* (New Haven: Yale University Press, 1913), 401.

7. Anne Swardson, "Aren't Fountain Pens Fun? No; They're Too Hard," *Washington Post,* October 5, 1997; Bruno Delmas, "Révolution Industrielle et Mutation Administrative: l'Innovation dans l'Administration Française aux XIXème Siècle," *Histoire, Économie et Société,* vol. 4, no. 2 (1985), 210.

8. See Henry Petroski, *The Pencil: A History of Design and Circumstance* (New York: Alfred A. Knopf, 1990).

9. Tamara Plakins Thornton, *Handwriting: A Cultural History* (New Haven: Yale University Press, 1996), 46–66.

10. Ibid., 66–71, 163–64.

11. Samuel Solly, "Scriveners' Palsy, or the Paralysis of Writers," *Lancet,* December 24, 1864, 709–11.

12. Richard N. Current, *The Typewriter and the Men Who Made It* (Urbana: University of Illinois Press, 1954), 22–26.

13. See Paul Israel, *From Machine Shop to Industrial Laboratory: Telegraphy and the Changing Context of American Invention, 1830–1920* (Baltimore: Johns Hopkins University Press, 1992).

14. G. C. Mares, *The History of the Typewriter: Successor to the Pen* (Arcadia, Calif.: Post-Era Books, 1985 [1909]), 28–32; A. Baggenstos, *Von der Bilderschrift zur Schreibmaschine* (Zurich: n.p., 1977), 37–39.

15. Cited in Current, *Typewriter,* 11; Cynthia Monaco, "The Difficult Birth of the Typewriter," *American Heritage of Invention and Technology,* vol. 4, no. 21 (Spring/Summer 1988), 12.

16. Mares, *Typewriter,* 140–44; Current, *Typewriter,* 42–43.

17. Current, *Typewriter,* 37–38.

18. Monaco, "Difficult Birth," 17–18.

19. On the effects of organizational growth and education on office information technology, see the brilliant work of Delmas, "Révolution Industrielle," 205–32.

20. Chas. Howard Montague, cited in Bates Torrey, *Practical Typewriting: By the All-Finger Method,* . . . (New York: Fowler & Wells, 1889), [4].

21. Christopher Keep, "The Cultural Work of the Type-Writer Girl," *Victorian Studies,* vol. 40, no. 3 (Spring 1997), 405; Current, *Typewriter,* 120; Jack Quinan, *Frank Lloyd Wright's Larkin Building: Myth and Fact* (Cambridge, Mass.: MIT Press, 1987), 48–49, 53.

22. Israel, *Machine Shop,* 134–35, 163–65; Basil Kahan, *Ottmar Mergenthaler: The Man and His Machine* (New Castle, Del.: Oak Knoll Press, 2000), 179–81, 282–84; Ava Baron, "Contested Terrain Revisited: Technology and Gender Definitions of Work in the Printing Industry, 1850–1920," in Barbara Drygulski Wright et al., eds., *Women, Work, and Technology: Transformations* (Ann Arbor: University of Michigan Press, 1987), 58–83.

23. *The History of Touch Typing* (New York: Wyckoff, Seamans, and Benedict, n.d.), 6–7.

24. Ibid., 9–11.

25. Ibid., 13–15.

26. Ibid., 17–21; Torrey, *Practical Typewriting,* 3–20.

27. Current, *Typewriter,* 55–58.

28. Mares, *Typewriter,* 145–51; Paul Lippman, *American Typewriters: A Collector's Encyclopedia* (Hoboken, N.J.: Original & Copy, 1992), 33–40.

29. See "Clio and the Economics of QWERTY," *American Economic Review,* vol. 75, no. 2 (May 1985), 332–37.

30. August Dvorak et al., *Typewriting Behavior: Psychology Applied to Teaching and Learning Typewriting* (New York: American Book Company, 1936), 267–75; Delphine Gardey, "The Standardization of a Technical Practice: Typing (1883–1930)," *History and Technology,* vol. 15 (1999), 326.

31. Dvorak et al., *Typewriting Behavior,* esp. 81–101, 284–301.

32. Ibid., 208–16.

33. Ibid., 217–28; William E. Cooper, "Introduction," in William E. Cooper, ed., *Cognitive Aspects of Skilled Typewriting* (New York: Springer-Verlag, 1983), 6.

34. Cooper, "Introduction," 7; Dvorak et al., *Typewriting Behavior,* 237.

35. Gardey, "Standardization," 323–39; Cooper, "Introduction," 8.

36. Lippman, *American Typewriters,* 41–43.

37. Allan G. Bromley, "Analog Computing Devices," in William Aspray, ed., *Computing Before Computers* (Ames: Iowa State University Press, 1990), 156–85.

38. Jay Hersh, "The Tyranny of the Keyboard," in http://www.tifaq.org/articles; "DEC Terminals," in http://www.cs.utk.edu/~shuford.

39. See Edward Tenner, *Why Things Bite Back: Technology and the Revenge of Unintended Consequences* (New York: Vintage Books, 1997), 222–29.

40. Joy M. Ebben, "It's Not Just with Keyboards," *Occupational Health & Safety,* vol. 70, no. 4 (April 2001), 65 ff.; Pamela Mendels, "Protecting Little Hands," *New York Times,* January 9, 2000.

41. Donald A. Norman and Diane Fisher, "Why Alphabetic Keyboards Are Not Easy to Use: Keyboard Layout Doesn't Much Matter," *Human Factors,* vol. 24, no. 5 (October 1982), 509–19; Donald R. Gentner and Donald A. Norman, "The Typist's Touch," *Psychology Today,* vol. 18, no. 3 (March 1984), 66–72; S. J. Liebowitz and

Stephen E. Margolis, "The Fable of the Keys," *Journal of Law and Economics,* vol. 33, no. 1 (April 1990), 1–26.

42. Scot Ober, "Review of Research on the Dvorak Simplified Keyboard," *Delta Pi Epsilon Journal,* vol. 34, no. 4 (Fall 1992), 167–82; Scot Ober, "Relative Efficiencies of the Standard and Dvorak Simplified Keyboards," *Delta Pi Epsilon Journal,* vol. 35, no. 1 (Winter 1993), 1–13, and references cited in each article; Peter J. Howe, "Different Strokes Catch On," *Boston Globe,* January 15, 1996; Jennifer B. Lee, "Keyboards Stuck in the Age of NumLock; Defunct Keys and Odd Commands Still Bedevil Today's PC User," *New York Times,* August 12, 1999.

43. Neal Taslitz, telephone interview, August 17, 2002.

44. John Markoff, "Microsoft Sees New Software Based on Pens," *New York Times,* November 9, 2000; "Newtonian Marketing Lessons," *Advertising Age,* August 15, 1994, 20.

45. Frank R. Wilson, *The Hand: How Its Use Shapes the Brain, Language, and Human Culture* (New York: Pantheon Books, 1998), 309–10; Lisa Guernsey, "For Those Who Would Click and Cheat," *New York Times,* April 26, 2001.

46. Monaco, "Difficult Birth," 15.

47. Lucy C. Bull, "Being a Typewriter," *Atlantic Monthly,* vol. 76, no. 458 (December 1895), 822–31.

CHAPTER NINE

1. Vincent Canby, "Nobody's Spared in Dame Edna's One-Man Show," *New York Times,* October 24, 1992; Christopher Lehmann-Haupt, "Behind the Superman, an All-Too-Human Thinker," *New York Times,* November 23, 1998; Marc Winoto, "Ways of Walking," *New Scientist,* vol. 164, no. 2218 (December 25, 1999), 87.

2. "The Adventure of the Golden Pince-nez," in *The Annotated Sherlock Holmes,* ed. William S. Baring-Gould, 2 vols. (New York: Potter, 1967), vol. 2, 356, 364; Frank DeCaro, "Dame Edna Speaks Her Mind, and Yours," *New York Times,* October 3, 1999.

3. Hugh R. Taylor, "Racial Variations in Vision," *American Journal of Epidemiology,* vol. 133, no. 1 (January 1981), 62–80; Seang-Mei Saw et al., "Epidemiology of Myopia," *Epidemiologic Reviews,* vol. 18, no. 2 (1996), 176–77; Richard A. Post, "Population Differences in Visual Acuity," *Social Biology,* vol. 29, no. 3–4 (Fall–Winter 1982), 319–43; Jane E. Brody, "In Debate on Myopia's Origins the Winner Is: Both Sides?" *New York Times,* June 1, 1994.

4. Douglas Fox, "Blinded by Bread," *New Scientist,* vol. 174, no. 2337 (April 6, 2002), 9; Loren Cordain et al., "An Evolutionary Analysis of the Aetiology and Pathogenesis of Juvenile-Onset Myopia," *Acta Ophthalmologica Scandinavica,* vol. 80, no. 2 (April 2002), 125–35.

5. Anneliese A. Pontius, "In Similarity Judgments Hunter-Gatherers Prefer Shapes over Spatial Relations in Contrast to Literate Groups," *Perceptual and Motor Skills,* vol. 81, no. 3, pt. 1 (December 1995), 1027–41.

6. Jah M. Enoch, "The Enigma of Early Lens Use," *Technology and Culture,* vol. 39, no. 2 (April 1998), 278–91.

7. "Lenses and Eyeglasses," *Encyclopedia of the Middle Ages,* vol. 7, 538; Nicholas Horsfall, "Rome Without Spectacles," *Greece & Rome,* vol. 42, no. 1 (April 1998), 49–56.

8. Dora Jane Hamblin, "What a Spectacle! Eyeglasses, and How They Evolved," *Smithsonian,* vol. 13, no. 12 (March 1983), 102–3; Vincent Ilardi, "Renaissance Flo-

rence: The Optical Capital of the World," *Journal of European Economic History*, vol. 22, no. 3 (Winter 1993), 508–10.

9. Ilardi, "Renaissance Florence," 512–36.

10. Dennis Simms, electronic mail, August 25, 2002; Vincent Ilardi, telephone interview, August 24, 2002. See D. J. Bryden and D. L. Simms, "Spectacles Improved to Perfection and Approved by the Royal Society," *Annals of Science* (U.K.), vol. 50, no. 1 (1993), 1–32.

11. Henri Jean Martin, *The History and Power of Writing*, trans. Lydia G. Cochrane (Chicago: University of Chicago Press, 1994), 32; John Dreyfus, "The Invention of Spectacles and the Advent of Printing," *The Library*, 6th ser., vol. 10, no. 2 (June 1988), 94–106; Lynn White, Jr., "Technology Assessment from the Stance of a Medieval Historian," *American Historical Review*, vol. 79, no. 1 (February 1974), 1–13.

12. Heinz Herbert Mann, *Augenglas und Perspektive: Studien zur Ikonographie zweier Bildmotive* (Berlin: Gebr. Mann, 1992), 43–57.

13. Ibid., 30–32, 35–39; Jean-Claude Margolin, "Des Lunettes et des Hommes ou la Satire des Mal-voyants au XVIe Siècle," *Annales E.S.C.*, vol. 30, no. 2–3 (March–June 1975), figs. V–IX.

14. Margolin, "Des Lunettes et des Hommes," 375–93.

15. German text: "Was heissen Fackeln / Liecht oder Brilln / Wann die Leute nich sehen wölln." Claude K. Deischer and Joseph L. Rabinowitz, "The Owl of Heinrich Khunrath: Its Origin and Significance," *Chymia*, vol. 3 (1950), 243–50; Mann, *Augenglas*, 92–94, 99–100.

16. Jill Bepler, "Cultural Life at the Wolfenbüttel Court, 1635–1666," *A Treasure House of Books: The Library of Duke August of Brunswick-Wolfenbüttel* (Wolfenbüttel: Herzog August Bibliothek, 1998), 140–46.

17. Michael Scholz-Hänsel, *El Greco, Der Großinquisitor: Neues Licht auf die Schwarze Legende* (Frankfurt am Main: Fischer Taschenbuch Verlag, 1991), 47–60.

18. Ibid., 67–72.

19. Richard Corson, *Fashions in Eyeglasses* (Chester Springs, Pa.: Dufour, 1967), 63.

20. Ibid., 69–74.

21. Ibid., 69–70; Rita Reif, "Fashions in Glasses, Sights for Sore Eyes," *New York Times*, December 21, 1997; "On the Nose: Spectacles and Other Optical Fashions," exhibition text, New-York Historical Society (I am grateful to Margaret K. Hofer of the Society for providing a copy); Billie J. A. Follensbee, "Rubens Peale's Spectacles: An Optical Illusion?" *History of Opththalmology*, vol. 41, no. 5 (March–April 1997), 417–24.

22. James R. Gregg, *The Story of Optometry* (New York: Ronald Press, 1965), 130–31; "Durchs Glas den Rehbock Geniessen," *Frankfurter Allgemeine Zeitung*, March 12, 2001; Deutsche Presse-Agentur (DPA)—Europadienst, release March 8, 2001.

23. Gregg, *Story of Optometry*, 132–33, 179–201; Hermann von Helmholtz, *Popular Lectures on Scientific Subjects* (New York: Appleton, 1873), 197–226.

24. "Education," *Encyclopaedia Britannica*, 14th ed.; David Vincent, *The Rise of Mass Literacy: Reading and Writing in Modern Europe* (Cambridge, Eng.: Polity, 2000), 1–26.

25. Roy Porter, "Reading Is Bad for Your Health," *History Today*, vol. 48, no. 3 (March 1998), 11–16.

26. Suzanne Hahn, "Die Schulhygiene zwischen naturwissenschaftlicher Erkenntnis, sozialer Verantwortung und 'vaterländischem Dienst': Das Beispiel der Myopie in

der zweiten Hälfte des 19. Jahrhunderts," *Medizinhistorisches Journal,* vol. 29, no. 1 (1994), 23–38; Emil Ludwig, *Gifts of Life: A Retrospect* (Boston: Little, Brown, 1931), 3–19.

27. Hermann Cohn, *The Hygiene of the Eyes in Schools,* trans. W. P. Turnbull (London: Simpkin, Marshall, 1886), 54–83; James C. Albisetti, *Secondary School Reform in Imperial Germany* (Princeton: Princeton University Press, 1983), 123–24.

28. Albisetti, *Secondary School Reform,* 130–39; Roy Porter, *Companion Encyclopedia of the History of Medicine* (London: Routledge, 1993), 585–600.

29. Hermann Cohn, *Lehrbuch der Hygiene des Auges* (Vienna: Urban & Schwarzenberg, 1892), 252–68.

30. Hahn, "Schulhygiene," 33–38.

31. Ernest Hart, "Spectacled Schoolboys," *Atlantic Monthly,* vol. 72, no. 433 (November 1893), 681–84.

32. *Oxford English Dictionary,* 2nd ed., s.v. "four"; *Random House Historical Dictionary of American Slang,* s.v. "four-eyes" and "four-eyed."

33. "Spectacles," *Saturday Review,* vol. 50, no. 1295 (August 21, 1880), 234; John F. Kasson, *Houdini, Tarzan, and the Perfect Man: The White Male Body and the Challenge of Modernity in America* (New York: Hill & Wang, 2001), 30–31.

34. "Spectacles," 235; L. D. Bronson, *Early American Specs: An Exciting Collectible* (Glendale, Calif.: Occidental Publishing, 1974), 34; Corson, *Fashion in Spectacles,* 215; Lyned Isaac, "Speculating Through Spectacles: A Material Culture Study of Eyeglasses in Late Nineteenth Century Australia" (M.A. thesis, Department of History, Monash University, 1993), 40–59, 69–72. I am grateful to the School of Historical Studies, Monash University, for providing a copy of this excellent thesis.

35. Corson, *Fashions in Eyeglasses,* 106; "Gli Occhiali di Corbu," *Abitare,* no. 307 (May 1992), 244–46.

36. Quoted in Richard Bernstein, "Isaac Babel May Yet Have the Last Word," *New York Times,* July 11, 2001.

37. Henry Allen, "Everything You Always Wanted to Know About Specs," *Washington Post,* December 16, 1991.

38. Nick Ravo, "Altina Schinasi Miranda, 92, Designer of Harlequin Glasses," *New York Times,* April 21, 1999; "Take Another Look at Glasses," *Glamour,* September 1992, 298–301; David Shields, "Optical Illusions," *Vogue,* April 1993, 120 ff.

39. Joseph L. Bruneni, *Looking Back: An Illustrated History of the American Ophthalmic Industry* (Torrance, Calif.: Ophthalmic Laboratories Association, 1994), 78–82, 138.

40. Ibid., 42.

41. Anahad O'Connor, "Harold Ridley, Eye Doctor, 94, Early Developer of Lens Implants," *New York Times,* June 6, 2001.

42. Susan Ferraro, "Patients Are Blindsided," *New York Daily News,* February 25, 2001; Bob Garfield, "The Focus Thing," *Washington Post,* January 7, 2001.

43. Laura Johannes, "Is End of Reading Glasses in Sight?" *Wall Street Journal,* March 29, 2001.

44. Saw et al., "Epidemiology of Myopia," 175; Jane Gwiazda and Lynn Marran, "The Many Facets of the Myopic Eye: A Review of Genetic and Environmental Factors," *Vision Science and Its Applications,* vol. 35 (2000), 393.

45. Gwiazda and Marran, "Myopic Eye," 394–95; Naomi Lee, "A Close Look at the Cause of Myopia," *South China Morning Post,* April 16, 2000; Roger Dobson, "The Future Is Blurred," *The Independent* (London), May 20, 1999; Saw et al., "Epidemiology of Myopia," 177.

46. Joshua Wallman, "Nature and Nurture of Myopia," *Nature,* vol. 371, no. 6494 (September 15, 1994), 201–2; John Schwartz, "In Sharpening Children's Focus, Glasses May Fuzz the Future," *Washington Post,* August 1, 1995.

CHAPTER TEN

1. Edward B. Becker, "Helmet Development and Standards," excerpt from N. Yoganandan et al., eds., *Frontiers in Head and Neck Trauma: Clinical and Biomedical* (London: IOS Press, 1998), 1–2.

2. M. D. W. Jeffreys, "Ibo Warfare," *Man,* vol. 56 (June 1956), 77–79; E. W. Gudger, "Helmets from Skins of the Porcupine-Fish," *Scientific Monthly,* vol. 30, no. 5 (May 1930), 432–42.

3. "Helme," *Der neue Pauly,* vol. 5, 327; Richard A. Gabriel and Karen S. Metz, *From Sumer to Rome: The Military Capabilities of Ancient Armies* (New York: Greenwood Press, 1991), 54–58.

4. Gabriel and Metz, *From Sumer to Rome,* 51, 60–63; Yigael Yadin, *The Art of Warfare in Biblical Lands in the Light of Archaeological Discovery,* trans. M. Pearlman (London: Weidenfeld & Nicolson, 1963), 40–44, 49.

5. Walter Mayer, *Politik und Kriegskunst der Assyrier* (Münster: Ugarit-Verlag, 1995), 431, 461–63; Gabriel and Metz, *From Sumer to Rome,* 29–30.

6. A. M. Snodgrass, *Arms and Armour of the Greeks* (Ithaca: Cornell University Press, 1967), 48–51.

7. Victor Davis Hanson, *The Western Way of War: Infantry Battle in Classical Greece* (New York: Alfred A. Knopf, 1989), 71–75, 213–14.

8. See A. K. Goldsworthy, "The Othismos, Myths and Heresies: The Nature of Hoplite Battle," *War in History,* vol. 4, no. 1 (January 1997), 1–26; "Alternative Agonies: Hoplite Martial and Combat Experiences Beyond the Phalanx," in Hans van Wees, ed., *War and Violence in Ancient Greece* (London: Duckworth and the Classical Press of Wales, 2000), 233–59; Hanson, *Western Way of War,* 58–59.

9. Hanson, *Western Way of War,* 57–58; Snodgrass, *Arms and Armour,* 93–95.

10. Jacques Ellul, *The Technological Society* (New York: Vintage Books, 1964), 29; Victor Duruy, *History of Rome and of the Roman People,* trans. M. M. Ripley (Boston: Dana Estes, 1883–86), vol. 5, sec. 2, 320; see the fascinating evolutionary diagrams prepared by Marcus Junkelmann in *Römische Helme* (Mainz: Sammlung Guttmann bei Philipp von Zabern, 2000), n.p.

11. Adrian Goldsworth, *Roman Warfare* (London: Cassell, 2000), 53; Peter Connolly, "The Roman Fighting Technique Deduced from Armour and Weaponry," *Roman Frontier Studies 1989* (Exeter: University of Exeter Press, 1991), 358–63; Junkelmann, *Römische Helme,* 24–30.

12. Junkelmann, *Römische Helme,* 32, 85.

13. Ibid., 33.

14. T. Philip D. Blackburn et al., "Head Protection in England Before the First World War," *Neurosurgery,* vol. 47, no. 6 (December 2000), 1268, 1281; Christopher Knüsel and Anthea Boylston, "How Has the Towton Project Contributed to Our Knowledge of Medieval and Later Warfare?" in Veronica Fiorata, Anthea Boylston, and Christopher Knüsel, eds., *Blood Red Roses: The Archaeology of a Mass Grave from the Battle of Towton AD 1461* (Oxford: Oxbow Books, 2000), 169–88.

15. Bashford Dean, *Helmets and Body Armor in Modern Warfare,* 2nd ed. (Tuckahoe, N.Y.: Carl J. Pugliese, 1977), 30, 34.

16. Claude Blair, *European Armour, Circa 1066 to Circa 1700* (London: B. T. Bats-

ford, 1958), 31–32; A. V. B. Norman and Don Pottinger, *English Weapons & Warfare, 449–1660* (London: Arms and Armour Press, 1979), 46–47.

17. Blair, *European Armour,* 105–7; A. W. Boardman, *The Medieval Soldier in the Wars of the Roses* (Stroud, Gloucester: Sutton Publishing, 1998), 129–31; Donald La Rocca, electronic mail, February 7, 2003.

18. Blackburn et al., "Head Protection in England," 1280–81; Dean, *Helmets and Body Armor,* 42–43; Blair, *European Armour,* 191–92.

19. Richard A. Preston, Alex Roland, and Sydney F. Wise, *Men in Arms: A History of Warfare and Its Interrelationships with Western Society,* 3rd ed. (Fort Worth: Harcourt Brace Jovanovich, 1991), 98.

20. Dean, *Helmets and Body Armor,* 50–53; Blair, *European Armour,* 296–97.

21. Ivor Noël Hume and Audrey Noël Hume, *The Archaeology of Martin's Hundred,* 2 parts (Philadelphia: University of Pennsylvania Museum of Archaeology and Anthropology, and Williamsburg, Va.: The Colonial Williamsburg Foundation, 2001), part 1, 158; Harold L. Peterson, *Arms and Armor in Colonial America, 1526–1783* (New York: Bramhall House, 1956), 103–51.

22. Dean, *Helmets and Body Armor,* 56–57; R. D. Stiot, "Les Armures de Tranchées en Usage durant la Révolution et l'Empire et l'Armure de Sapeurs Modèle 1833," and "Les Armures de Sapeurs Modèle 1836 et Fabrication 1838," *Armes Blanches Militaires Françaises,* vol. 28 (1983), n.p.; "Equestrian," "Heavy Cavalry Helmets" (letter), *Times* (London), March 31, 1865.

23. John W. Waterer, *Leather and the Warrior* (Northampton, Eng.: The Museum of Leathercraft, 1981), 32–43; Günter Gall, *Leder im Europäischen Kunsthandwerk* (Braunschweig: Klinkhardt & Biermann, 1965), 182–87; Dean, *Helmets and Body Armor,* 56.

24. Kenneth Holcombe Dunsee, *Engine! Engine! A Story of Fire Protection* (New York: Harold Vincent Smith for the Home Insurance Company, 1939), 47–51. Dennis Smith, *Dennis Smith's History of Firefighting in America: 300 Years of Courage* (New York: Dial Press, 1978), 34.

25. Francis Bertin, Pascal Courault, and Joan Deville, *Le Feu Sacré* (Rennes: Éditions Ouest-France, 1994), passim; Jean-Claude Demory, *Pompiers Militaires de France* (Boulogne: E.T.A.I., 1997), 10–13, 16–17; W. Eric Jackson, *London's Fire Brigades* (London: Longmans, 1966), 34, 46–48, 80, 82; Ludwig Baer, *The History of the German Steel Helmet, 1916–1945,* trans. K. Daniel Dahl (San Jose, Calif.: R. James Bender, 1985), 267; Robyn Cooper, "The Fireman: Immaculate Manhood," *Journal of Popular Culture,* vol. 28, no. 4 (1985), 139–70.

26. Dean, *Helmets and Body Armor,* 58–65.

27. See the essays in Roger Cooter and Bill Luckin, eds., *Accidents in History: Injuries, Fatalities, and Social Relations* (Amsterdam: Rodopi, 1997).

28. David G. Herrmann, *The Arming of Europe and the Making of the First World War* (Princeton: Princeton University Press, 1996), 17–21; John Ellis, *Eye-Deep in Hell: Trench Warfare in World War I* (Baltimore: Johns Hopkins University Press, 1989), 61–62; Dean, *Helmets and Body Armor,* 68–70.

29. Dean, *Helmets and Body Armor,* 74–75.

30. Petros Dintsis, *Hellenistische Helme,* 2 vols. (Rome: G. Bretschneider, 1986), vol. 1, 113–33, and vol. 2, supp. 9; Dean, *Helmets and Body Armor,* 9, 74–83.

31. Dean, *Helmets and Body Armor,* 128–31, 193–96, 314.

32. Baer, *German Steel Helmet,* 7–24.

33. Dean, *Helmets and Body Armor,* 138.

34. Baer, *German Steel Helmet,* 85–89.

35. Dean, *Helmets and Body Armor,* 208–17; Stephen V. Grancsay, "Helmets and Body Armor in Modern Warfare," in *Arms & Armor: Essays from the Metropolitan Museum of Art Bulletin, 1920–1964* (New York: Metropolitan Museum of Art, 1986), 286–87; "Helmets Presented a Challenge to Metal Workers," *Scientific American,* vol. 171, no. 4 (October 1994), 153; Robert K. Southee, "The Steel Pot," *American Legion,* vol. 119, no. 2 (February 1982), 24 ff.

36. Mike Mason, "So Long, Ol' Pot," *Soldiers,* November 1982, 21–22.

37. Baer, *German Steel Helmet,* 390–402; Dean, *Helmets and Body Armor,* 208–13. Dean's Model 2 had a visor and side/rear apron like the original *Stahlhelm,* but with a more gentle curve.

38. "Army Hails New Helmets' Value in Grenada," *New York Times,* December 8, 1982; "PASGT, Mon Armor," *Washington Post,* August 14, 1985.

39. "Trench Helmets in Mines," *New York Times,* August 12, 1923.

40. "From the Hard-Boiled Hat to Today's Skull Bucket," *California Builder and Engineer,* February 16, 1998, 16–22.

41. General Detroit Corp., *The Buyers' Encyclopedia,* Catalogue No. 20.

42. Earl W. Murphy, "Hardhats Become Heroes of New Novel by Idaho Author," *Idaho Voter,* January 26, 1956; Thomas M. Pryor, "Wald Will Film 'The Hard Hats,' " *New York Times,* September 3, 1956; "Who's Wearing That Helmet—and Why?" *Popular Science,* vol. 173, no. 5 (November 1958), 122–23.

43. Homer Bigart, "City Hall Is Stormed," *New York Times,* May 9, 1970; Richard Rogin, "Why the Construction Workers Holler, 'U.S.A., All the Way!' " *New York Times Magazine,* June 28, 1970, 7 ff.

44. "A New Hat in Political Ring," *New York Times,* October 27, 1970.

45. "Police Here Start to Wear Helmets for Hydrant Duty," *New York Times,* July 5, 1961; John Kifner, "300 in S.D.S. Clash with Chicago Police," *New York Times,* October 9, 1969; Steven Mikulan, "A Wintry Discontent," *LA Weekly,* February 8, 2002.

46. Andy Beckett, "Spot the Difference," *The Guardian,* July 12, 2002; Tom Utley, "The Policeman's Helmet May Be Impractical—But It Still Works," *Daily Telegraph,* April 6, 2002.

47. Becker, "Helmet Development and Standards," 2–3; Hugh Cairns, "Head Injuries in Motor-Cyclists," *British Medical Journal,* no. 4213 (October 4, 1941), 465–71; Hugh Cairns and H. Holbourn, "Head Injuries in Motor-Cyclists," *British Medical Journal,* no. 4297 (May 15, 1942), 591–97.

48. "Jockeys Must Wear Helmets in Steeplechases, Is Ruling," *New York Times,* January 9, 1925; Henry R. Isley, "Cap Saves M'Atee as Mount Falls," *New York Times,* June 3, 1924.

49. J. Nadine Gelberg, "The Lethal Weapon: How the Plastic Football Helmet Transformed the Game of Football, 1939–1994," *Bulletin of Science, Technology, and Society,* vol. 15, no. 5–6 (1995), 304; Manny Topol, "Evolution of a Football Helmet," *Newsday* (Long Island), July 8, 1996.

50. Edward Tenner, *Why Things Bite Back: Technology and the Revenge of Unintended Consequences* (New York: Vintage Books, 1997), 277–78; Gelberg, "Lethal Weapon," 304–8.

51. Dean Fisher, "History, Helmets and Standards: 40 Years of Advancement in Head Protection," *ASTM Standardization News,* vol. 20, no. 6 (June 1992), 34–37; Ed Hinton, "Head Impact Isn't Necessary in Fatal Crashes," *Los Angeles Times,* February 11, 2001; Dick Teresi, "The Case for No Helmets," *New York Times,* June 17, 1995.

52. Matt Walker, "In the Firing Line," *New Scientist,* vol. 162, no. 2191 (June 19, 1999), 25.

53. Amber Smith, "Heads Up on Headgear: Safety Outweighs Similarities When It Comes to Buying a Helmet," *Newark Star-Ledger,* May 7, 2002, 33; Sarah Collins and Lisa Gubernick, "The Riskiest Sports," *Wall Street Journal,* July 6, 2001.

54. Phil Patton, "Hard Hats," *I.D.,* vol. 41, no. 3 (May–June 1994), 62–67.

55. Julian F. Barnes, "A Bicycling Mystery: Head Injuries Piling Up," *New York Times,* July 29, 2001; Gary D. Mower, "Bicycle-Helmet Article Misled," *Deseret News* (Salt Lake City), August 6, 2001.

56. Douglass Carnall, "Cycle Helmets Should Not Be Compulsory," *British Medical Journal,* vol. 318 (June 5, 1999), 2005; Mayer Hillman, *Cycle Helmets: The Case For and Against* (London: Policy Studies Institute, 1993); Gerald J. S. Wilde, *Target Risk* (Toronto: PDE Publications, 1994).

57. Gregory Bayan, "Reflections on a Hockey Helmet," *Newsweek,* March 12, 1984, 13.

58. Judith VandeWater, "Babies Can Get a Better Head Start When Helmets Round Out Flat Spots," *St. Louis Post-Dispatch,* March 28, 2002; Geoffrey A. Fowler, "Baby May Have a Flat Head, but Parents Shouldn't Get Bent Out of Shape," *U.S. News & World Report,* August 28, 2000, 57.

EPILOGUE

1. Donna Haraway, "A Cyborg Manifesto," in *Simians, Cyborgs and Women: The Reinvention of Nature* (New York: Routledge, 1991), 149–81; Chris Hables Gray, *Cyborg Citizen* (New York: Routledge, 2001).

2. Reidar F. Sognnaes, "America's Most Famous Teeth," *Smithsonian,* vol. 2, no. 11 (February 1973), 47–51; Robert Cohen and J. Scott Orr, "Often the Patient Is the Last to Know," *Newark Star-Ledger,* August 11, 2002; Dean E. Murphy, "Betraying John Hancock," *New York Times,* February 4, 2001.

3. Justin Bachman, "Boy's Attached Fin Roils the Waters," *Newark Star-Ledger,* August 16, 2002; Clare Wilson, "Cutting Edge," *New Scientist,* vol. 175, no. 2355 (August 10, 2002), 32–35.

4. Steve Mann with Hal Niedzviecki, *Cyborg: Digital Destiny and Human Possibility in the Age of the Wearable Computer* (Toronto: Doubleday Canada, 2001), 4–5, 149–54.

5. Phil Patton, "New Mouse Takes Shoulder off the Wheel," *New York Times,* April 12, 2001; Ann Carrns, "Hard Hit by Imports, American Pencil Icon Tries to Get a Grip," *Wall Street Journal,* November 24, 2000.

6. *Tactical Display for Soldiers* (Washington, D.C.: National Academy Press, 1997); Geoffrey A. Fowler, "In the Digital Age, 'All Thumbs' Is Term of Highest Praise," *Wall Street Journal,* April 17, 2002; Libby Copeland, "Thumbs Up: After Eons Spent in Its Siblings' Shadow, the Dumpy Digit Finally Counts," *Washington Post,* June 24, 2002; Frank Wilson, *The Hand: How Its Use Shapes the Brain, Language, and Human Culture* (New York: Pantheon Books, 1998), 146.

7. Eric von Hippel, *The Sources of Innovation* (New York: Oxford University Press, 1988), 1–25, 102–16; Kim J. Vicente, *Cognitive Work Analysis: Towards Safe, Productive, and Healthy Computer-Based Work* (Mahwah, N.J.: Erlbaum), 109–38.

8. John Markoff, "Kristen Nygaard, 75, Who Built Framework for Modern Computer Languages," *New York Times,* August 14, 2001.

Suggestions for Further Reading

On human culture in comparative perspective, the best survey that I have found is Tim Ingold, ed., *Companion Encyclopedia of Anthropology* (London: Routledge, 1994), with a number of important articles by leading figures on themes such as tools and tool behavior, technology, and artifacts. For biologists' views of the human/nonhuman divide, if there is one, see W. C. McGrew, *Chimpanzee Material Culture: Implications for Human Evolution* (Cambridge, Eng.: Cambridge University Press, 1992), Donald R. Griffin, *Animal Minds: Beyond Cognition to Consciousness* (Chicago: University of Chicago Press, 2001), and Frans de Waal, *The Ape and the Sushi Master: Cultural Reflections of a Primatologist* (New York: Basic Books, 2001).

Frank R. Wilson's *The Hand: How Its Use Changes the Brain, Language, and Human Culture* (New York: Pantheon Books, 1998) is a brilliant look at the relationship between human mental and physical skills. John Napier's *Hands,* revised by Russell H. Tuttle (Princeton: Princeton University Press, 1993), remains a classic of biology and culture. Marcel Mauss's short essays on body techniques in his *Sociology and Psychology: Essays,* trans. Ben Brewster (London: Routledge & Kegan Paul, 1979) are indispensable. William H. McNeill's *Keeping Together in Time: Dance and Drill in Human History* (Cambridge, Mass.: Harvard University Press, 1995) is a pioneering study of body techniques in history.

Jacques Ellul's *The Technological Society,* trans. John Wilkinson (New York: Alfred A. Knopf, 1964), was one of the twentieth century's truly prophetic works. As a study of the mechanical and the organic, Siegfried Giedion's *Mechanization Takes Command: A Contribution to Anonymous History* (New York: Norton, 1969) is also a masterpiece. Excellent recent studies of the body and technology are Anson Rabinbach's *The Human Motor: Energy, Fatigue, and the Origins of Modernity* (New York: Basic Books, 1990) and the papers, including a reprint of Mauss's essay on body techniques, collected in Jonathan Crary and Sanford Kwinter, eds., *Zone 6* (New York: Zone, 1992).

On material culture and history, one of the most delightful books remains Asa Briggs, *Victorian Things* (Chicago: University of Chicago Press, 1988), a social historian's survey of the nineteenth-century proliferation of Stuff. Its counterpart from an engineer's perspective, and even more absorbing, is Henry Petroski's *The Evolution of Useful Things* (New York: Alfred A. Knopf, 1995). George Basalla, *The Evolution of Technology* (Cambridge, Eng.: Cambridge University Press, 1988) and Wiebe E. Bijker, *Of Bicycles, Bakelite, and Bulbs* (Cambridge, Mass.: MIT Press, 1995) mix theory and example agreeably. Bruno Jacomy's *L'Âge du Plip: Chroniques de l'Innovation Technique* (Paris: Éditions du Seuil, 2002) examines human-machine interaction from a French perspective that should become more familiar in Anglo-American history of technology.

On breast-feeding and the bottle, the best academic survey is Patricia Stuart-Macadam and Katherine A. Dettwyler, eds., *Breastfeeding: Biocultural Perspectives* (New York: Aldine de Gruyter, 1995).

Much of the history of footwear is preserved in trade publications, catalogues, and specialized textbooks. There is no book on zori in any Western language, but a wonderful cultural background exists in Susan B. Hanley's *Everyday Things in Premodern Japan: The Hidden Legacy of Material Culture* (Berkeley: University of California Press, 1997), which also is relevant to Japan's retention of mat-level sitting and living. Most writing about athletic shoes deals more with marketing strategy than with technical design. An enjoyable exception is Tom Vanderbilt's *The Sneaker Book: Anatomy of an Industry and an Icon* (New York: New Press, 1998).

Of hundreds of studies of seating, Galen Cranz's *The Chair: Rethinking Culture, Body, and Design* (New York: Norton, 1998) stands out as a historical critique of conventional design. Katherine C. Grier's *Culture and Comfort: Parlor Making and Middle-Class Identity, 1850–1930* (Washington, D.C.: Smithsonian Institution Press, 1997), Kenneth L. Ames, *Death in the Dining Room and Other Tales of Victorian Culture* (Philadelphia: Temple University Press, 1992), and Leora Auslander, *Taste and Power: Furnishing Modern France* (Berkeley: University of California Press, 1996) are equally readable social histories of furniture. Clive D. Edwards, *Victorian Furniture: Technology and Design* (Manchester, Eng.: Manchester University Press, 1993) links industry and form.

A superb study of piano making is Edwin L. Good, *Giraffes, Black Dragons, and Other Pianos: A Technological History from Cristofori to the Modern Concert Grand*, 2nd ed. (Stanford: Stanford University Press, 2001). The best multiauthor survey of piano history is James Parakilas, ed., *Piano Roles: Three Hundred Years of Life with the Piano* (New Haven: Yale University Press, 1999). Beside the breezy but learned *Men, Women, and Pianos* by Arthur Loesser (New York: Simon & Schuster, 1954) is Craig H. Roell's more focused *The Piano in America, 1890–1940* (Chapel Hill: University of North Carolina Press, 1989).

August Dvorak et al., *Typewriting Behavior: Psychology Applied to Teaching and Learning Typewriting* (New York: American Book Company, 1936) is about much more than keyboard arrangements. It is a window on the industrial psychology of the 1930s. (One revelation: since a learning curve charts the growth of performance over time, a steep learning curve represents rapid rather than slow mastery.) In contemporary media studies, Friedrich A. Kittler's *Gramophone, Film, Typewriter,* trans. Geoffrey Winthrop-Young and Michael Wutz, and Lisa Gitelman, *Scripts, Grooves, and Writing Machines: Representing Technology in the Edison Era* (both Stanford: Stanford University Press, 1999) stand out.

Among contemporary books on eyeglasses, the best historical survey is Joseph L. Bruneni's *Looking Back: An Illustrated History of the American Ophthalmic Industry* (Torrance, Cal.: Optical Laboratories Association, 1994). The best European-oriented reference is Richard Corson, *Fashions in Eyeglasses* (Chester Springs, Pa.: Dufour Editions, 1967). Both are unfortunately hard to find.

On ancient helmets the most accessible book is probably A. M. Snodgrass, *Arms and Armour of the Greeks* (Ithaca: Cornell University Press, 1967). Bashford Dean's *Helmets and Body Armor in Modern Warfare,* originally published by Yale University Press in 1920, was reprinted with an appendix on World War II armor in 1977

(Tuckahoe, N.Y.: C. J. Pugliese). It remains the best all-around survey. On the German steel helmet, Ludwig Baer's *The History of the German Steel Helmet, 1916–1945,* trans. K. Daniel Dahl (San Jose, Calif.: R. J. Bender, 1985) is based on original documents.

Donald A. Norman's *The Psychology of Everyday Things* (New York: Basic Books, 1988), now reprinted as *The Design of Everyday Things,* emphasizes the mental side of physical objects. The most important recent study of the body in today's workplace is Shoshana Zuboff, *In the Age of the Smart Machine: The Future of Work and Power* (New York: Basic Books, 1988). For the history of the visionary side of technology, mind, and body, there is Thierry Bardini's *Bootstrapping: Douglas Engelbart, Coevolution, and the Origins of Personal Computing* (Stanford: Stanford University Press, 2000).

In cyborg anthropology, the starting point is Donna Haraway, *Simians, Cyborgs, and Women: The Reinvention of Nature* (New York: Routledge, 1991). Taking the theme into the Web era are Chris Hables Gray, *Cyborg Citizen: Politics in the Posthuman Age* (New York: Routledge, 2001), and N. Katherine Hailes, *How We Became Posthuman: Virtual Bodies in Cybernetics, Literature, and Informatics* (Chicago: University of Chicago Press, 1999).

Finally, for two complementary views of the human future, I suggest Christopher Wills, *Children of Prometheus: The Accelerating Pace of Human Evolution* (Reading, Mass.: Perseus Books, 1998), and Rodney R. Brooks, *Flesh and Machines: How Robots Will Change Us* (New York: Pantheon Books, 2002).

Index

Italicized page numbers refer to illustrations.

ABOUT THE AUTHOR

Edward Tenner has been an executive editor of Princeton University Press and a visiting scholar in the university's Departments of Geosciences and English, and in the School of Social Science of The Institute for Advanced Study. He has been a John Simon Guggenheim Memorial Fellow and a Fellow of the Woodrow Wilson International Center for Scholars. He is now a senior research associate of the Jerome and Dorothy Lemelson Center for the Study of Invention and Innovation, National Museum of American History. He lives in Plainsboro, New Jersey.

A NOTE ON THE TYPE

The text of this book was set in Berkeley Old Style, a typeface designed by Tony Stan based on a face originally developed by Frederic Goudy in 1938 for the University of California Press at Berkeley.

Composed by North Market Street Graphics, Lancaster, Pennsylvania
Printed and bound by Berryville Graphics, Berryville, West Virginia
Designed by Robert C. Olsson